Engineering Illumination Optics

Case Studies and Exercises

Henning Rehn and Julius Muschaweck

WILEY-VCH

Authors

Henning Rehn
Illuminatio Solutions GmbH
Waltharistrasse 3
9400 Rorschach
Switzerland

Julius Muschaweck
JMO Illumination Optics
Zugspitzstr. 66
82131 Gauting
Germany

Cover Design: Wiley
Cover Image: © Chris Rogers/Getty Images

Typesetting: Lumina Datamatics, Madurai, India

Printing and Binding: CPI Group (UK) Ltd, Croydon, CR0 4YY

Library of Congress Card No.: applied for

British Library Cataloguing-in-Publication Data A catalogue record for this book is available from the British Library.

Bibliographic information published by the Deutsche Nationalbibliothek The Deutsche Nationalbibliothek lists this publication in the Deutsche Nationalbibliografie; detailed bibliographic data are available on the Internet at <http://dnb.d-nb.de>

Print ISBN: 9783527414635
ePDF ISBN: 9783527853137
ePub ISBN: 9783527853120
oBook ISBN: 9783527853144

C9783527414635_140526

Contents

List of Exercises

List of Knowledge Corner Topics

List of Simulations

Preface

To become a good optical designer for illumination, an engineer must acquire a set of technical skills: He or she must learn to understand the underlying theory, must know the toolbox of optical elements and how to combine them, and become an expert user of the optical design software of choice. Besides reading books, experiencing projects, and getting advice from senior colleagues, you, as an engineer on your way to grow your skills, must think through problems on your own, use your own mind, stumble over obstacles yourself, and find your own way to a solution, all by yourself. This is hard, sometimes exhausting, sometimes even embarrassing. Designing good illumination optics requires creative thinking, juggling, combining and (rarely) inventing optical elements. But still quite early in the process, this creative thinking must be guided and fenced in by numbers, which come from calculations. This book deals with the calculation part. Many simple calculations in illumination optics can be of great help to guide the optical designer. We aim to show you how. Calculate before you embark on concrete design work. If you don't, you will likely end up with inefficient solutions (not knowing they are inefficient) or failed attempts to violate fundamental laws (not knowing why you failed). Accordingly, this book also serves as a handbook of illumination pre-design.

This book's primary goal is to let you, the reader, learn by studying typical problems, solving them yourself. Some books avoid formulas ("each formula will cut your number of readers in half", as the saying goes) and explain topics by pictures and descriptions. To gain understanding about how something works in principle, such a presentation style, i.e., a style that plays to our visualization capabilities, is great. However, solving concrete engineering problems also requires skill and experience in using formulas and numbers. Hence, the book contains a curated selection of exercise problems, inspired by real world problems we had to solve and whose solutions we found helpful to know. We introduce and discuss the topics with many, many calculations.

Our book as presented here is a result of our own continuing journey and originates from the tasks we presented in a large number of live courses and online trainings. Writing this book took us about forty years: From the first optics lectures in university, to where we stand now, in our long careers in the optical industry.

During those years, many persons supported our journey, and we are deeply grateful to all of them, even though we can only name very few here.

First of all, our deepest thanks, for everything and more, go to our spouses, Liby and Gisela, and to our parents. University professors like Richard Kowarschik (for HR) and Rudolf Sizmann (for JM) shared their insights about the theory of light: a theory that is deep, beautiful, and immensely useful in daily engineering work, all at the same time. They made us realize that knowing how to calculate useful quantities, based on fundamental first principles, is essential to the deep satisfaction we get from a job well done.

We also had the great luck to be supported by several illuminated managers who gave us the freedom, and even encouraged us, to develop our current knowledge of illumination optics.

And we appreciate the many challenges brought to us by customers and colleagues which we could not always solve immediately – the type of problems that keep lingering at the back our heads, and help us to grow insights while solving them eventually.

Last, but not least, we thank our test readers, Bernhard Michel, Dirk Döring, Gokce Toprak, Martin Daniels, Matthias Walser, Oliver Dross, Rich Pfisterer, and Ronian Siew: Their valuable comments and criticism greatly helped to bring this book into its final form.

Rorschach and Gauting, March 2025 *Henning Rehn* and *Julius Muschaweck*

Acronyms

ANSI	American National Standards Institute
AR	Anti-reflection
ARRI	Arnold & Richter Cine Technik GmbH & Co, a provider of cine cameras and lighting fixtures
ASAP	Advanced Systems Analysis Program, one of the first non-sequential software packages
BEF	Brightness Enhancement Film
CAD	Computer Aided Design
CBCP	Central Beam Candle Power, luminous intensity on axis
CCD	Charge Coupled Device
CCT	Correlated Color Temperature
CEC	Compound Elliptical Concentrator
CHC	Compound Hyperbolic Concentrator
CIE	Commission Internationale de l'Eclairage, the international commission on illumination
CMOS	Complementary Metal Oxide Semiconductor
CPC	Compound Parabolic Concentrator
CPV	Concentrated Photovoltaics
CRI	Color Rendering Index
CTE	Coefficient of Thermal Expansion
CVRL	Color & Vision Research Laboratory
DGaO	Deutsche Gesellschaft für Angewandte Optik
DIN	Deutsches Institut für Normung
DLP	Digital Light Processing
DMD	Digital Micromirror Device
ECE	Economic Commission for Europe
EMVA	European Machine Vision Association
EOS	European Optical Society
ESA	European Space Agency
ESO	European Southern Observatory
FRED	an optical engineering software
FWHM	Full Width at Half Maximum
GRIN	Gradient Index
HBO	a high-pressure short-arc mercury lamp technology
HCR	High Concentration Region
HDR	High Dynamic Range
HDTV	High Definition Television
HID	High Intensity Discharge

HIL	Homogeneous, Isotropic, Linear
HLX	a high-pressure short-arc mercury lamp technology
HPL	High-performance Lamp
HPS	High Pressure Sodium
HUD	Head-up Display
IEC	International Electrotechnical Commission
IEEE	Institute of Electrical and Electronics Engineers
IES	Illuminating Engineering Society
IODC	International Optical Design Conference
IRC	Infrared Reflective Coating
ISO	International Organization for Standardization
LARP	Laser Activated Remote Phosphor
LCD	Liquid Crystal Display
LCR	Low Concentration Region
LDC	Light Distribution Curve
LED	Light Emitting Diode
LES	Light Emitting Surface
LightTools	a non-sequential illumination optics software
LM	Lighting Measurement
LPD	Limiting Photometric Distance
MR	Multifaceted Reflector
NEMA	National Electrical Manufacturers Association
NIR	Near Infrared
NSP	Narrow Spot
OLED	Organic Light Emitting Diode
OPD	Optical Path Difference
OSLO	Optics Software for Layout and Optimization
PAR	Parabolic Reflector
PBS	Polarizing Beam Splitter
PCB	Printed Circuit Board
PCX	Plano Convex
PFAS	Per- and polyfluoroalkyl Substances
PMMA	Poly(methyl methacrylate)
POF	Plastic Optical Fiber
POP	Physical Optics Propagation
PPF	Photosynthetic Photon Flux
PPFD	Photosynthetic Photon Flux Density
PRTE	Paraxial Ray Tracing Equation
PSA	Projected Solid Angle
SAE	Society of Automotive Engineers
SI	International System of Units
SID	Society of Information Display
SNR	Signal-to-Noise Ratio
SPIE	Society of Photographic Instrumentation Engineers, later the Society of Photo-Optical Instrumentation Engineers (now the International Society of Optical Engineering)
SSL	Solid State Lighting
TIR	Total Internal Reflection
TO	Transistor Outline
UCS	Uniform Color Space
UGR	Unified Glare Rating
UHP	Ultra High Pressure
WPE	Wall Plug Efficacy
XBO	a Xenon high-pressure short-arc lamp technology
ZEMAX	a sequential/non-sequential optical design software

1

Introduction

To become a good optical designer for illumination applications, an engineer must acquire a very specific set of technical skills: he or she must understand the underlying theory, must know the toolbox of optical elements and know how to combine them, and must be a proficient user of (at least one) optical design software. In addition to reading books, experiencing projects, and learning from senior colleagues, as an engineer on your way to expanding your skills, you must also think through problems on your own, use your own mind, stumble over obstacles, and find your way to a solution – all by yourself. This may be hard, sometimes exhausting or even embarrassing. However, solving problems by yourself is an indispensable part of the journey.

For an introduction to approaches, processes and methods, we recommend our book *Designing Illumination Optics* [1]. In addition, we are also planning to publish a tutorial with the tentative title *Design Elements of Illumination Optics* intended to help readers to understand the function and design parameters of individual optical elements. Other books that we find useful and frequently consult ourselves are Julio Chaves' *Introduction to Nonimaging Optics* [2] for the nonimaging part of illumination optics, Michael Kidger's two books on *Fundamental* and *Intermediate Optical Design* [3, 4] for the imaging aspects, the SPIE *Field Guide on Illumination* [5] for short, concise definitions, and John Koshel's great edition of selected chapters on *Illumination Engineering* [6].

The book you hold in your hands is the result of our own continuing journey and has its roots in a large number of live and online training courses we held over the years. Hence, the book contains a curated selection of exercises and problems, inspired by real problems we faced and the solutions of which we found instructive. Our goal was to touch upon numerous facets and fields of illumination design such as lighting design [7], imaging optics (e.g. to calculate image brightness [8, 9]), medical optics [10] or horticulture [11], to name just a few examples.

1.1 Goals, or How to Use This Book

This text is a technical book aimed at professionals from many fields who want to dive deeper into optics or, specifically, practical illumination design. It is a collection of knowledge – much of it in terms of formulas – but it is *not* a textbook. Rather, it is intended as a *resource* and a *guide* for the many lateral entrants to the field of illumination optics as well as for those studying optics theory, optical design, and the function of optical elements or struggling to understand the details and inner workings of an optical design software. While the book does follow a story line (as laid out in Section 1.2), it is perfectly fine to scan the lists of exercises, knowledge items and simulations provided throughout the text and to jump back and forth between places of particular interest to you.

This book's primary goal is to let you learn by studying typical problems, solving them by yourself. We provide solutions to all exercises, but **we strongly encourage you to *not* peek ahead** to the solution before trying really hard to think through the problem yourself. What is printed in this book is mere *information*, but the ultimate goal is to build *knowledge*. Information is the score, but knowledge is the music [12]. It takes hard work, and time, to learn how to play.

On the way, we aim to expand your theoretical knowledge, to acquaint you with illumination laws and quantities, and to develop your intuitive feeling for units and numbers. You will expand your understanding of the laws by applying them repeatedly. After completing the exercises in this book, you will be able to deal with a wide variety of situations and to apply the appropriate fundamental laws with confidence.

Since this book is a practical guide and not a textbook, many basic laws are presented without proof (but with appropriate citations). We do not aim to replace a university course: we are simply showing the way toward solutions and designs, trying to adequately balance the level of detail to avoid overwhelming one half of the readers while boring the other half.

Designing good illumination optical systems requires creative thinking, juggling, combining and (rarely) inventing optical elements. But still quite early in the process, this creative thinking must be guided and fenced in by numbers that result from calculations. This book deals with the calculation part. In illumination optics, simple calculations can often be of great help to guide the optical designer. We aim to show you how. Always calculate before you embark on specific design work. If you fail to do this, you will likely end up with inefficient solutions (while not knowing they are inefficient) or failed attempts to violate fundamental laws (not knowing why you failed). Accordingly, this book also serves as a handbook of illumination *pre-design*.

Some books try to avoid formulas ("each formula will cut your number of readers in half", as the saying goes) and explain topics merely by pictures and descriptions. Such a presentation style, i.e. a style that invokes the readers' visualization capabilities, is very helpful to understand how things work *in principle*. However, solving specific engineering problems also requires skill and experience in using formulas and numbers. An engineer must be able to perform precise calculations, often with the help of computers. However, such calculations are always prone to errors.

To avoid such errors, we found two approaches extremely beneficial in our work as optical designers. First, it is useful to develop some skill in "keeping a mental tab" on results by using approximations, quick estimates and mental arithmetic to

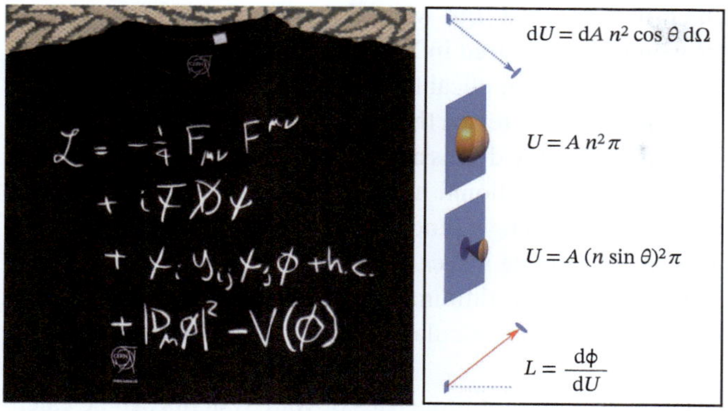

Figure 1.1 Left: Lagrange density with bosonic and fermionic parts and the Higgs field, on a T-shirt from CERN. Right: Important formulas for etendue and radiance.

judge whether a result can be correct. The most dangerous results are just a little too good to be true! Second, simulations may help. Ray tracing software just traces the propagation of rays, knowing nothing about our beautiful calculations, and this is exactly what we want: Confirmation or refutation by an independent referee.

This is the reason why this book introduces and discusses all topics with the help of many (really many!) calculations.[1]

To help you develop the "third eye" that comes with mental tab-keeping, we present many formulas that are simple enough to be used regularly (in contrast to, e.g. the world formula on the left side of Figure 1.1, which includes everything but is quite impossible to apply to daily work), but still profound enough to guide you in the world of illumination optics (e.g. the etendue and radiance formulas on the right side of Figure 1.1).

Talking about a "third eye": It is generally not easy to visualize the flow of light in an optical system just by looking at it from the outside. One approach for helping ourselves in this regard is to think of ourselves as being tiny – just a few μm –, but endowed with superpowers: We are able to fly, neither extreme temperatures nor immense radiation doses afflict us, we can swim through glass like it was water, and we have keen hyperspectral eyes with a sense for quantitatively correct radiometric values. Thus endowed, we take a mental virtual tour through the optical system and imagine what we see when the book, employing our "Lightman" character (see Section 1.4).

You will learn that many parameters of commercial light sources and luminaires can be calculated with some knowledge of the source at hand and a basic knowledge of the optics inside. *Vice versa*, you will learn how to interpret datasheet values, to fill the gaps in the datasheets, and to derive information about components and technologies.

Occasionally, we feel puzzled by little inconsistencies that are commonly glossed over. For example, we all (should) know that according to the celebrated inverse square law, irradiance E (flux per area) and intensity I (flux per solid angle) are related by $E = I/r^2$, where r is the distance of the irradiance sensor to a point-

---

1 As Henning's high school math teacher Dr. Schiemann put it: "Formeln sind klüger als Menschen" (Formulas are smarter than people).

like source, toward which it is oriented. But what about the units? I is measured in W/sr. Therefore, I/r^2 would be measured in W/m^2 sr (which would be the physical unit of radiance L – which is not applicable here), whereas the physical units of irradiance E must be W/m^2. We will reveal the mystery of the appearing and disappearing solid angle unit sr during the discussion of the *spherical excess* (Eq. 2.26), and we make it a somewhat unconventional habit to write, for example, 4π sr as opposed to $4r^2\pi$ when we mean solid angle (and not surface area!) of a sphere. Similar puzzles will be solved at appropriate places.

In summary: To become a proficient illumination optics designer, you will have to get your feet wet with formulas and calculations. Let's leave the beach and go swimming!

In illumination optics, many important properties of your systems can be calculated, and we show how to achieve this. Sometimes, you can analytically derive results such as a light distribution to an incredible degree of correspondence to the simulation (cf. Exercise 3.13 and Simulation 3.6). In other cases, you need the help of a spreadsheet or a Python/Matlab®/Mathematica® notebook.

Calculations are complemented by simulations – both are indispensable. We routinely use simple ray tracing simulations to test and (hopefully!) confirm our calculations (a habit we strongly recommend!), and we use ray tracing simulations of more complex systems in situations where calculations are too hard to apply (you will find many examples of this in the book, e.g. in Simulation 3.3). This book should help you to develop a feeling to distinguish those cases.

1.2 Structure

This book presents a colorful mixture of knowledge and formulas from various parts of optics that play a role in illumination. We begin by introducing some basics of optics and geometry that are crucial for proper handling of design tasks. We then introduce various aspects of phase space mathematics and etendue, fill phase space with light (radiometry), and extend the calculus to spectral concepts and colorimetry as far as needed for illumination problems.

At this point, you should be well prepared for the subsequent "practical" chapters. There, we consider ideal and real light sources, discuss lighting situations for various purposes, and evaluate how various quantities evolve as light propagates in optical systems. In the final chapter, we will demonstrate how to lay the groundwork for conceptual design, how to extract information from product sheets and how to select suitable light sources for various applications.

Sometimes, it is not obvious into which category (practice or design) a topic should fall. And since things are inevitably connected with each other, it is sometimes impossible to arrange topics in a linear sequence. In such cases we tried to ensure narrative continuity and we hope that our frequent cross-references may help.

1.2.1 Exercises and Solutions

Exercises play a central role in our approach. In our classroom and online courses, we found that exercises are crucial: Attendees are easily persuaded that something is so and so, but only when they apply the freshly presented information to actual problems, they convince themselves that it is indeed true. Their feedback told us time and again that exercises are particularly effective in converting information to knowledge, teaching us to transfer what we learned in an exercise to similar, but different, situations in real life. Accordingly, we greatly expand the selection of course exercises in this book.

Frequent insets about theory, formulas, simulation examples, and of course detailed solutions complement the exercise problems.

We teach technical content as a spin-off of the actual problem solving (see below for simulations and visits to the Knowledge Corner café). In many instances, we provide auxiliary materials, such as spreadsheet and optical simulation files for download, both with a clickable link for the electronic version of this book, and a corresponding QR code in the margin for using a mobile device with the paper version of this book.

Each exercise problem is described in an *Exercise* box:

Exercise 1.1: Snell's law applied to the mirage phenomenon

When looking along a hot summer road, we can sometimes see the sky above the horizon reflected by the road below, creating the illusion of water puddles; this phenomenon is known as a *mirage* (Figure 1.2).

Figure 1.2 Top: A Highway Mirage – the road appears wet, caused by a layer of hot air above the road surface that leads to a redirection of rays by total internal reflection (TIR). The slightly irregular density profile of the hot air layer causes the wavy distortion of the mirror image. *Source:* Yuri Khristich/Wikimedia Commons/Public Domain. Bottom: A simplified model of the hot air layer in a cross section along the driving direction, here represented by two homogeneous layers of air with a sharp boundary between them. The orange ray from the right is refracted into the hotter layer and eventually absorbed by the road; the green ray with a direction slightly closer to the road suffers total internal reflection at the same interface. The image shows rays with higher angles and layers with larger index differences than for real air; otherwise, the angles could not be discerned in this image.

(Continued)

Exercise 1.1: (Continued)

Assume that a thin planar layer of air above a perfectly horizontal hot road has a temperature of 60 °C, with a sudden sharp transition to a temperature of 30 °C above, and that the air has a refractive index of $n_{\text{air},0\,°\text{C}} = 1.000293$ at 0 °C. Further assume that the refractive index n of air depends on its density ϱ as $n^2 - 1 = \text{const} \cdot \varrho$, and that the air is an ideal gas ($\varrho = \text{const}/T$ with absolute temperature T).

Since the hotter air has a slightly lower refractive index than the colder air above, total internal reflection can occur at the interface between the layers at large incidence angles, causing the mirage phenomenon.

Compute the critical angle for total internal reflection.

Most exercises are immediately followed by their *Solution*, often followed by a comment or advice providing additional information.

Solution 1.1:

First, we calculate the densities of air at 30 °C and 60 °C relative to that at 0 °C = 273.15 K:

$$\varrho_{\text{rel, 30 °C}} = \frac{0 + 273.15}{30 + 273.15} = 0.9010$$

$$\varrho_{\text{rel, 60 °C,rel}} = \frac{0 + 273.15}{60 + 273.15} = 0.8199$$

Next, we calculate $n^2 - 1$ for all three temperatures:

$$n_{0\,°\text{C}} = 1.000293$$

$$n^2_{0\,°\text{C}} - 1 = 5.8609 \cdot 10^{-4}$$

$$n^2_{30\,°\text{C}} - 1 = (n^2_{0\,°\text{C}} - 1) \cdot \varrho_{\text{rel, 30 °C}} = 5.2809 \cdot 10^{-4}$$

$$n^2_{60\,°\text{C}} - 1 = (n^2_{0\,°\text{C}} - 1) \cdot \varrho_{\text{rel, 60 °C}} = 4.8053 \cdot 10^{-4}$$

These equations can be solved for the refractive indices:

$$n_{30\,°\text{C}} = \sqrt{1 + 5.2809 \cdot 10^{-4}} = 1.0002640$$

$$n_{60\,°\text{C}} = \sqrt{1 + 4.8053 \cdot 10^{-4}} = 1.0002402$$

The critical "mirage angle" is then:

$$\alpha_{\text{mirage}} = \arcsin\left(\frac{n_{60\,°\text{C}}}{n_{30\,°\text{C}}}\right) = 89.605°$$

You might as well get advice from experienced teachers (which will be introduced shortly, see Section 1.4) in special advice sections such as this one:

Dr. Wisenheimer:
I am aware that the authors did not introduce me and my cousins yet, but I still want to say: Try a web search for the term "mirage ship"! You will be presented a huge number of reports and photos of "hovering" ships.
And check out Les différents types de mirage.

1.2.2 Insights from the Knowledge Corner Café

In our daily professional lives, we keep encountering interesting questions to which we would like to find answers, such as "Where is this index of refraction coming from?" We just love to think such questions through, very often with help from the literature. Or, we might go to our favorite corner café to read a science book (yes, we are indeed nerdy enough to do that!) and discuss it with a friend. In this book, we call this place the "Knowledge Corner" science café (Figure 1.3).

From time to time, when such a question arises, we will invite you to follow us to the Knowledge Corner café for a discussion. There, we will introduce useful facts, theories – and occasionally trivia as well. Since this book is not a university course textbook, we do not even attempt to give full explanations or proofs; the Knowledge Corner conversations are intended to present information in a succinct, (hopefully) easy-to-understand way. We do, however, add many citations for further study.

Exercise 1.1, for example, uses a certain relation between density and refractive index that turns out to be quite useful for environmental tolerancing.

Figure 1.3 Our Knowledge Corner science café.

Knowledge Corner Topic 1.1: Lorenz–Lorentz equation

One way to understand the refractive index is by coherent scattering: The incoming electromagnetic disturbance makes polarizable atoms and molecules oscillate and therefore radiate with a phase delay. For refractive media such as glass or air, the superposition of all these dipole radiations with the incoming wave will result in a "drag", slowing the incoming wave down and thus shortening its wavelength [13]. Coherence ensures that in a homogeneous, isotropic, and linear medium (a so-called HIL medium), these dipole radiations have a combined effect only in the direction of propagation, while scattering to the side is suppressed by destructive interference. (Rayleigh scattering in air, which is responsible for the blue sky, is different because of the random density fluctuations in air that are absent in glass). If a material expands thermally, reducing its density but leaving the molecular structure unchanged, there will be fewer dipoles per volume and the refractive index n will decrease with density ϱ.

A detailed analysis [13] shows that

$$\frac{n^2 - 1}{n^2 + 2} = \text{const} \cdot \varrho \tag{1.1}$$

This relation is known as the Lorenz–Lorentz equation, named after Ludvig Lorenz and Hendrik Lorentz, who discovered it around 1870. For gases, the change of the numerator is negligible compared to the change of the denominator so that the relation $n^2 - 1 = \text{const} \cdot \varrho$ holds. For glasses and plastics, however, the full Lorenz–Lorentz relation from Eq. 1.1 should be used. The relation is valid for each wavelength individually. When the full dispersion curve $n(\lambda)$ is known for one temperature, the Lorenz–Lorentz equation gives us dispersion curves for a whole temperature range.

Why is this relation important for illumination optics? The reason is that particularly for plastics and silicones, the thermal expansion is so substantial that it cannot be neglected in many cases; one needs to take into account both the thermal shape change and the thermal refractive index change. However, while the refractive index at some standard wavelength and temperature is often known from datasheets, as well as the coefficient of thermal expansion (CTE), the refractive index as a function of temperature is rarely published. In such cases, the Lorenz–Lorentz equation is a good (and probably the only practical) way to calculate the refractive index at varying temperatures. However, beware of two pitfalls: (i) The CTE is sometimes listed as linear CTE and sometimes as volume CTE. We need the volume CTE here, which is three times the linear CTE (why three times? – Think about it!). (ii) The molecular polarizability changes at phase transitions – even phase transitions of higher order – such as, e.g., the glass transition of thermoplastics. This fact has been used to determine precise transition temperatures from refractive index measurements [14, 15]. Accordingly, you should make sure that no phase transition occurs in your temperature range: Apply the Lorenz–Lorentz equation only with great care when approaching, for example, the softening temperature of a thermoplastic.

Simulations often give additional insights beyond the mathematical results. For this mirage example, the sudden temperature change above the road is a crude, yet surprisingly valid model of the true, gradual temperature distribution. We show a gradient-index (GRIN) ray tracing simulation in a *Simulation* box, accompanied by further discussion and insights.

Simulation 1.1: GRIN ray trace

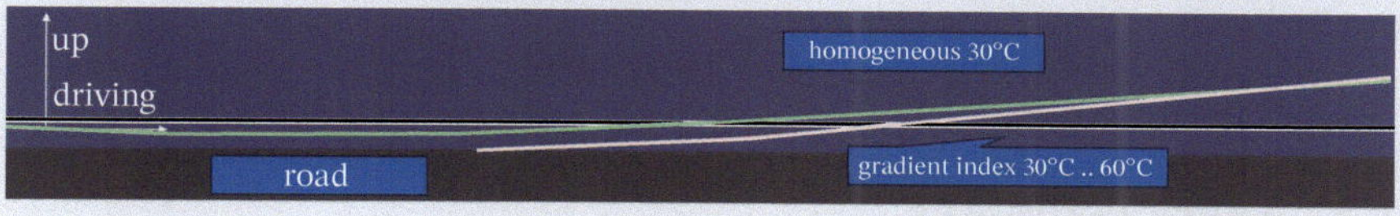

Figure 1.4 A similar simulation as in Figure 1.2, but now with a gradient-index layer. The same orange ray eventually hits the road; the same green ray enters into the GRIN layer, but returns eventually back up. The simulation file is provided here.

In reality, the air above the road should be described as a gradient-index (GRIN) material, since there is a temperature gradient, not a sudden temperature step. In Figure 1.4, the green ray enters the gradient-index section at a grazing incidence angle that would fulfil the condition for total internal reflection; it is bending its way back into the homogeneous air. The orange ray with a steeper incidence angle is also bending back, but not enough to avoid being absorbed (or scattered) by the road surface. To visualize the effect in this real ray trace with a linear GRIN profile, the rays are much steeper and the index gradient is much stronger than in the exercise; in the exercise situation, the rays travel quite long in the gradient layer before returning back, too long to show here.

Bonus question: As long as the temperature gradient and thus the index gradient is strictly vertical, so that you can think of the GRIN air as being composed of many thin planar parallel layers, but still with a minimum of 30 °C on top and a maximum of 60 °C at the bottom, do the details of the index profile influence whether the ray hits the road or is bending back into air, creating a mirage? The answer is no – the distribution details only influence the lateral shift and the precise path of bending, but not whether a ray hits the road or not, and also not its final angle back into homogeneous air. But why? To answer, apply Snell's law in the form of $n_i \sin\theta_i = n_{i+1} \sin\theta_{i+1}$ (Eq. 2.44) repeatedly for each interface. This result also tells you how the idea of using a multilayer coating to avoid TIR fails in terms of geometrical optics.

(The LightTools system file is available here.)

Besides a host of accompanying online materials, and eventually a list of errata, we provide supporting files (helpful or necessary for the solution) for download from our Illumination Optics website. We expect to add (hopefully much) additional material there in the coming months and years, beyond what we directly link to in this book. This is because sometimes it is adequate to use a spreadsheet or a Python notebook for numerical evaluation of a derived formula. We will also maintain a (hopefully short) list of errata there.

Finally, this book contains many useful links, which is adequate for an e-book, but also carries the risk of expiring content. We will try to add each link to the Wayback machine, to mitigate the expiring content issue at least to some degree. We apologize if some links may not be working anymore at a later date, and may provide updates on illuminationoptics.net.

We provide mainly LightTools® files for the simulations, apologizing to the users of ASAP®, CODE V®, FRED®, KostaCLOUD®, LucidShape®, OSLO®, Photopia®, Quadoa®, RayJack ONE®and other software packages. However, perhaps you accept the challenge and set the systems up your way.

For some simulations, you may need some ray files and download them yourself.

1.3 Quick Estimates

As we already mentioned in Section 1.1, we find it extremely useful to be able to keep a "mental tab" on results, to develop and nurture your ability to tell quickly if some quantitative statement may be correct or not. Again, the most dangerous results are just a little too good to be true, but still wrong. Quick estimates are the key. If possible, do them in your head, without a computer. This requires experience and mental arithmetic practice, but it allows you to follow the discussion, and it trains the "muscle".

Some quick estimates are useful, but dangerous – "dirty tricks" of the do-not-try-this-at-home variety: Rotational symmetry, for example, simplifies things greatly, but alas – square LED chips are not rotationally symmetric. As long as you know what you are doing, simply assuming rotational symmetry for the moment, and then taking "a quarter more" of the result will get you very close (actually, you need to multiply the result with the square-to-circle area ratio $4/\pi$, but $4/\pi \approx 1.273239 \approx 1.25 = 1 + 1/4$). We will tag such foolhardy approximations with a HANDLE WITH CARE sticker. While you develop such skills, it is a good idea to later simulate [16] the actual case and compare with your quick and dirty estimate.

Other quick estimates are beautifully simple, yet precise, for example the etendue of a telecentric, round beam, $U = A\,(n\,\sin\,\alpha)^2\pi$ sr (see Section 3.2). Many *specific* quick estimate formulas will appear in the exercises, solutions and knowledge boxes. Here in the introduction, we would like to introduce a few general quick estimate *techniques*.

Knowledge Corner Topic 1.2: Big *O* calculus

When dealing with the effects of small relative changes, big O calculus, also known as "big O notation", is very useful [17]. The idea is to approximate a complicated function $f(x)$ in a small region $x \approx x_0$ by a simple function $g\,(x)$ such that (i) $g\,(x_0) = f(x_0)$ and (ii) the approximation error $g\,(x) - f(x)$ vanishes faster than $x - x_0$. The small-angle approximation ($\alpha \approx \sin\,\alpha \approx \tan\,\alpha$, see Knowledge Corner Topic 2.3) is one example of this technique.

Another example: Recall the discussion of volume vs. linear coefficient of thermal expansion (CTE) in Knowledge Corner Topic 1.1, where we stated that the volume CTE_V equals three times the linear CTE_l. How did we know? Let us consider a cube at temperature T_0 with side lengths l_0 and thus volume $V_0 = l_0^3$, made of a material with known CTE_l. Then, by definition,

$$\begin{aligned} l(T_0 + \Delta T) &= l_0(1 + \Delta T \cdot \mathrm{CTE}_l) \\ V(T_0 + \Delta T) &= V_0(1 + \Delta T \cdot \mathrm{CTE}_V) \end{aligned} \tag{1.2}$$

By expanding $V = l^3$ and setting $\varepsilon = \Delta T \cdot \mathrm{CTE}_l$, we obtain

$$\begin{aligned} V(T_0 + \Delta T) &= (l(T_0 + \Delta T))^3 = l_0^3(1 + \varepsilon)^3 \\ &= V_0(1 + 3\varepsilon + 3\varepsilon^2 + \varepsilon^3) \\ &= V_0(1 + 3\varepsilon + O(\varepsilon^2)) \\ &\approx V_0(1 + 3\varepsilon) \end{aligned} \tag{1.3}$$

from which we see that $3\varepsilon = 3\,\mathrm{CTE}_l\Delta T = \mathrm{CTE}_V\Delta T$. Done!

More formally, "vanishing faster than ..." means to take a smooth function of a small quantity ε, to consider its Taylor expansion $\sum_{i=0}^{\infty} a_i\varepsilon^i$, and to ignore higher order terms, $O(n) = \sum_{i=n}^{\infty} a_i\varepsilon^i$, retaining only terms with $i < n$. In our case, we used $n = 2$, the most common choice, often called "linearizing around x_0."

Yet another example is

$$\sqrt{x_0 + \varepsilon} = \sqrt{x_0} + \frac{1}{2\sqrt{x_0}}\varepsilon - \frac{1}{8\sqrt{x_0^3}}\varepsilon^2 + O(\varepsilon^3) \tag{1.4}$$

The approximation

$$\sqrt{x_0 + \varepsilon} \approx \sqrt{x_0} + \frac{1}{2\sqrt{x_0}}\varepsilon \tag{1.5}$$

is extensively used in Knowledge Corner Topic 3.8.

Knowledge Corner Topic 1.3: Rounding, or how much precision is just right?

Most quick estimates are not entirely accurate. On the other hand, it is just as wrong to report a luminous flux of 12.46295 lm. To quote Jimmy Efird [18]:

> Rounding is a compromise strategy that involves replacing a true or more accurate value with one having less accuracy. The intent is to preserve the meaning and interpretation of the original result. It represents a balance between reporting too few significant digits and losing information versus failing to achieve parsimony (i.e. retaining too many inessential digits). In some cases, the numbers are rounded beyond the capacity of the

(Continued)

Knowledge Corner Topic 1.3: (Continued)

measurement device (spurious accuracy) or practical aspects of the problem at hand. The degree of rounding imposed in a situation depends on the desired accuracy and precision of the result. Additionally important is the need to minimize the error that may accumulate when performing complex computations on rounded values. Such sequential operations can lead to ill-conditioned and inaccurate findings.

In illumination optics, we mostly consider radiometric values such as flux, radiance, illuminance etc. It is extremely difficult, and often just not practical, to achieve a precision of 1 % in measurements! During production, LEDs are individually measured and then sorted into bins, such that LEDs with similar electro-optical properties are packed together. In a typical brightness bin, flux may vary by about 15 % from min to max. On top of this variation comes the measurement uncertainty during binning: In many of their LED data sheets, ams-OSRAM states: "Brightness values are measured ... with a tolerance of ±7 %". This uncertainty may appear large, but it is actually an achievement. Billions of LEDs must be binned each year, using many spectrometers, which are calibrated in a cascaded process against (ultimately!) a calibration standard lamp, which itself comes from national standardization institutes with uncertainties of about 1 %. And then, LED flux varies with current and temperature, neither of which are perfectly controlled in most applications.

Similar considerations apply to other light sources. We have seen a reflectorized Xenon discharge lamp, specified with "aperture lumens min. 1400 lm, max. 2300 lm".

In our experience, reporting radiometric values and etendue values with 1 % accuracy, wavelength values with 1 nm accuracy, and CIE x/y color coordinates with 0.001 accuracy is a good compromise: more precise than what can be measured, but allowing just enough additional precision for complex computations performed on rounded values.

1.4 Introducing the Wisenheimers

We have enlisted the assistance of four members of the international Wisenheimer family clan of scientists (Figure 1.5): Dr. Mary Wisenheimer, Dr. Linda Wisenheimer, Dr. John Wisenheimer, and Dr. Thomas Wisenheimer (left to right). (You have already experienced Dr. Mary Wisenheimer commenting on mirage images.) Their comments are always correct and helpful, but sometimes a little bit over the top, we're afraid.

From time to time, three other special characters will chime in as well (Figure 1.6):

First, we introduce Rube, a very kind but slightly confused inventor, kind of a Gyro Gearloose character. He will introduce what he thinks are bright ideas, but he lacks some respect for the fundamentals and gets carried away quite easily.

Figure 1.5 The Wisenheimers giving unsolicited advice.

Figure 1.6 Rube, the mad inventor, Ike, the stern boss, and the Lightman (from left to right).

Then there is Ike, a stern boss ruling the engineering department and single-handedly negotiating specifications with customers, while being oblivious to both the needs of his staff and the laws of nature. Similarities to a certain pointy-haired comic character are of course purely coincidental.

And finally, we sometimes call the Lightman for help when we don't understand how the light flows through a particular optic system. Lightman is really small, but he has superpowers: He can fly, even the highest temperatures and radiation doses do not affect him, he can swim through glass like a fish through through water, and he has keen hyperspectral eyes with a sense for quantitatively correct radiometric values.

Characters were produced by the authors using Stable Diffusion text to image AI.

References

1 Muschaweck, J. and Rehn, H. (2022). *Designing Illumination Optics*. Bellingham, WA: SPIE Press.

2 Chaves, J. (2016). *Introduction to Nonimaging Optics*, 2e. CRC Press.

3 Kidger, M.J. (2002). *Fundamental Optical Design*. Bellingham, WA: SPIE Press.

4 Kidger, M.J. (2013). *Intermediate Optical Design*. Bellingham, WA: SPIE Press.

5 Arecchi, A.V., Koshel, R.J., and Messadi, T. (2007). *Field Guide to Illumination*. Bellingham, WA: SPIE Press.

6 Koshel, R.J. (2013). *Illumination Engineering: Design with Nonimaging Optics*. Hoboken, NJ: Wiley–IEEE Press.

7 Bartenbach, C. (2021). *Licht: Meine Erkenntnisse*. Basel: Birkhäuser.

8 Czapski, S. (1904). *Grundzüge der Theorie der Optischen Instrumente nach Abbe*. Leipzig: Johann Ambrosius Barth.

9 Siew, R. (2024). *Modern Classical Optical System Design: Fundamentals, Techniques, Tips, and Tricks*. IOP Publishing.

10 Vreman, H.J., Wong, R.J., and Stevenson, D.K. (2004). Phototherapy: current methods and future directions. *Seminars in Perinatology* 28 (5): 326–333. https://doi.org/10.1053/j.semperi.2004.09.003.

11 Sloper, C. (2013). *The LED Grow Book: Better, Easier, Less Watts*. CreateSpace Independent Publ. Platform, S.l.

12 Barreca, G. (2021). What's the difference between knowledge and information? https://www.psychologytoday.com/ca/blog/snow-white-doesnt-live-here-anymore/202111/whats-the-difference-between-knowledge-and.

13 Born, M. and Wolf, E. (1999). *Principles of Optics*, 7e. Cambridge University Press.

14 Kasarova, S.N., Sultanova, N.G., and Nikolov, I.D. (2010). Temperature dependence of refractive characteristics of optical plastics. *Journal of Physics: Conference Series* 253: 012028. https://doi.org/10.1088/1742-6596/253/1/012028.

15 Michel, P., Dugas, J., Cariou, J.M. et al. (1986). Thermal variations of refractive index of PMMA, polystyrene, and poly (4-methyl-1 -pentene). *Journal of Macromolecular Science, Part B* 25 (4): 379–394. https://doi.org/10.1080/00222348608248046.

16 Rehn, H. and Muschaweck, J. (2020). Étendue estimation for a non-trivial geometry. In: *DGaO Proceedings*, Vol. 121. https://www.dgao-proceedings.de/download/121/121_b33.pdf.

17 Knuth, D. (1998). Teach calculus with big O. *Notices of the American Mathematical Society* 45 (6): 687.

18 Efird, J.T. (2021). Goldilocks rounding: achieving balance between accuracy and parsimony in the reporting of relative effect estimates. *Cancer Informatics* 20: 1176935120985132. https://doi.org/10.1177/1176935120985132.

2

Warm Up

To enable you to operate successfully in the world of illumination optics, we will need to lay some groundwork first. We start with geometry, introducing solid angles and some calculus around them. Then, we will practice Snell's law and basic imaging so that you will have such knowledge readily available for later illumination work. Finally, we will have a closer look at the optics of conic sections (and their beauty), which are widely used in illumination design and will appear in numerous exercises later.

2.1 Angles and Solid Angles

Knowledge Corner Topic 2.1: Angles

Plane angles are used in two different contexts in illumination optics.

First, *rotation*: A 30° incidence angle of a ray on a surface means that you would need to rotate the local surface normal by 30° within the plane of incidence to get the ray's direction (Figure 2.1). In optics, an angle of incidence (AOI) on a surface is measured with respect to the surface normal.

Second, a *contiguous range* of directions: A 30° beam means that there are two edge directions with an angle of 30° between them, and that the set of directions in between defines the angular range of that beam.

Plane angles are measured in degrees (°) or radians (rad), defined by an arc length s on a circle of radius r, or, equivalently, just the arc length when using a unit circle with radius 1 (Figure 2.1):

$$\theta = \frac{s}{r}\ \text{rad} = \frac{s}{r} \cdot \frac{360°}{2\pi} \tag{2.1}$$

(Continued)

Engineering Illumination Optics: Case Studies and Exercises, First Edition. Henning Rehn and Julius Muschaweck.
Published 2026 by Wiley-VCH GmbH.

Knowledge Corner Topic 2.1: (Continued)

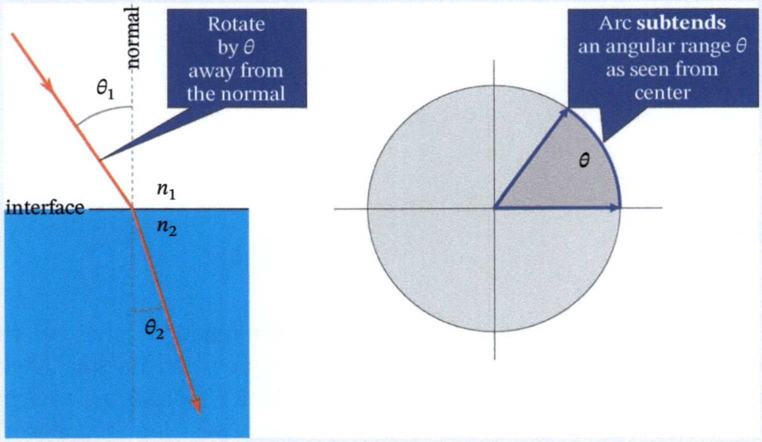

Figure 2.1 Plane angle definition by arc length and radius.

While the radian is the "natural" unit of an angle in a mathematical sense, the degree, representing 1/360 of a full circle, has been in use since thousands of years, probably dating back to Babylonian times. One advantage of the degree scale is that 360 is divisible by so many numbers (2, 3, 4, 5, 6, 8, 9, 10, 12, 15, 18, 20, 24, 30, 36, 40, 45, 60, 72, 90, 120, 180), so that almost all commonly used angles are described by integer values in degrees without rounding. Both ° and rad are here to stay. Exercise 2.1 helps you to get familiar with both.

Exercise 2.1: A table of common angles

As discussed in Section 1.3, it is almost a superpower to be able to quickly estimate quantities while thinking through or discussing problems. To this end, create a table with important, commonly occurring angles, both in degrees and radians, together with their sines and their cosines. Commit all values to memory!

Solution 2.1

We think that these angles are most important, and we know them by heart:

°	1	≈ 5.7	30	45	60	90	180
rad	≈ 0.017	0.1	$\frac{\pi}{6} \approx 0.52$	$\frac{\pi}{4} \approx 0.79$	$\frac{\pi}{6} \approx 1.05$	$\frac{\pi}{2} \approx 1.57$	π
sin	≈ 0.017	≈ 0.1	$\frac{\sqrt{1}}{2} = 0.5$	$\frac{\sqrt{2}}{2} \approx 0.71$	$\frac{\sqrt{3}}{2} \approx 0.87$	1	0
cos	0.99985	0.995	$\frac{\sqrt{3}}{2} \approx 0.87$	$\frac{\sqrt{2}}{2} \approx 0.71$	$\frac{\sqrt{1}}{2} = 0.5$	0	−1

Knowledge Corner Topic 2.2: Trigonometric identities

We summarize a few trigonometric identities that are used in this book.

Of course, we need the basic identity

$$\sin^2\theta + \cos^2\theta = 1 \tag{2.2}$$

for example in Exercise 4.12. We also use

$$\cos 2\theta = \cos^2\theta - \sin^2\theta \tag{2.3}$$

hence

$$1 - \cos\theta = 1 - \cos^2\frac{\theta}{2} + \sin^2\frac{\theta}{2} = 2\sin^2\frac{\theta}{2} \tag{2.4}$$

in Eq. 2.11.

Equation 3.50 uses the following expression for the sine by the tangent:

$$\sin^2\theta = \frac{\tan^2\theta}{1 + \tan^2\theta} \tag{2.5}$$

In Section 2.5.1, we use

$$\tan\frac{\theta}{2} = \frac{\sin\theta}{1 + \cos\theta} \tag{2.6}$$

For a complete presentation of such trigonometric identities we refer to an online summary.

Knowledge Corner Topic 2.3: Small-angle approximation

Both linear and solid angles play a central role in illumination optics. Unfortunately, they mostly appear in our equations as sines, cosines and tangents, the evaluation of which typically requires an electronic device. In imaging optics, the *paraxial approximation* is widely used to avoid trigonometric functions, to allow ray transfer matrix calculations to describe multi-lens systems, and to analyze imaging parameters like focal length and $F/\#$. In illumination optics, numerical apertures are often larger, but on the other hand, less accuracy is required. Therefore, the small-angle approximation given by

$$\sin(\theta) \approx \tan(\theta) \approx \theta, \qquad \cos(\theta) \approx 1 - \frac{\theta^2}{2} \approx 1 \tag{2.7}$$

can be used up to surprisingly large angles. We will need it for paraxial optics (Knowledge Corner Topic 2.10).

(Continued)

Knowledge Corner Topic 2.3: (Continued)

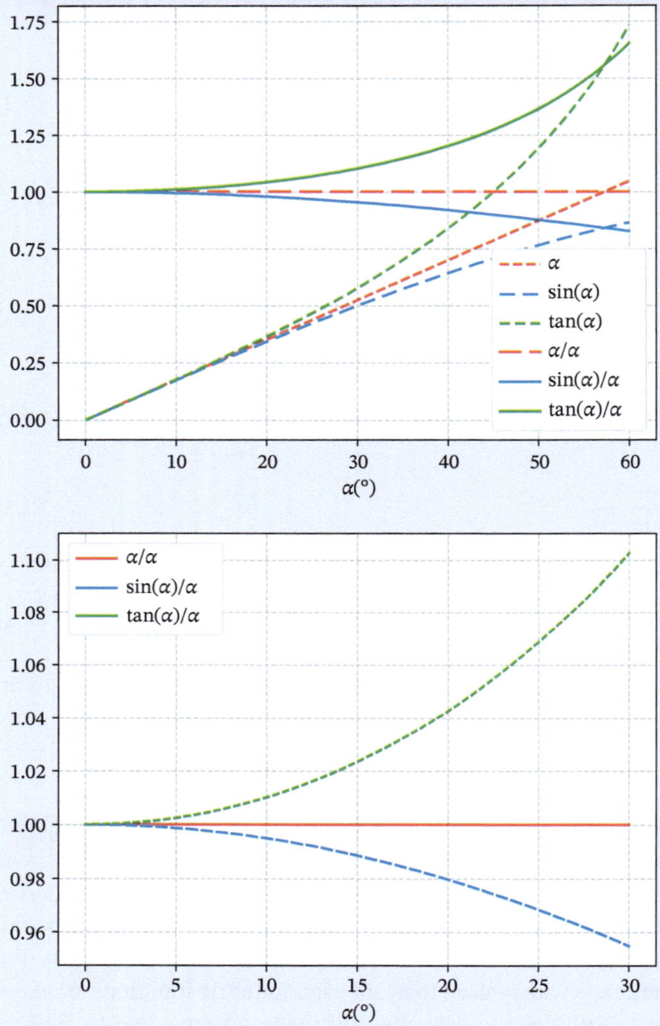

Figure 2.2 Illustrating the small angle approximation.

Figure 2.2 shows angles along with their sines and tangents as dashed lines and their ratios as solid lines ("what is my relative error if ...";) at the top. At the bottom, it shows zoomed-in ratios for small values of θ, demonstrating that up to 20° for the tangent and up to 30° for the sine, the error introduced by the small-angle approximation is only a few percent, good enough for most quick estimates. Similar considerations apply to the solid angle, see Knowledge Corner Topic 2.4. See also Knowledge Corner Topic 1.2 about how the small-angle approximation is just one example of big O notation.

Dr. Wisenheimer:
When using the approximation from Eq. 2.7, always use the unit radians (rad) for the angle!

Knowledge Corner Topic 2.4: Solid angle and projected solid angle

While plane angles play dual roles (rotation vs. direction range), *solid angles* are only used to measure an extended range of directions in 3D space. A point on a sphere defines a direction from the sphere's center. Thus, an area on a sphere defines a range of directions (Figure 2.3). Analogous to the planar angle being defined by an arc length s on a circle with radius r, the solid angle Ω defined by some area A on a sphere with radius r is measured by

$$\Omega = \frac{A}{r^2}\ \text{sr} \tag{2.8}$$

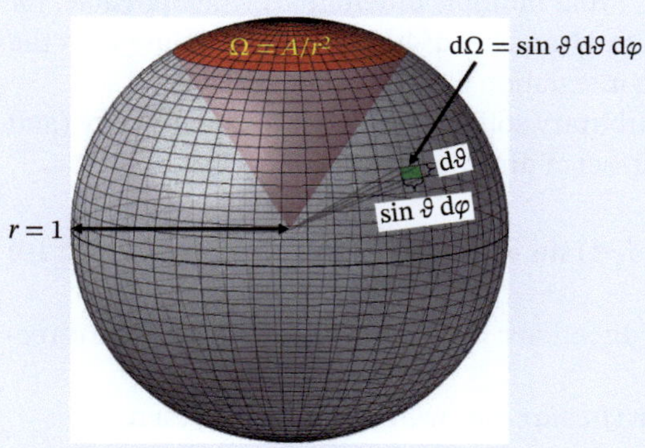

Figure 2.3 Visualization of solid angle Ω (in this case, that of a circular cone) and its differential element.

The unit of solid angle is the *steradian* (abbreviated as sr); there is no such thing as a "solid angle degree". The radian, degree and steradian are special: They are dimensionless units. While this may sound like an oxymoron, angular units are very useful to define densities: An amount of X per steradian tells us how directionally concentrated some quantity X is. Prime examples, discussed in detail in Section 3.4, are, of course, radiant intensity $I = \Phi/\Omega$, measured in W/sr, and luminous intensity $I_v = \Phi_v/\Omega$, measured in cd = lm/sr.

Carrying the dimensionless angular units with us, we can check consistency in our equations – not just in optics, but in science and engineering in general [1]. After all, a cd is quite different from a lm, and a solid angle of π sr is quite different from a plane angle of π rad. An extreme example is presented in Eq. 2.25 which may be hard to understand without explicitly mentioning sr and rad.

(Continued)

Knowledge Corner Topic 2.4: (Continued)

Frequently, the area that defines a solid angle is a spherical cap, i.e. the solid angle describes a circular cone of directions. (For a common counterexample, see Exercise 2.5). A ±30° beam then means that any ray that has been rotated away from a central direction by an angle of ≤30° belongs to that conical beam. In contrast, a 60° FWHM (full width at half maximum) beam means that you consider a ray bundle whose luminous or radiant intensity distribution is concentrated around a central direction such that directions with half the peak intensity or more form a cone with full tip angle of 60°. These two definitions of the same beam width are easily confused; the $\pm\alpha$ notation lends itself to etendue calculations, where the sine of the half angle plays a prominent role, while β FWHM is common in the general lighting world, in stage lighting, and sometimes in lasers. We strive, in speaking and in writing, to always add "plus-minus", "half angle", or "FWHM" to avoid this trap. See also Knowledge Corner Topic 3.15.

The differential solid angle element (see Figure 2.3) is given in spherical coordinates (ϑ, φ) by

$$\mathrm{d}\Omega = \mathrm{d}\vartheta\,(\sin\vartheta\,\mathrm{d}\varphi) = \sin\vartheta\,\mathrm{d}\vartheta\,\mathrm{d}\varphi \tag{2.9}$$

(Throughout the book, we will avoid multiple differentials when possible; for simplicity and readability, we write $\mathrm{d}\Phi$ and not $\mathrm{d}^2\Phi$. In integrals, we use ϑ for the integration variable and θ for the integration limit of polar angles.)

In the most general case, an arbitrary solid angle, given by an arbitrary (and not necessarily simply connected) set of directions is given by

$$\Omega = \int \mathrm{d}\Omega = \int_{\varphi=0}^{2\pi}\int_{\vartheta=0}^{\pi} \Gamma(\vartheta, \varphi)\sin\vartheta\,\mathrm{d}\vartheta\,\mathrm{d}\varphi \tag{2.10}$$

where Γ defines the set of desired directions: $\Gamma(\vartheta, \varphi) = 1$ if (ϑ, φ) is a desired direction, and zero otherwise.

The solid angle of the common *circular cone with half angle* θ is therefore

$$\Omega = \int_{\varphi=0}^{2\pi}\int_{\vartheta=0}^{\theta} \sin\vartheta\,\mathrm{d}\vartheta\,\mathrm{d}\varphi = 2\pi\,\mathrm{sr}\cdot(1-\cos\theta) = 4\pi\,\mathrm{sr}\cdot\sin^2\frac{\theta}{2} \tag{2.11}$$

Projected Solid Angle

Later, in Section 3.2, we will introduce and discuss etendue as a geometrical quantity of fundamental importance in illumination optics. Etendue measures the extent of a beam, considering both spatial and angular aspects. As it turns out, it is not the solid angle, but a related quantity known as the *projected solid angle* (PSA for short), that is needed to define the angular extent of a beam. We define PSA by

$$\Omega_\mathrm{p} = \int \cos\vartheta\,\mathrm{d}\Omega \tag{2.12}$$

The $\cos\vartheta$ term is the obliquity factor (the scalar product between the directions of the optical axis and the direction toward the differential $\mathrm{d}\vartheta$, cf. Section 3.2).

Geometrically, applying the obliquity factor means to project each infinitesimal solid angle element on a circular disk through the sphere's center (Figure 2.4).

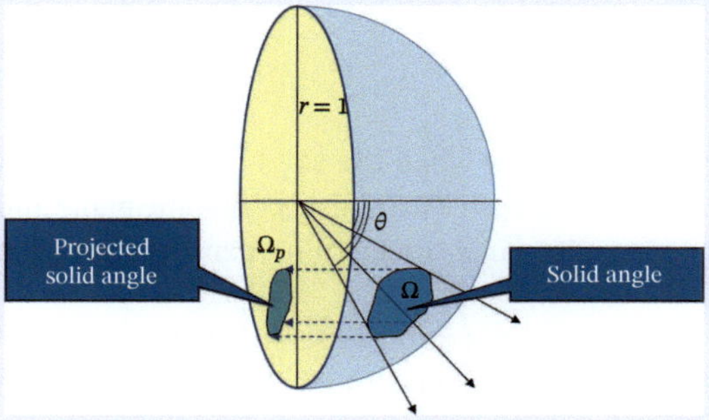

Figure 2.4 The projected solid angle is obtained by projecting the corresponding area on the unit sphere on to a circular disk.

Hence, the PSA of a centered circular cone with half opening angle θ is (Figure 2.5)

$$\Omega_{\text{p, centered cone}} = \int_{\vartheta=0}^{\theta} \int_{\varphi=0}^{2\pi} \sin\vartheta \cos\vartheta \, \mathrm{d}\vartheta \, \mathrm{d}\varphi = \pi \, \text{sr} \sin^2\theta \tag{2.13}$$

This relation will us allow to derive Eq. 3.10, which – in our opinion – is probably the single most important etendue formula.

The differential projected solid angle element is

$$\mathrm{d}\Omega_{\text{p}} = \sin\vartheta \cos\vartheta \, \mathrm{d}\vartheta \, \mathrm{d}\varphi = \cos\vartheta \, \mathrm{d}\Omega \tag{2.14}$$

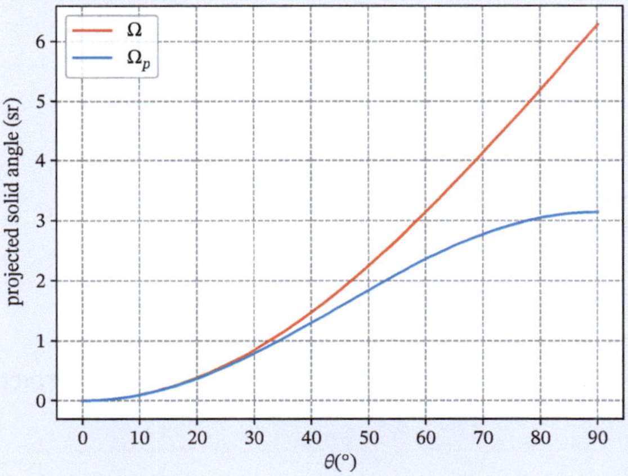

Figure 2.5 Solid angle and PSA (in sr) of a cone as function of half cone angle θ.

What happens when the $\pm\theta$ circular cone is not centered, but tilted by an angle α? Surprisingly, the same cosine factor as for the differential solid angle

(Continued)

Knowledge Corner Topic 2.4: (Continued)

element (Eq. 2.14) applies to the whole cone! (Thanks to Bernhard Michel for pointing this out to us!)

$$\Omega_{\text{p,tiltedcone}} = \pi\text{sr}\ \sin^2\theta\ \cos\alpha \tag{2.15}$$

More generally, the same $\cos\alpha$ factor applies when the solid angle itself is not cone-shaped, but only possesses mirror symmetry with respect to some plane and is rotated away from the axis by an angle α in a direction perpendicular to that plane.

As for the small-angle approximation, there is no big difference between a solid angle and a projected solid angle for small cone angles (3% at 20°, 7% at 30° half angle), but for the full hemisphere, the error amounts to a factor of two (Figure 2.6).

For small cone angles, we can use the simple approximation (θ in radians)

$$\Omega \approx \Omega_{\text{p}} \approx \theta^2\pi\text{sr} \tag{2.16}$$

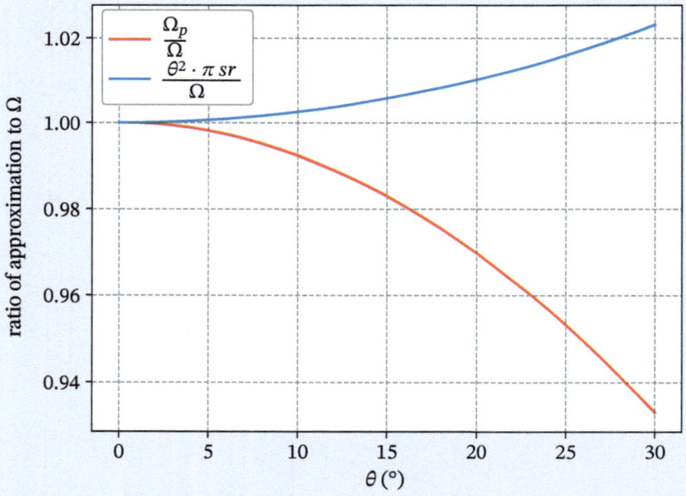

Figure 2.6 Solid angle vs. PSA approximation errors for a cone with a half angle θ.

In Knowledge Corner Topic 2.9, we introduce a closely related quantity termed the *angular extent*.

In illumination optics, the solid angle is closely related to *luminous intensity* $I_{\text{v}} = \text{d}\Phi_{\text{v}}/\text{d}\Omega$ (see Section 3.4), defined as flux per solid angle, emitted by a small, point-like source. For intensity, only the angular distribution is considered; the spatial distribution across the source aperture is disregarded.

Exercise 2.2: Solid angle

What is the cone half angle corresponding to a solid angle and to a projected solid angle of 1 sr each?

Solution 2.2

With $\Omega = 2\pi\,\mathrm{sr}\,(1 - \cos\theta)$ from Eq. 2.11 and $\Omega = 1$ sr, we solve for θ:

$$\theta = \arccos\left(1 - \frac{1\,\mathrm{sr}}{2\pi\,\mathrm{sr}}\right) \approx 0.572\,\mathrm{rad} \approx 32.8° \tag{2.17}$$

With $\Omega_\mathrm{p} = \pi\,\mathrm{sr}\sin^2\theta$ from Eq. 2.13, we find for $\Omega_\mathrm{p} = 1$ sr that

$$\theta = \arcsin\sqrt{\frac{1\,\mathrm{sr}}{\pi\,\mathrm{sr}}} \approx 0.599\,\mathrm{rad} \approx 34.3° \tag{2.18}$$

confirming that below 30°, there is no big difference between Ω and Ω_p (cf. Figures 2.5 and 2.6).

Exercise 2.3: Observing a sphere

You observe a sphere (radius R, center distance D, see Figure 2.7). Calculate its projected solid angle. (As we shall see in Section 3.4, we need the PSA subtended by a homogeneous source to compute local irradiance.)

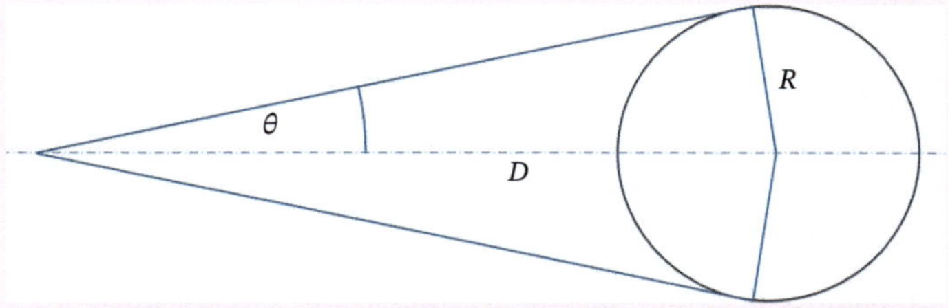

Figure 2.7 Projected solid angle of a sphere.

Solution 2.3

Equation 2.13 yields

$$\Omega_\mathrm{p} = \pi\,\mathrm{sr}\,\sin^2\theta = \pi\,\mathrm{sr}\cdot\frac{R^2}{D^2} \tag{2.19}$$

Combining Eqs. (2.15) and (2.19) allows to compute the PSA under which a reflector area element from Exercise 4.3 would see the bulb.

Exercise 2.4: Incident cone

A receiver accepts radiation up to a half angle of θ_0 from its normal direction (e.g., a fiber with corresponding NA). Now, a beam of homogeneous luminance within half cone angle θ_1 (zero luminance outside) arrives in, whose chief ray is tilted by an angle γ to the normal (e.g., when fibers in a coupler were tilted with respect to each other, see Figure 2.8).

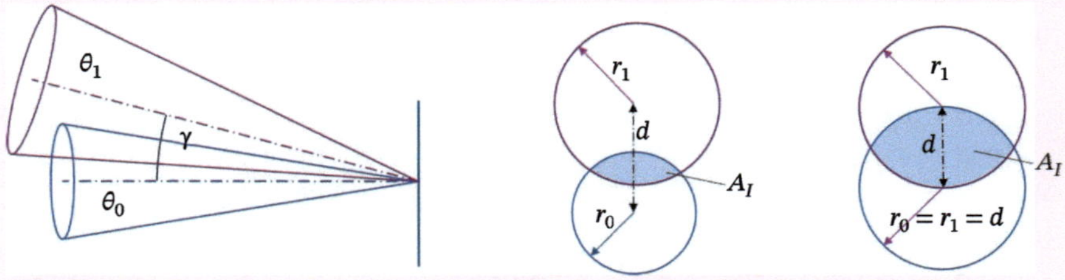

(a) Overlapping cones (b) Overlapping circles (c) $r_0 = r_1 = d$

Figure 2.8 Overlapping cones and circles. (a) The actual situation in 3D. (b) General overlapping cones. (c) The situation $r_0 = r_1 = d$ to be analyzed in this exercise.

Estimate the collection efficiency for the case $\theta_0 = \theta_1 = \gamma$ (Figure 2.8c), assuming homogeneous luminance!

Solution 2.4

We borrow an approximate solution HANDLE WITH CARE from the Circle–Circle Intersection on a plane (Figure 2.8b):

$$A_\mathrm{I} = r_0^2 \arccos\left(\frac{d^2 + r_0^2 - r_1^2}{2\,d\,r_0}\right) + r_1^2 \arccos\left(\frac{d^2 - r_0^2 + r_1^2}{2\,d\,r_1}\right) \\ -\frac{1}{2}\sqrt{(-d + r_0 + r_1)(d + r_0 - r_1)(d - r_0 + r_1)(d + r_0 + r_1)} \tag{2.20}$$

(radii r_0, r_1, center distance d, assuming actually intersecting circles, since other cases such as separate circles or one circle completely inside the other are trivial).

In the task, the circles are situated on a sphere representing cones, but we simply replace radii by angles as if they were situated on a plane (small-angle approximation):

$$r_0 = \tan\theta_0 \tag{2.21}$$

$$r_1 = \tan\theta_1 \tag{2.22}$$

$$d = \tan\gamma \tag{2.23}$$

For $r_0 = r_1 = d$ we get an intersection area of

$$A_\mathrm{I} = 2r_0^2 \arccos\left(\frac{1}{2}\right) - \frac{\sqrt{3}}{2}r_0^2 = r_0^2\left(\frac{2\pi}{3} - \frac{\sqrt{3}}{2}\right) \tag{2.24}$$

The overlap of the circles and thus the cones is

$$\eta = \frac{\Omega_\mathrm{I}}{\Omega_0} = \frac{A_\mathrm{I}}{\pi r_0^2} = \frac{2}{3} - \frac{\sqrt{3}}{2\pi} \approx 40\% \tag{2.25}$$

It would have been better to use projected solid angles for this analysis, but our solution is based on a crude approximation anyway.

When the angles are more complicated or become too large for this approximation to be applicable, we would set up the problem in ray tracing simulation software, effectively computing the overlap by Monte Carlo integration. For a LightTools® example, see OverlappingCircles.1.lts.

Knowledge Corner Topic 2.5: Spherical excess

What follows is an advanced topic in solid angle calculus, a helpful tool, and some sort of mathematical magic.

Curvature is a fascinating notion in differential geometry. Intuitively, when you consider points on a surface and the normal vectors at these points, then curvature is related to how the normal vector "wiggles" as you move from point to point. If there is no change of normal vector direction, you are (locally) on a plane. But can you measure curvature without surface normals, by "intrinsic" quantities only, i.e. quantities that do not require embedding the surface into 3D space? The answer is yes – you can. Without going into details (for a great, geometry-guided introduction, see [2]), we introduce an intrinsic-only theorem of ancient origin (Hariot and Euler) that is related to solid angle: The solid angle of a triangle on a sphere (Figure 2.9) is equal to the *spherical excess* (sum of the angles beyond 180°):

$$\Omega\frac{\mathrm{rad}}{\mathrm{sr}} = \alpha + \beta + \gamma - \pi\mathrm{rad} \tag{2.26}$$

(Mathematicians tend to omit the units, which may lead to confusion: It is perfectly admissible to take the sine of $\alpha + \beta + \gamma - \pi$ rad, but not of Ω!)

In references [3–6], this result is used to derive several simple vector formulas to compute the solid angle subtended by a triangle, e.g.

$$\Omega = 2\ \arctan\left(\frac{\mathbf{a}\cdot(\mathbf{b}\times\mathbf{c})}{1+\mathbf{b}\cdot\mathbf{c}+\mathbf{c}\cdot\mathbf{a}+\mathbf{a}\cdot\mathbf{b}}\right)\frac{\mathrm{sr}}{\mathrm{rad}} \tag{2.27}$$

which, in turn, allows easy computation of the solid angle subtended by an object that can be well approximated by a small number of triangles, not necessarily coplanar.

(Continued)

Knowledge Corner Topic 2.5: (Continued)

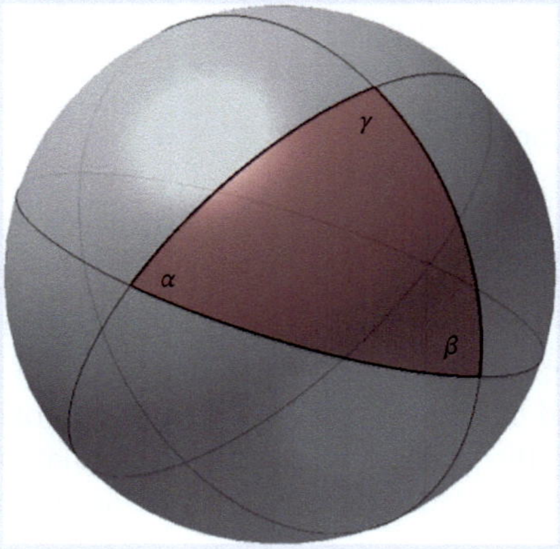

Figure 2.9 A spherical triangle, where each side is part of a great circle. The solid angle subtended by an arbitrary plane triangle as seen from an arbitrary point in space is always such a spherical triangle.

Knowledge Corner Topic 2.6: Rectangular limits

We consider the particularly annoying case of the rectangular far field ("annoying" since it is much harder than it seems, "particularly" since it is so common). The task is to calculate the solid angles of the two areas depicted in red and blue, respectively, in Figure 2.10.

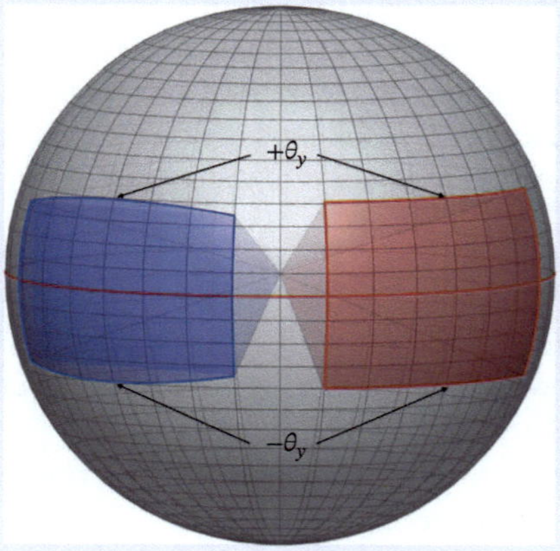

Figure 2.10 Rectangular solid angle limits. On the left, the blue area shows the solid angle subtended by a distant planar rectangle. All boundaries are great circles (why?); therefore the spherical excess formula, Eq. 2.26, could be readily applied (the corner angles are all >90°). On the right, the red area shows the solid angle for all directions with $-\theta_x < \varphi < +\theta_x$ and $-\theta_y < \vartheta < +\theta_y$. Here, the red horizontal lines at $\vartheta = \pm\theta_y$ are *not* great circles, and their projection to a distant plane yields quasi-hyperbolic curves.

For the "blue" situation (solid angle subtended by a distant rectangle, centered at $\theta = \varphi = 0$), we have [7] (a derivation, and an extension to non-centered rectangles is given in [8])

$$\Omega = 4 \arctan \frac{\tan\theta_x \tan\theta_y}{1 + \tan^2\theta_x + \tan^2\theta_y} \cdot \frac{\text{sr}}{\text{rad}} \tag{2.28}$$

via the spherical excess (Knowledge Corner Topic 2.5).

For the "red" situation (longitude/latitude rectangle) we find by integration

$$\Omega = \int_{\vartheta=-\theta_y}^{\theta_y} \int_{\varphi=-\theta_x}^{\theta_x} \sin(\pi/2 + \vartheta)\, d\vartheta\, d\varphi = 4\theta_x \cdot \sin\theta_y \frac{\text{sr}}{\text{rad}} \tag{2.29}$$

This is all very nice, but in real life – under pressure, when discussing whether a goal can be reached or not – there may be no time for the beauty of spherical trigonometry.

So let us estimate both situations in (slightly) dirty ways (Figure 2.11).

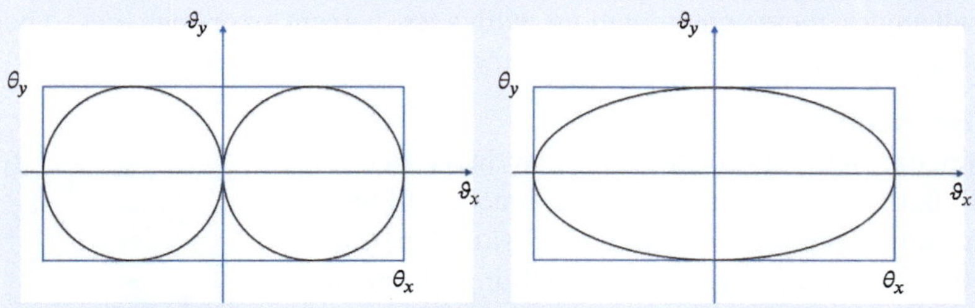

Figure 2.11 Approximations for rectangular solid angle limits.

In case of a 2:1 aspect ratio we may approximate the rectangle by two circles as their solid angle is easily estimated using Eq. 2.11 and correct the results by a factor of $4/\pi$ to get to the squares.

$$\Omega = 2 \cdot 2\pi\,\text{sr}\,(1 - \cos\theta_y) \cdot \frac{4}{\pi} \tag{2.30}$$

We can also use the projected solid angle Eq. 2.12 for the solid angle (as we have small angles, we may do this without introducing big deviations) and modify it:

$$\Omega \approx \pi\,\text{sr} \cdot \sin\theta_x \sin\theta_y \cdot \frac{4}{\pi} \tag{2.31}$$

(the approach is later used in Eq. 5.98).

Exercise 2.5: Rectangular limits

For rectangular limits of $(\pm 5°) \cdot (\pm 10°)$ – which may stem from a problem in the automotive or search light sector – calculate the solid angle using the exact equations and compare! Do the approximations work reasonably well?

Solution 2.5

The exact results obtained from Eq. 2.28 are

$$\Omega = 0.0594\,\text{sr} \tag{2.32}$$

for the "blue" situation and

$$\Omega = 0.0608\,\text{sr} \tag{2.33}$$

for the "red" one – a little larger than "blue", as already suspected from Figure 2.10.

The two-circle approximation yields

$$\Omega = 0.0609\,\text{sr} \tag{2.34}$$

and the ellipse

$$\Omega = 0.0605\,\text{sr} \tag{2.35}$$

For small angles, these approximations, with a few percent error, may indeed be good enough. We provide a file for the numerics.

Dr. Wisenheimer:

In summary, these approximations may indeed be good enough. Both situations are relevant in practice. Illuminating a distant, planar rectangle (the "blue" situation) is ubiquitous (e.g. video projection, museum lighting, illuminating the field of view of a CMOS camera). Rectangular angular limits (the "red" situation), on the other hand, appear in the SAE (Society of Automotive Engineers) automotive lighting standards, and in IES (Illuminating Engineering Society) type A and type B photometry [9].

2.2 Refraction

Knowledge Corner Topic 2.7: Snell's law and TIR

We assume that you are familiar with Snell's law of refraction,

$$n \sin\theta = \text{const} \tag{2.36}$$

across an interface, with incidence angle θ and refractive index n (Figure 2.12). Still, we think it is worthwhile to add some information you may not know yet.

The law has been discovered and rediscovered several times. Willebrord Snellius, the Dutch astronomer, indeed discovered the law, but it should be rightly attributed to Ibn Sahl, a Persian scientist, who found it about 600 years earlier [10].

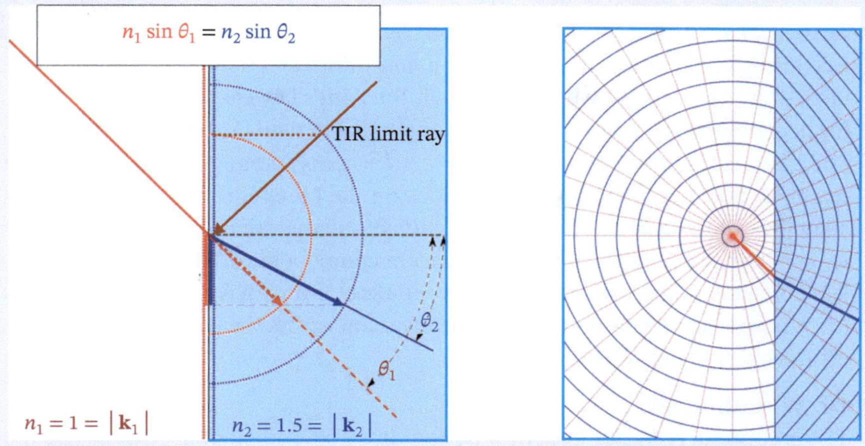

Figure 2.12 Ibn Sahl's (better known as Snell's) law. On the left, a ray from air (red) with incidence angle θ_1 is refracted (blue) toward the surface normal with an exit angle θ_2 in glass with a refractive index $n = 1.5$. The invariance of $n \sin\theta$ can be visualized geometrically by representing the sines in an unusual way: Instead of the common *unit* vectors for ray directions **k**, we assign them a length $|\mathbf{k}| = n$. Thus we see that the tangential component of **k**, i.e. $n \sin\theta$, is invariant. The orange ray, coming from inside the glass, is the *TIR limit ray,* whose **k** tangential component is just getting too large to have room on the other side. Rays with even larger incidence angles from glass cannot propagate into air. On the right, the same ray is shown as if originating from a nearby point source emitting spherical wavefronts. After refraction, the wavefronts are no more spherical; in glass, their wavefront profiles appear somewhat hyperbolic due to the requirement that phases are matched across an interface.

The law can be derived in several ways. (i) Solving Maxwell's equations tells us how wavefronts propagate across an interface. Figure 2.12 (right) shows how the slower propagation in glass with its shorter wavelength leads to the hyperbolic-looking wavefronts due to the requirement of phase continuity across the interface [11]. (ii) Fermat's variational principle (rays follow paths of extremal time) leads to the same law, but does not explain why. A deeper explanation is given by (iii) quantum electrodynamics (QED), which states that photons "sniff out" all possible paths, but that the complex amplitudes for most of these cancel, except for those close to the ones with extremal lengths [12]. If you now think that you still do not really understand the propagation of light, especially the QED part – don't worry, you are in excellent company. To quote Feynman again [12]:

> It is my task to convince you *not* to turn away because you do not understand it. You see, my physics students don't understand it either. That is because *I* don't understand it. Nobody does ... The theory of quantum electrodynamics describes Nature as absurd from the point of view of common sense. And it agrees fully with experiment. So I hope you can accept Nature as it is – absurd.

(Continued)

Knowledge Corner Topic 2.7: (Continued)

According to Feynman, theories explain *how* things are, but not necessarily *why*. Snell's law (grudgingly, we will not call it Ibn Sahl's law elsewhere in this book), immediately explains another phenomenon: that of *total internal reflection (TIR)*. For large incidence angles $\theta > \theta_{\mathrm{TIR}} = \arcsin(n_1/n_2)$ from the optically denser material, no corresponding ray directions on the other side exist. Algebraically, such rays would correspond to angles with sines larger than 1. Geometrically, as shown in Figure 2.12 on the left, the tangent components of the direction vectors $\mathbf{k}$ become too large. Physically, our direction vector $\mathbf{k}$ with $|\mathbf{k}| = n$ is the *wave vector*, whose tangent component must be invariant across the interface as required, ultimately, by Maxwell's laws.

Dr. Wisenheimer:

TIR is supremely useful in optical design. It is 100% efficient at no additional cost!

However, avoid rays too close to θ_{TIR} in your designs: If you want TIR, then tolerances easily cause light leaks of such rays; if you want refraction, then rays close to θ_{TIR} can suffer from large partial (Fresnel) reflections. And when the rays come from the optically denser material, their directions in air may become essentially uncontrollable due to tolerances.

Attention has to be paid to the evanescent field outside the light guide, in particular for rays that are incident at an angle only slightly larger than the TIR angle, which show the largest penetration depth. The fact that any material or body placed there may absorb some flux imposes a rule for the design of light guide mounts: reduce the contact area to a minimum and avoid any surface contamination [13].

Exercise 2.6: Light guide

Which refractive index requirement must a transparent material meet so that a solid, linear, rectangular block light guide in air will guide *all* incident rays without leakage (see Figure 2.13)?

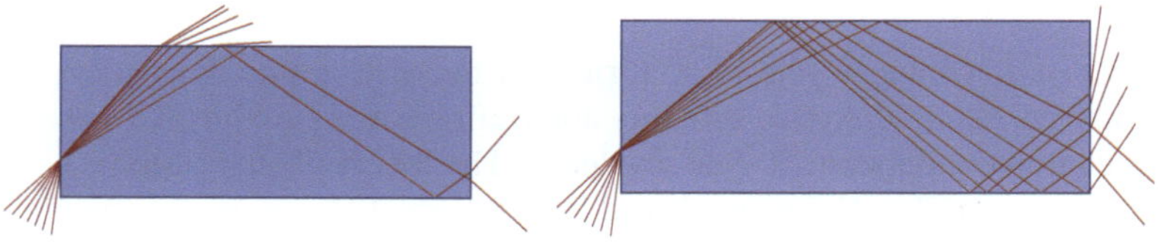

Figure 2.13 Left: A linear light guide made from a hypothetical, unsuitable material that does not guide all incident rays. Right: Same rays, same geometry, but now with a suitable material where no rays can leak (based on real ray tracing simulations).

Solution 2.6

Total internal reflection (TIR) at the side wall occurs if

$$\sin \gamma > \frac{1}{n} \tag{2.37}$$

with index of refraction n (see Figure 2.14). Thus,

$$\frac{1}{n} < \sin \gamma = \cos \beta = \sqrt{1 - \sin^2 \beta} = \sqrt{1 - \frac{1}{n^2} \sin^2 \alpha} \tag{2.38}$$

implies

$$n > \sqrt{1 + \sin^2 \alpha} \tag{2.39}$$

to ensure TIR. A ray of grazing incidence ($\alpha \to 90° \Rightarrow \sin \alpha \to 1$) at the entrance face produces the maximum of the right side. Hence, TIR and thus no leakage is ensured for

$$n > \sqrt{1 + (\sin 90°)^2} = \sqrt{2} \approx 1.41 \tag{2.40}$$

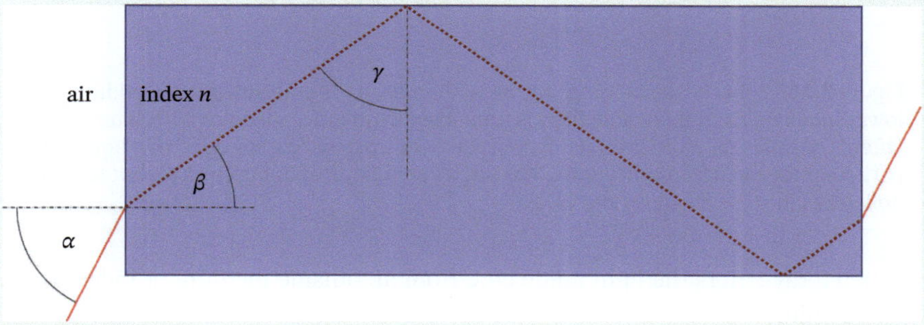

Figure 2.14 Light guide geometry with incidence and exit angles in air and in a specific material.

Dr. Wisenheimer:

This is fortunately ensured by all common glass and plastic materials available in the optical engineer's toolbox: Light that is incident onto the front face of a cuboid light guide in air cannot escape via its side faces [14].

Knowledge Corner Topic 2.8: Numerical aperture

When we embed a cylindrical light guide core in a cladding, as in an optical fiber, the TIR condition changes (Figure 2.15). With core index n_C and cladding index n_{Cl}, we obtain TIR when

$$\cos\beta = \sin\gamma > \frac{n_{Cl}}{n_C} \tag{2.41}$$

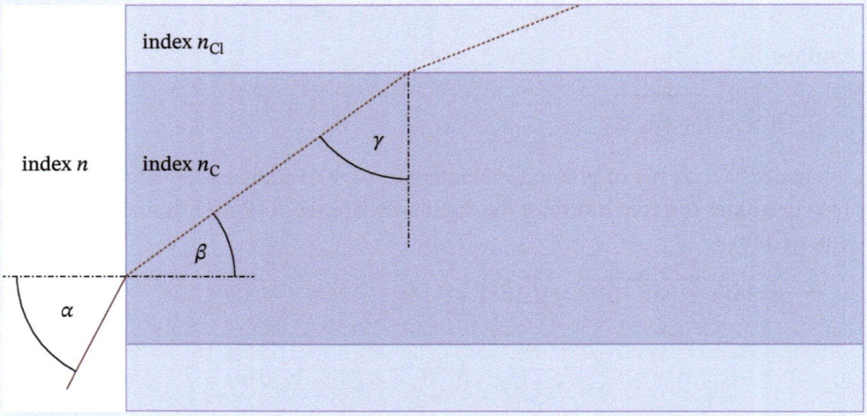

Figure 2.15 Fiber with cladding: A core with index n_C is embedded in a cladding with lower index n_{Cl}, and the whole fiber is immersed in an outer medium with index n. Often, the outer medium is air. However, the fiber tip may be embedded in a liquid, in particular for medical applications. The angles are identical to Figure 2.14, but the ray now leaks into the cladding and is lost.

When a ray enters the light guide core from an outside medium of index n, the TIR condition is

$$n\sin\alpha = n_C\sin\beta = n_C\sqrt{1-\cos^2\beta} < n_C\sqrt{1-\frac{n_{Cl}^2}{n_C^2}} = \sqrt{n_C^2 - n_{Cl}^2} \tag{2.42}$$

Rays at the TIR limit angle or rays that touch the edge of an aperture (or, more generally: rays that separate different light paths) are known as *edge rays* in illumination optics [15, 16]. When they form an angle α with a surface normal or with an optical axis, the quantity $n\sin\alpha$ is so ubiquitous and important that it has acquired its own name: Numerical aperture (NA) [17]. For the cladded fiber,

$$\text{NA} = n\sin\alpha = \sqrt{n_C^2 - n_{Cl}^2} \tag{2.43}$$

Accordingly, NA = 1 for a light guide with $n_C > \sqrt{2}$ in air (no cladding, $n_{Cl} = 1$), which means that all incoming ray angles will be guided.

For a high-index core, Eq. 2.43 yields NA > 1. In this case, a ±90° beam incident onto the light guide from air would not completely fill the ±90° range of propagation directions inside the medium. To guide the maximum amount of light, the entrance face then needs to be *immersed* in a material of index $n \geq$ NA to accommodate rays with the highest guided angles inside the light guide. Conversely, fibers with NA > 1 can be used effectively for wide-angle illumination in endoscope applications, where the fiber end is immersed in a liquid (e.g. inside the human eye).

Dr. Wisenheimer:
Commercial high-NA fibers often use claddings made of fluoropolymers [18] with refractive indices of $n_{\mathrm{Cl}} < 1.4$. These materials may potentially be restricted in Europe, because there is a proposal regarding the future use of PFAS (per- and polyfluoroalkyl substances) from the ECHA (European Chemicals Agency), which may also affect optical fibers.

Exercise 2.7: Solid angle during refraction

The law of refraction (Knowledge Corner Topic 2.7) may be written as

$$n \sin \theta = \text{const} \tag{2.44}$$

for a ray before and after passing an interface. Obviously, something is invariant here! Now, what about a whole beam described by a solid angle Ω or a projected solid angle Ω_{p}? Compare Ω and Ω_{p} before and after refraction and try to identify an invariant for the whole beam.

Solution 2.7

For a centered cone with half angle θ, Eq. 2.13 implies that

$$\Omega_{\mathrm{p,\ centered\ cone}} = \pi \,\mathrm{sr}\, \sin^2 \theta$$

Combining this with Eq. 2.43, we find that

$$n^2\Omega_{\mathrm{p}} = \pi \mathrm{sr}\,(n \sin \theta)^2 = \text{const} \tag{2.45}$$

across refraction and reflection. Thus, we have found a conservation law not for Ω_{p} itself, but for the quantity $n^2\Omega_{\mathrm{p}}$, at least for centered cones. Can we generalize this result?

Take a look again at Figure 2.12 (left): The equally long thick red and blue lines parallel to the interface are just a 2D representation of the projected solid angle. Refraction changes ray directions as well as the shape and size of the incoming and exiting solid angle, but $n^2\Omega_{\mathrm{p}}$ remains identical in size and in shape for the incoming and outgoing rays.

Just like Ω_{p}, the solid angle Ω shrinks when refracted into a denser medium. However, there is no similarly simple relation for Ω.

Knowledge Corner Topic 2.9: Angular extent X

Now that we found an invariant quantity (i.e. $n^2\Omega_p$), should we not give it proper recognition? We introduce

$$X = n^2\Omega_p \tag{2.46}$$

which we term the *angular extent* of a beam. This is not yet a commonly accepted symbol and name, but we think it should be: In phase space of geometrical optics, angular extent and area are two sides of the same coin, just like momentum and position in the phase space of classical or quantum mechanics. Moreover, we have seen in Exercise 2.7 that X is conserved during refraction, and we will see (beginning with Eq. 3.6) how X will blend itself into the definition of etendue. It is common good practice in science to assign proper names and symbols to conserved important quantities; we follow this practice here and hope that the word will spread.

Dr. Wisenheimer:

We conclude that the relationship between refractive index and angle is a rather close one. We should group the components of upcoming long formulas accordingly.

Exercise 2.8: Plane parallel plate

A convergent beam is modified by inserting a plane parallel plate (optical flat, index n) of thickness t tilted at an angle α (Figure 2.16).

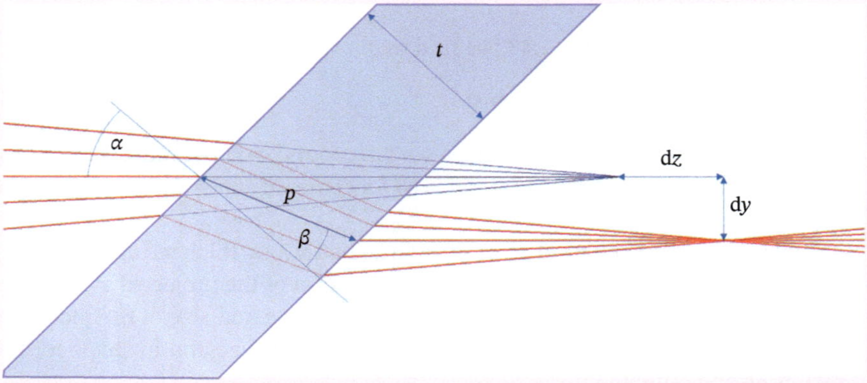

Figure 2.16 Focus displacement by a plane parallel plate. The focus of a convergent beam (blue, incident from the left) is shifted when a plane is inserted (red rays).

When looking from the left through the tilted plane (see Figure 2.16) at a point source (bottom right, red), its virtual image (blue) is displaced in the top left direction (and can be located inside the plate when it is sufficiently thick).

Calculate dy and dz.

Solution 2.8

The lateral displacement is

$$\mathrm{d}y = p \sin(\alpha - \beta) \tag{2.47}$$

with the geometrical path length p within the plate, and

$$p = \frac{t}{\cos \beta} \tag{2.48}$$

If you like, print Figure 2.16 and draw the corresponding triangles. Snell's law tells us that

$$\sin \beta = \frac{1}{n} \sin \alpha \tag{2.49}$$

Together, these equations result in [19]

$$\mathrm{d}y = t(\sin \alpha - \cos \alpha \tan \beta) = t \sin \alpha \left(1 - \frac{\sqrt{1 - \sin^2 \alpha}}{\sqrt{n^2 - \sin^2 \alpha}}\right) \tag{2.50}$$

The longitudinal displacement is

$$\mathrm{d}z = \frac{\mathrm{d}y}{\tan \alpha} = t \cos \alpha \left(1 - \frac{\sqrt{1 - \sin^2 \alpha}}{\sqrt{n^2 - \sin^2 \alpha}}\right) \tag{2.51}$$

which for perpendicular incidence ($\alpha = 0$) yields

$$\mathrm{d}z = t\left(1 - \frac{1}{n}\right) = t\left(\frac{n-1}{n}\right) \tag{2.52}$$

(cf. [20, Ch. 16]).

Dr. Wisenheimer:

Despite being just flat, a plane parallel plate introduces aberrations. For example, Figure 2.17 (left) shows a point source behind a parallel plate with its aberrated virtual image inside the plate. The spherical aberration can then be compensated rather easily using a plano-convex (PCX) lens (same figure, right).

(Continued)

Solution 2.8 (Continued)

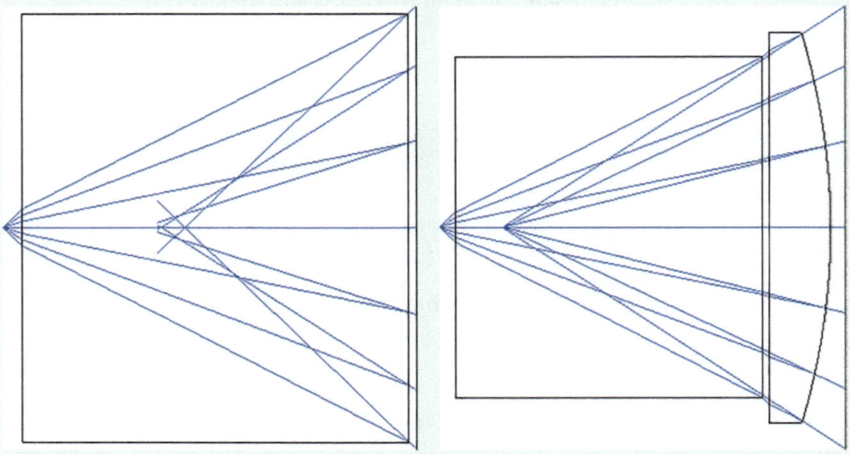

Figure 2.17 Parallel plate causing spherical aberration. When observing a microscopic object (left of the plate) from the right, one will see an aberrated virtual image of it (left). The aberration can be compensated with a lens (right).

The effect can be severe for microscopes. Most microscope objective lenses feature a "nominal cover glass thickness" as a part of their optical design, which is usually indicated on the label of the lens.

Cover glasses play their role in illumination optics as well, for example on top of an LED. This discussion will be continued in Knowledge Corner Topic 3.9.

The next exercise will introduce aspects of tolerancing – a step that is often neglected in illumination design. The art is to understand what is going on and which degree of tolerancing is necessary. This will depend on the optical elements in the beam path, the effects these rely on (refraction or reflection) and on surface slopes.

Exercise 2.9: Surface slope error

Surface slopes are subject to tolerances: refracted and reflected ray directions will deviate from their ideal paths. Compare by how much the surface slope deviation $\Delta\psi$ of an optical element changes the direction θ of the outgoing ray, i.e. compute $\Delta\theta$ after a refracting and after a reflecting surface. Also, consider how an internally reflected ray changes its final direction in air.

Solution 2.9

For *refraction*, a ray perpendicularly incident at an exit surface from the inside experiences a deflection according to Snell's law (Figure 2.18):

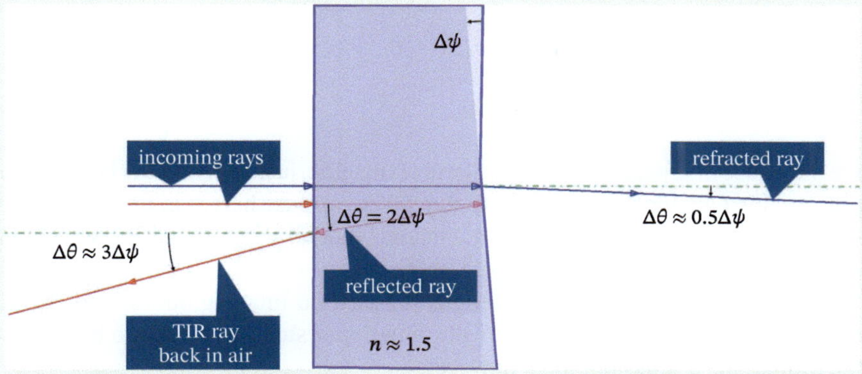

Figure 2.18 Ray deflection $\Delta\theta$ by a surface with slope error $\Delta\psi$, for refraction (solid blue ray), reflection (dotted red ray), and TIR back in air (solid red ray).

$$\sin(\Delta\theta + \Delta\psi) = n \sin \Delta\psi \tag{2.53}$$

which for small $\Delta\psi$ yields HANDLE WITH CARE

$$\Delta\theta = (n-1)\,\Delta\psi \tag{2.54}$$

i.e. $\approx 0.5\,\Delta\psi$ for typical materials.

For *reflection*, the incidence and exit angles w.r.t. the surface normal are equal. The ray directions relative to the surface normal change by the same $\Delta\psi$; therefore the outgoing ray deviation is

$$\Delta\theta = 2\,\Delta\psi \tag{2.55}$$

When a ray is internally reflected (via TIR or partial Fresnel reflection), its internal $2\,\Delta\psi$ deviation is amplified by subsequent refraction into air. When the final surface is planar, we have

$$\Delta\theta = 2n\,\Delta\psi \tag{2.56}$$

i.e. $\approx 3\,\Delta\psi$ for typical materials. When the final surface is curved, there is no general answer.

Dr. Wisenheimer:

Reflection is more sensitive to surface slope errors than refraction by a factor of four; for TIR, that factor is six! The same surface roughness, by the way, also causes much more scattering for reflective surfaces than for refractive surfaces. Spend your surface quality budget on reflective surfaces first.

(Continued)

Solution 2.9 (Continued)

In the ISO 10110 lens drawings standard, slope errors can be described as part of the “Surface form and figure” tolerances (part 10110-5, “3/ΔS” or “3/RMSΔS”) [21].

2.3 Paraxial Optics

We introduce or repeat the very basic laws of imaging, just applying the law of refraction and introducing some approximations. The purpose here is to understand how such approximations work and to understand whether you can still apply them in a later exercise.

A certain part of illumination designs is based on imaging optics, and paraxial equations may lead the way to your illumination system. You need to be aware of the limitations of this approach, however.

Knowledge Corner Topic 2.10: The paraxial ray tracing equation

Consider the following meridional section of an optical arrangement: A ray intersects a lens surface and is refracted (Figure 2.19).

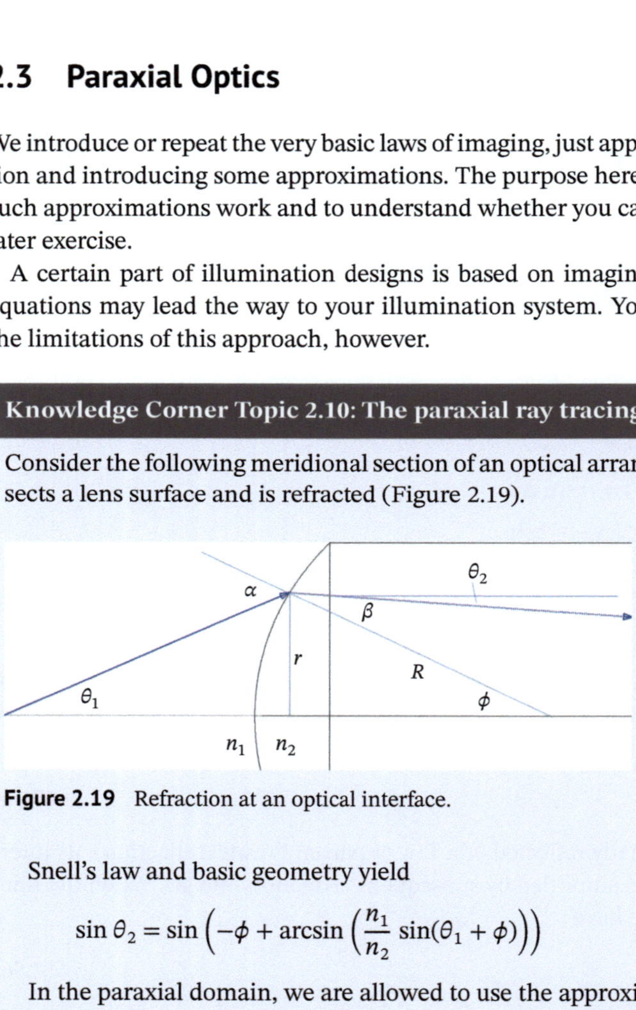

Figure 2.19 Refraction at an optical interface.

Snell’s law and basic geometry yield

$$\sin\theta_2 = \sin\left(-\phi + \arcsin\left(\frac{n_1}{n_2}\sin(\theta_1 + \phi)\right)\right) \tag{2.57}$$

In the paraxial domain, we are allowed to use the approximations

- $\sin\theta \approx \theta$ (thus $r = r_0 + \theta z$) and
- $r \ll R$ (thus ray height $r = \phi R$)

and thus obtain the *paraxial ray-tracing equation* (PRTE, [22])

$$n_2\theta_2 = n_1\theta_1 + \frac{r}{R}(n_1 - n_2) \tag{2.58}$$

Be aware that in the literature of imaging optics, formulas may differ from our notation as very specific sign rules are sometimes applied. While the latter are helpful in complex setups, they do not convey any additional insights in our case.

The single surface focal length f_1 of the surface and its inverse, the surface's refractive power P, are defined by

$$P = \frac{1}{f_1} = \frac{n_2 - n_1}{R} \tag{2.59}$$

For a thin lens (radii R_1 and R_2, but negligible path length inside, thus the same r at both surfaces), the powers simply add, and we have

$$n_3\theta_3 = n_1\theta_1 + \frac{r}{R_1}(n_1 - n_2) + \frac{r}{R_2}(n_2 - n_1) \tag{2.60}$$

which in air ($n_1 = n_3 = 1$) simplifies to

$$\theta_3 = \theta_1 - \frac{r}{f} \tag{2.61}$$

with the thin-lens focal length

$$\frac{1}{f} = (n_2 - 1)\left(\frac{1}{R_1} - \frac{1}{R_2}\right) \tag{2.62}$$

which is termed the *lens maker equation*, for obvious reasons.

For convenience, we present the same content in a different way in the following Knowledge Corner Topic.

Knowledge Corner Topic 2.11: Paraxial matrix formalism

The laws of paraxial optics may be written as matrix multiplication

$$\begin{pmatrix} r_2 \\ \theta_2 \end{pmatrix} = \mathbf{M} \cdot \begin{pmatrix} r_1 \\ \theta_1 \end{pmatrix} \tag{2.63}$$

with

$$\mathbf{M}_{\text{free space}} = \begin{pmatrix} 1 & z \\ 0 & 1 \end{pmatrix} \tag{2.64}$$

and

$$\mathbf{M}_{\text{lens}} = \begin{pmatrix} 1 & 0 \\ -\frac{1}{f} & 1 \end{pmatrix} \tag{2.65}$$

which allows for easy evaluation of paraxial ray propagation through a multi-lens system.

Exercise 2.10: Imaging equations

An axial point has a distance g from an ideal lens (focal length f, object and image in air, see Figure 2.20).

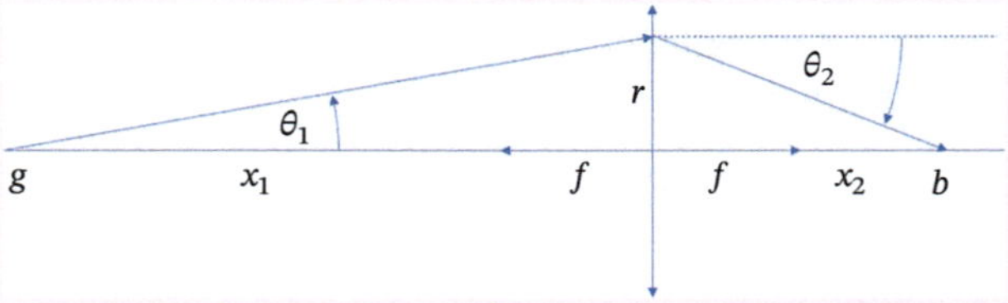

Figure 2.20 Imaging of an axial point with an ideal lens.

Using the PRTE, calculate the image distance b.

Solution 2.10

With $\theta_1 \approx r/g$ and $\theta_2 \approx -r/b$, Eq. 2.61 yields

$$-\frac{r}{b} = \frac{r}{g} - \frac{r}{f} \tag{2.66}$$

and thus

$$\frac{1}{g} + \frac{1}{b} = \frac{1}{f} \tag{2.67}$$

which is the *thin-lens imaging formula*, valid for the above assumptions.

Dr. Wisenheimer:
Setting $g = x_1 + f$ and $b = x_2 + f$, we obtain

$$f^2 = x_1 x_2 \tag{2.68}$$

which is *Newton's imaging equation*.

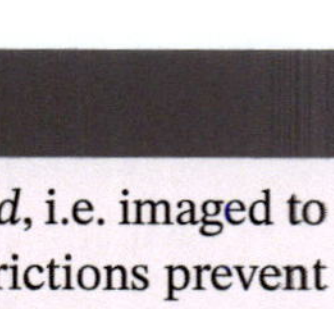

Exercise 2.11: LED collimation (1)

An LED (bare chip in air, emitting amber light) shall be *collimated*, i.e. imaged to infinity, by using a lens. In your laboratory, however, space restrictions prevent you from making the necessary arrangements. To find the focal plane for the LED, you intend to adjust the distance between the LED and the lens until you achieve a sharp image on the screen, and then to correct the LED-to-lens distance. This is your setup:

Quantity	Value
Chip area	$A = 1\ \text{mm}^2$
Wavelength	$\lambda \approx 590\ \text{nm}$
Focal length	$f = 50\ \text{mm}$
Maximum screen distance	$D = 1\ \text{m}$

Calculate the necessary distance correction. Does the $F/\#$ of the lens affect the result?

Solution 2.11

We start with an ideal lens with a large $F/\#$ (we are in the paraxial world and can use Eq. 2.66). In our case of an image at infinity ($b \to \infty$), we obviously have $g = f$ (object in the focal plane). An image on the wall, $b = D = 1$ m is obtained for

$$\frac{1}{g} = \frac{1}{50\ \text{mm}} - \frac{1}{1000\ \text{mm}} \tag{2.69}$$

which yields $g = 52.63$ mm. In the experiment, we adjust the LED position until we get a sharp (and square) image on the wall. Then, by decreasing the LED–lens distance by $\Delta g = 2.63$ mm we obtain the desired collimating arrangement.

This procedure (adjusting for sharp image on screen without measurements) nicely takes care of the fact that our real lens will not be thin.

However, any possible effect of the $F/\#$ (i.e. real rays, large angles, aberrations) is not covered by these formulas. So let us check these effects by simulation.

Simulation 2.1: LED collimation (1)

This is a rather simple example to begin with. If you have access to optical software, we recommend setting up a system by yourself.

We use a sequential or non-sequential system with a sole wavelength of 587.56 nm. For a thick (10 mm) plano-convex lens, $f = 50$ mm, $d = 20$ mm, made of N-SF6 ($n = 1.80518$ at 587.56 nm) we use Eq. 2.61 and obtain a radius of curvature of 40.259 mm. The planar side of the lens should face the source to get nearly equal refraction at both faces – this is a basic principle to reduce aberrations (Figure 2.21). For a thick lens, we have to measure object and image distances from the respective principal planes. Our PCX lens features the object side principal plane at a position of 5.54 mm from the flat surface inside the lens. Your software might tell you where the principal planes are; equivalently, it might provide the front focal length (the distance from the front focal plane to the first lens surface vertex). The image side principal plane of the lens is situated at the vertex of the curved face. Thus, for collimation, we need a source to lens working

(Continued)

Simulation 2.1: (Continued)

distance of g = 44.46 mm. Set it up and check the collimation (look at far-field intensity).

Figure 2.21 Single lens as an LED collimator.

You will see that collimation is acceptable (full aperture rays with larger angles will be refracted a little stronger than necessary), but not perfect due to spherical aberration.

We provide an LED and Lens LightTools system.

Now, increase the distance from LED to lens by the value of $\Delta g = 2.63$ mm calculated above, to 47.09 mm and check the beam at an image distance of $b =$ 1 m: Image a 1 mm^2 source and check the illuminance distribution at the wall as a function of the lens aperture (Figure 2.22). (Make sure that the source will always fill the entire lens area with light.)

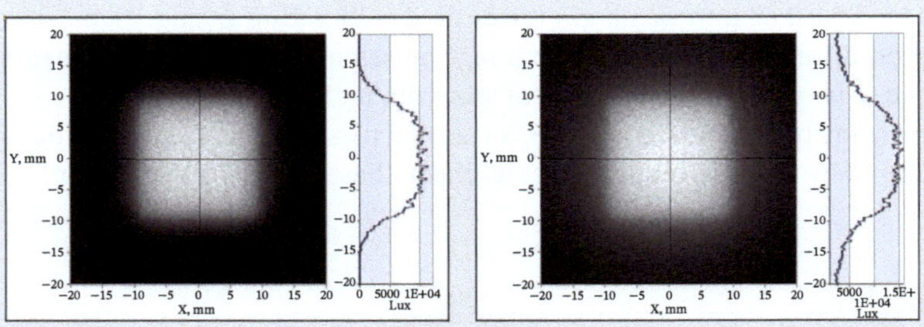

Figure 2.22 Source image at the wall for lens diameters of 20 mm and 50 mm, the latter case showing considerable stray light outside the source image as a result of more spherical aberration.

The aberrations of a large plano-spherical lens cause image degradation. Do they also affect the optimal LED–lens distance for imaging to the wall and for collimation? We answer this question with a *parametric study*, driving the simulation software from an external program; the result is shown in Figure 2.23.

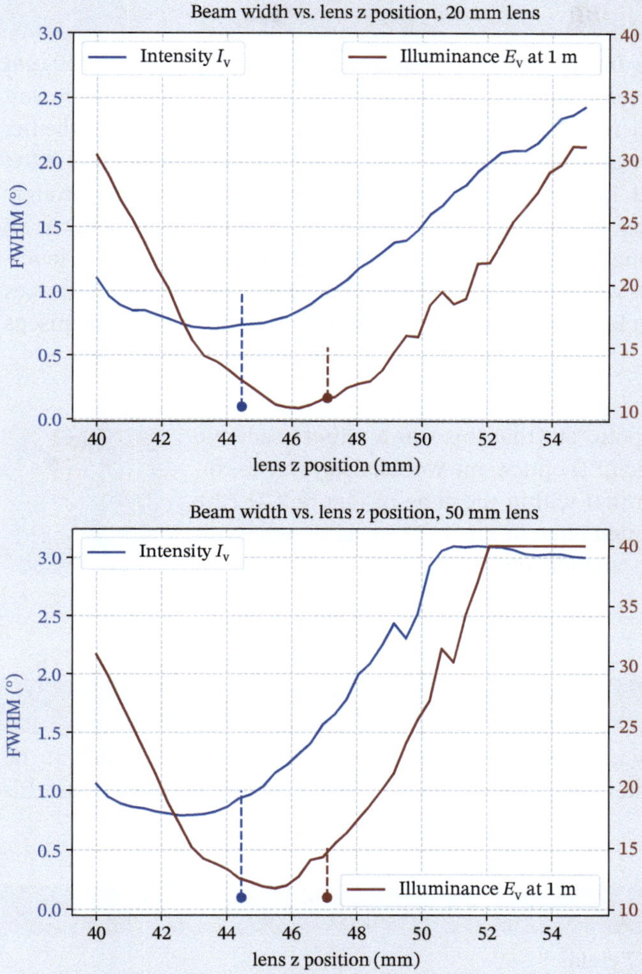

Figure 2.23 LED collimation as function of LED–lens distance, for two different lens apertures (20 mm, left and 50mm, right), evaluated in non-sequential software. Shown are the beam semidiameters at infinity (blue) and at 1 m (red), at 25% of peak value. The circles with the dotted lines show the paraxial values.

For the smaller 20 mm lens, there is negligible offset, while for the larger 50 mm lens, the optimum position to minimize beam width is about 2 mm closer to the lens than indicated by the paraxial calculations. A general conclusion is that an optimum illumination design may well differ from paraxial estimates, sometimes to a much larger extent than in this example. You may, and should, use the calculus and the approximations taught in this book for *predesign*, but be prepared to *fine tune* the system by hand or with a suitable optimizer.

However, there is a serious problem with the larger lens for the purpose of collimation. Can you spot it, studying the results and the plots above? (Do not peek ahead – think first!)

(Continued)

Simulation 2.1: (Continued)

The problem is the unexpectedly low peak intensity/illuminance for the 50 mm lens. When you increase the lens diameter by our factor $f = 50/20 = 2.5$, you would hope for an increase in on-axis intensity by $f^2 = 6.25$. Figure 2.23 (the lux scale) shows an increase by only a factor of about 1.5. Spherical aberration is the culprit: Looking back into the lens from afar, you would hope to see the strongly magnified central spot of the LED, but in reality, you cannot see the LED at all through the outer ring of the large lens. Use your simulation software to investigate further. Which working distance maximizes peak intensity, and how does the beam shape then look like? Would an aspheric lens help? Would a meniscus lens help?

Dr. Wisenheimer:

Please allow me to point out that this was a rather academic example. For realistic LED optics, one would always try to collect more light than that within the cone of Figure 2.21 (the reason will be presented later, in Eq. 3.10), i.e. to design a collimator with a much smaller source side $F/\#$.

Exercise 2.12: Glass sphere

An object at infinity is focused by a glass sphere, with its image located at the rear surface. Calculate: Which properties must the glass sphere have to meet this requirement?

Solution 2.12

For our case, Eq. 2.57 yields

$$n_2\theta_2 = \theta_1 - \frac{r}{R}(n_2 - 1) \tag{2.70}$$

The situation described here means $\theta_1 = 0$ (object at infinity) and $r = -\theta_2 \cdot 2R$. We easily obtain $n_2 = 2$. Such a high index of refraction is not typical, but suitable glasses do exist (although they are not cheap), and corresponding ball lenses are available from stock. The size of the sphere is not relevant – everything scales with size.

Dr. Wisenheimer:

We assumed paraxial conditions. For a large entrance pupil, the image would become blurry as a result of spherical aberration. This can be mitigated by an aspherical (in particular: elliptical) shape of the refracting surface, as will be shown in Exercise 2.19.

Let's invent something!

When you cover part of the back surface of such a sphere with a reflective coating, you create a *retroreflector*. This works good enough even for spheres of smaller refractive index. We can use this principle to make large, affordable retroreflecting foils such as those used in road signs, which appear very bright at night when illuminated by car headlamps. There, transparent spheres are embedded in a white matrix, peeking out a little bit, thus exposing their clear, spherical caps. They focus the headlamp light to a bright small spot on the sphere's back surface. The same spherical cap then images the bright small spot back into a narrow beam around the incoming headlamp light.

Dr. Wisenheimer:

Optical designers know that the effect of some spherical aberration can be mitigated by a little defocus. Here, we use it the opposite way and compensate the effect of defocus by spherical aberration. For our inventor's retroreflective foils, however, the combination of slightly smaller index (caused by the lack of cheap $n = 2$ materials) and spherical aberration is a feature, making the defocused spot just the right size for a narrow beam retroreflection (see Figure 2.24).

2.4 Laser Beams

Lasers do have their roles in illumination projects, either as a necessary part of the solution or as a competition to LEDs. For example, a blue laser diode can pump a phosphor with higher irradiance than possible with an LED, and any laser can help minimize the optical system due to its small etendue (to be introduced in Section 3.2).

On the one hand, laser beams follow their own laws (which can also help for design estimates), but often they can be handled in the same way as other (incoherent) light sources. You will find *ray files* for diode lasers at the website of some laser suppliers.

We restrict our considerations to the most relevant rotationally symmetric case. This means that we do not cover higher laser modes, but in exchange can offer analytical solutions.

Simulation 2.2: Refraction at a sphere

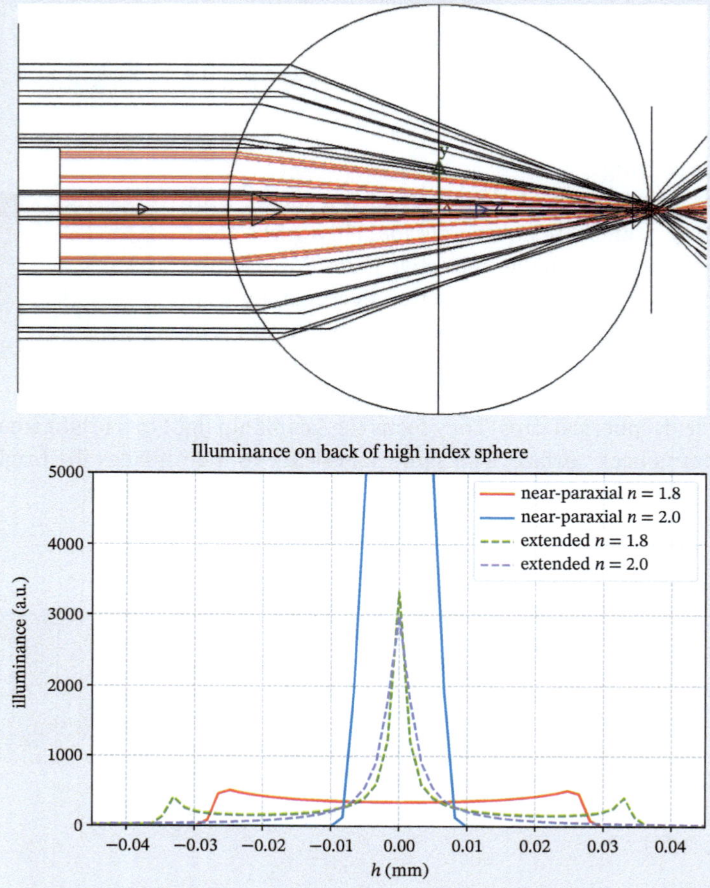

Figure 2.24 Top: While imaging of a distance point to the rear of a glass sphere works for $n = 2$ with paraxial perfection, a somewhat lower index of $n = 1.8$ causes defocus. The red rays come from a narrow, near-paraxial source; the black rays from an extended source. Note how both ray bundles have approximately same size focal spots. Bottom: Illuminance distributions in the focal region. The defocus due to $n = 1.8$ spreads the paraxial focus, strongly decreasing peak illuminance, but not the focus of an extended source.

We provide the design file.

Knowledge Corner Topic 2.12: Gaussian distributions

A Gaussian *irradiance* distribution on a screen is

$$E(r) = E_C \exp\left(-2\frac{r^2}{w^2}\right) \tag{2.71}$$

with peak irradiance E_C and waist w (see Figure 2.25). The standard deviation of this function is $\sigma = w/2$. Throughout this book, the Gaussian waist w and angle θ (Eq. 2.74) refer to the *half width of the* $1/e^2$ *value* of the irradiance or intensity peak.

In the laser community, the irradiance in Eq. 2.71 is often termed intensity [23], measured in W/mm^2, whereas *we* measure intensity in W/sr. Always be aware which community you are communicating with!

The Excel Normdist function is convenient to compute values of a Gaussian distribution and/or a Gaussian beam, but be aware of different factors as compared to Gaussian optics.

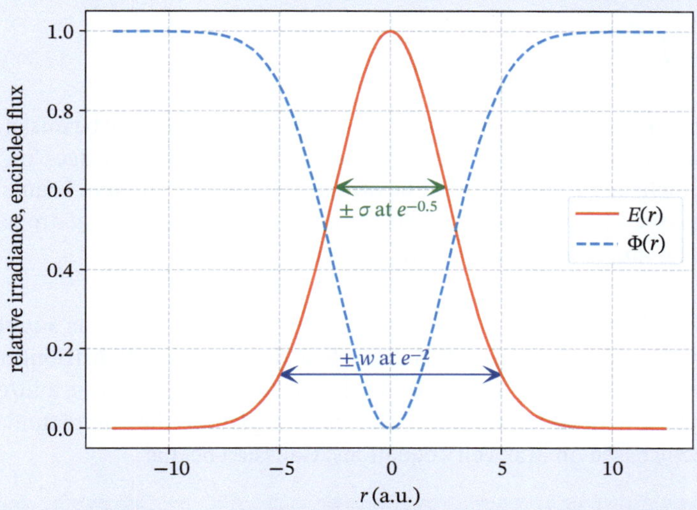

Figure 2.25 Solid red line: Cross section of a Gaussian irradiance profile, Eq. 2.71, with peak irradiance $E_C = 1E_C$, waist $w = 5$ and therefore standard deviation $\sigma = w/2 = 2.5$. Dashed blue line: Relative flux of the same distribution contained in a circle with radius r.

Since $\mathrm{d}A = r\,\mathrm{d}\varphi\,\mathrm{d}r$ in polar coordinates (r, φ), and

$$\frac{\mathrm{d}}{\mathrm{d}r}\exp\left(-2\frac{r^2}{w^2}\right) = -\frac{4\,r}{w^2}\exp\left(-2\frac{r^2}{w^2}\right) \tag{2.72}$$

integration gives the radiant flux (optical power) contained in a disk of radius r as

$$\Phi(r) = \int_{r'=0}^{R}\int_{\varphi=0}^{2\pi} E(r')\,r'\,\mathrm{d}\varphi\,\mathrm{d}r' = E_C\frac{\pi w^2}{2}\left(1-\exp\left(-2\frac{r^2}{w^2}\right)\right) \tag{2.73}$$

The relative encircled flux varies like the same bell curve inverted (see Figure 2.25)! The total flux of the beam ($R \to \infty$) is

$$\Phi = E_C\frac{\pi w^2}{2} \tag{2.74}$$

(Continued)

Knowledge Corner Topic 2.12: (Continued)

In the angular domain, a Gaussian *intensity* (flux per solid angle) distribution is given by

$$I(\vartheta) = I_0 \exp\left(-2\,\frac{\sin^2\vartheta}{\sin^2\theta}\right) \tag{2.75}$$

or, in the paraxial approximation, by

$$I(\vartheta) = I_0 \exp\left(-2\,\frac{\vartheta^2}{\theta^2}\right) \tag{2.76}$$

The total power in the beam is

$$\Phi = 2\pi\ \mathrm{sr} \int_0^{90^\circ} I(\vartheta)\sin\vartheta\,\mathrm{d}\vartheta \tag{2.77}$$

which can be integrated only numerically. For a practical description of beams in the angular domain we recommend to use the $\cos^n$ beam model (Knowledge Corner Topic 3.15), which, for not-too-wide beams, approximates Gaussian beams almost perfectly (see Figure 3.34) *and* is easy to integrate analytically *and* drops to zero at 90°, as it must for planar emitting apertures.

Gaussian distributions of irradiance or intensity are commonly used as simple models, or as optimization goals, of "soft" beams. Exactly how this soft distribution is achieved may vary; methods include optimized reflector and lens shapes, and/or use of diffuser sheets. With lasers, Gaussian distributions play another, more fundamental role, directly based on Maxwell's equations: Gaussian beams.

Knowledge Corner Topic 2.13: Gaussian beams

The propagation of laser beams is usually (and successfully) modeled using the Gaussian beam concept. A Gaussian beam is an approximate localized solution of Maxwell's equations (more specifically: the paraxial Helmholtz equation) describing the electric field $\mathcal{E}$ in the form [24, 25]

$$\mathcal{E}(r,z) = \mathcal{E}_0\frac{w_0}{w(z)}\exp\left[-\frac{r^2}{(w(z))^2}\right]\exp\left[-\mathrm{i}\left(kz + k\frac{r^2}{2R(z)} + \Psi(z)\right)\right] \tag{2.78}$$

with the waist w_0 as the minimum semi-diameter along the beam, and located at $z = 0$ without loss of generality. The first fraction is an integration constant, the first exponential describes the radial Gaussian amplitude distribution, and the second exponential is the wavefront-defining phase term, for whose details we refer you to the literature [25, 26].

During propagation, the beam's lateral extent expands to

$$w(z) = w_0\sqrt{1 + \left(\frac{\lambda z}{\pi w_0^2}\right)^2} \tag{2.79}$$

which describes a hyperbola (cf. Eq. 2.156) in the z-w plane with an asymptote of (cf. Figure 2.26)

$$\tan\theta = \frac{\lambda}{\pi w_0} \approx \sin\theta \tag{2.80}$$

Tacitly replacing the tangent by the sine, we become compatible with the calculus in other parts of optics, and avoid the pure paraxial notation: For diode laser optics, angles may be a little larger than paraxial.

Diffraction at a small aperture ties waist and asymptotic angle together [24]:

$$w_0 \sin\theta = \frac{\lambda}{\pi} \tag{2.81}$$

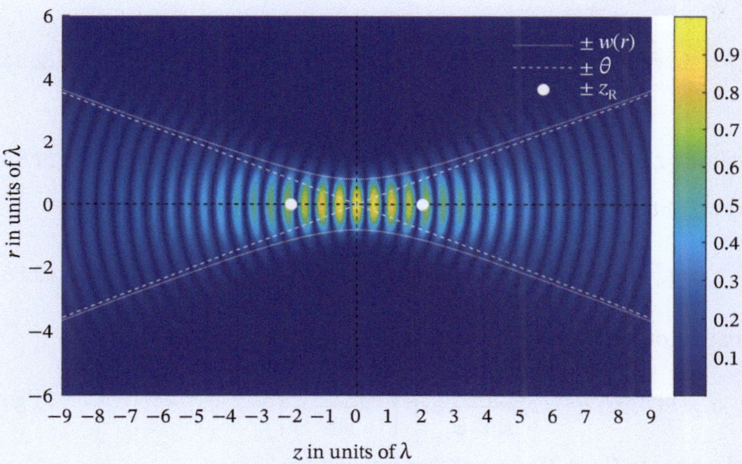

Figure 2.26 The field distribution (normalized absolute value of the electric field at time $t = 0$) of a Gaussian beam shows how the beam converges to its waist and then re-expands symmetrically. The waist radius is chosen as $w_0 = 0.8\lambda 1/e^2$. The dotted white line ($w(z)$) is the hyperbola where the field has dropped laterally to 1/e (when the field is 1/e, the irradiance/intensity is $1/e^2$). The dashed white lines denote the asymptotic beam width with half angle θ, and the white dots show the Rayleigh length z_R.

just like for the Airy disk. But this $w_0 \sin\theta$ product, also known as the *beam parameter product* (BPP), can also be viewed as a 2D etendue expression (cf. Section 3.3) of the fundamental Gaussian beam [25], connecting lateral and angular extent. The full 4D etendue of the beam is, as usual, cross-section area times projected solid angle; Eq. 3.11 yields

$$U_{\text{Gaussian beam}} = A\,\Omega_p = w_0^2\pi \cdot \sin^2\theta\,\pi = \lambda^2 \tag{2.82}$$

which is yet another manifestation of the fact [25] that the etendue of a single electromagnetic mode is λ^2. This is also termed the etendue *quantum* U_Q. See below (Knowledge Corner Topic 3.5) for further discussion.

(Continued)

Knowledge Corner Topic 2.13: (Continued)

Irradiance

By taking the square of the electric field in Eq. 2.78, we obtain an expression for irradiance in a $z = \text{const}$ plane:

$$E(r,z) = \varepsilon(r,z)\,\varepsilon^*(r,z) = \varepsilon_0^2 \frac{w_0^2}{w^2(z)} \exp\left[-2\left(\frac{r}{w(z)}\right)^2\right] \tag{2.83}$$

and conclude that Eq. 2.71 does indeed describe the lateral irradiance profile for a Gaussian beam.

Intensity

But what about far-field intensity? For large z, we can write on the basis of Eqs. 2.79 and 2.80

$$w(z) = \sin\theta \cdot z \tag{2.84}$$

and with

$$r = \sin\vartheta \cdot z \tag{2.85}$$

($\vartheta = 0\ldots\theta$) we obtain

$$E(\vartheta,z) = \varepsilon_0^2 \frac{w_0^2}{\sin^2\theta z^2} \exp\left[-2\frac{\sin^2\vartheta}{\sin^2\theta}\right] \tag{2.86}$$

The application of the inverse square law (Eq. 3.142) leads us to

$$I(\vartheta)\,\text{sr} = E(\vartheta,z)\cdot z^2 = \varepsilon_0^2 \frac{w_0^2}{\sin^2\theta} \exp\left[-2\frac{\sin^2\vartheta}{\sin^2\theta}\right] \tag{2.87}$$

which is identical to Eq. 2.75: a Gaussian intensity profile.

Rayleigh length

The z distance where peak irradiance drops to 50% is found to be (cf. Figure 2.26)

$$z_R = \frac{w_0}{\sin\theta} = \frac{\pi}{\lambda} w_0^2 \tag{2.88}$$

In contrast to the product $w \sin\theta$ (the 2D etendue of the beam), the ratio $w_0/\sin\theta$ (the Rayleigh length of the beam) is a measure of the shape of the beam's etendue in phase space and may be modified by imaging the beam with magnification.

Mixed-mode beams

Multi-mode or mixed-mode beams have a more complicated field distribution and are characterized by a BPP which is larger than the diffraction-limited value of λ/π by a factor of $M^2 > 1$ [27, 28]. When we view the beam as having a larger waist $W = Mw$ and larger angle $\sin\Theta = M\sin\theta$, compared to the fundamental Gaussian mode with waist w, which constitutes a larger 2D etendue of

$$W\sin\Theta = M^2\frac{\lambda}{\pi} \tag{2.89}$$

showing that under these assumptions, the Rayleigh length is independent of M^2.

However, it is also common to view the mixed-mode beam as having a certain (measured) waist w_0, and a divergence angle $\Theta = M^2\theta$; then, the Rayleigh length is inversely proportional to M^2.

When you send a Gaussian beam through a (series of) coaxial lenses and/or mirrors, it turns out that the image of an incoming coaxial Gaussian beam is another Gaussian beam. You can therefore equate waist locations with axial object and image points. Fundamental relations still hold: The paraxial thin-lens imaging formula (Eq. 2.66) carries over with a correction term that vanishes with the Rayleigh length, and the paraxial ABCD matrix calculus (see Knowledge Corner Topic 2.11) can still be applied. Consult Brooker's book [25] and Self's paper ([29]) for a detailed discussion.

Knowledge Corner Topic 2.14: Imaging Gaussian beams

Consider a laser beam refracted by an ideal, paraxial lens with focal length f and with the waist at a distance x_0 to the front focus of the lens. We term w_0 and θ_0 waist and angle in the object space and w_1 and θ_1 waist and angle after imaging. Imaging is performed by the *transformation parameter*

$$\Gamma = \frac{f^2}{x_0^2 + z_{\mathrm{R0}}^2} \tag{2.90}$$

For $x_0 = 0$ we have collimation

$$\Gamma = \frac{f^2}{z_{\mathrm{R0}}^2} \tag{2.91}$$

and the collimated beam has a waist of

$$w_1 = w_0\sqrt{\Gamma} = f \sin \theta_0 \tag{2.92}$$

at

$$x_1 = \Gamma x_0 = 0 \tag{2.93}$$

which means that the waist is in the rear focal plane.

The new Rayleigh length is

$$z_{\mathrm{R1}} = z_{\mathrm{R0}}\Gamma \tag{2.94}$$

After collimation, the divergence is

$$\sin \theta_1 = \frac{\lambda}{\pi w_1} = \frac{\lambda}{\pi f \sin \theta_0} = \frac{w_0}{f} \tag{2.95}$$

(corresponding to a situation with a disk source in front of a lens with focal length f).

Exercise 2.13: Imaging a Gaussian waist

Derive an equation similar to the paraxial thin-lens imaging formula Eq. 2.66 for the image distance of a laser waist.

Solution 2.13

We use the same symbols as in Eq. 2.66.

Using Eqs. 2.93 and 2.90, we obtain

$$b = f + x_1 = f + \Gamma x_0 = f + \frac{f^2(g-f)}{(g-f)^2 + z_{\mathrm{R0}}^2} \tag{2.96}$$

$$(b-f)(g-f)^2 + (b-f)z_{\mathrm{R0}}^2 = f^2(g-f) \tag{2.97}$$

$$bg - bf - fg + \frac{b-f}{g-f} z_{\mathrm{R0}}^2 = 0 \tag{2.98}$$

$$\frac{1}{f} = \frac{1}{g} + \frac{1}{b} - \frac{b-f}{g-f} \cdot \frac{z_{\mathrm{R0}}^2}{bgf} \tag{2.99}$$

This is the imaging equation with a correction term that is proportional to the square of the Rayleigh length.

Dr. Wisenheimer:

The location of the image of a small waist is almost exactly where we would expect it according to the paraxial formula Eq. 2.66.

Deviations from paraxial imaging may occur for large waists, i.e. expanded beams with a large Rayleigh length.

2.5 Conic Sections

A *conic section* is a curve obtained by intersecting the surface of a cone with a plane (Figure 2.27). Depending on the orientation of the plane, we may obtain a circle, an ellipse, a parabola or a hyperbola (plus the special cases of a point and one or two lines).

Conic sections pop up at a myriad of occasions in nature, geometry, physics, and optics. Mathematicians of all times have investigated them with great interest [30–32]. Some fundamental properties of conic sections lead to construction methods on the basis of strings (attributed to Anthemius of Tralles (474–558) [33]), compasses [34] or the method by Proclus [35] and de la Hire [36] using the circles through opposite vertices.

Their use in optics is mostly based on *stigmatic* (point to point) imaging properties (reflective and refractive) as explained in detail in [16, Ch. 4.2]: Revolved conic sections perfectly image one special point to another special point, both termed *foci*. When used to image *extended* objects such as light sources, the performance of conic

Figure 2.27 Planes intersecting the surface of a cone at different tilt angles: circles (red, plane perpendicular to cone axis), ellipses (orange, plane somewhat tilted), parabolas (yellow, plane parallel to mantle) and hyperbolas (green, plane even more tilted).

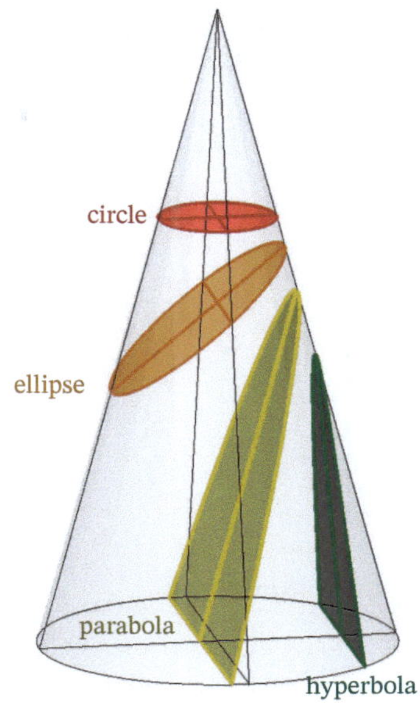

sections is fundamentally affected by *coma*, also known as *angular-dependent magnification*, as discussed, for example, in Knowledge Corner Topic 2.16. Mitigation is possible by either using multiple conics in the optical system [37–40] or by adding an aspheric lens to a conical reflector [41, 42]. Moreover, conic sections are the basis for the thermodynamically ideal *compound concentrator* designs (such as the CPC, CEC, and CHC, introduced later in this chapter, see Knowledge Corner Topic 2.17).

Let us take a closer look at these all-important optical surface shapes.

Knowledge Corner Topic 2.15: Properties common to all conics

It is hard to pick a single defining property of conics from the plethora of equivalent relations, of which Wikipedia provides a great overview. We will present three, and as you may already guess, we will do so in terms of formulas for later use in illumination design.

The first is particularly simple, and coordinate-free; it will later help us to understand the perfect refractive focusing property of an ellipsoidal lens surface (Exercise 2.20):

> A conic section is that set of points whose distances to a given point (focus) and a given line (directrix) have a constant ratio e, known as the numerical eccentricity (Figure 2.28):
>
> $$\frac{r_1}{l} = e = \text{const} \tag{2.100}$$

This is known as the focus–directrix property of conic sections [43]. Note that the humble circle cannot be described this way: it is the limiting case of $e \to 0$, with the directrix moving to infinity.

(Continued)

Knowledge Corner Topic 2.15: (Continued)

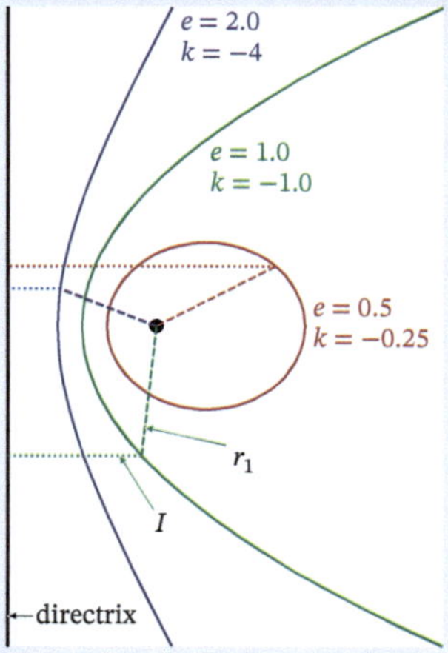

Figure 2.28 Focus–directrix property of conic sections. For any point on a conic section, the eccentricity e is the ratio of r_1 (distance to focus) over l (distance to directrix).

We have sneaked in the notion of *focus point*, or simply focus (plural: foci). For a given conic section, a mathematical way to define its focus (or, in general, one of its two foci) is to say that it is the point on the axis where the radial distance from that axis point to the conic section equals the radius of curvature at the vertex (see Figure 2.29). More important, optically, is the fact that all rays first emitted from the focus and then reflected from the conic section converge on another focus. For ellipses and hyperbolas, the second focus is a distinct point; for circles, both foci happen to be at the center, and for parabolas, the second focus is at infinity.

Second, the *vertex equation* for all conic sections, with one vertex at $r = z = 0$, is

$$(1 + k)z^2 - 2\rho z + r^2 = 0 \tag{2.101}$$

with the shape defined by the conic constant k as shown in Figure 2.28. This equation is immensely important, because describing conic section shapes via the conic constant k is ubiquitous in optical theory and software. It turns out that for negative k, the simple relation

$$k = -e^2 \tag{2.102}$$

holds, describing a hyperbola, a parabola and an ellipse for $k < -1$, $k = -1$ and $-1 < k < 0$, respectively. For $k = 0$, we obtain a circle with radius ρ, and for $k > 0$, the vertex equation yields so-called oblate ellipses, which are perfectly normal ellipses, just rotated by 90°. In the vertex equation, z is the coordinate along the optical axis, r is the radius from the optical axis, and ρ is a scaling constant, the *semilatus rectum* (the distance from the focus to the conic section along a line perpendicular to the optical axis, which also happens to be the curvature radius at the vertex).

Dr. Wisenheimer:

Take care when comparing different equations of conic sections. There are equations with the origin ($z = 0$) located in the vertex, in the focus or in the center of the conic. Do not try to combine them (for example, Eqs. 2.100 and 2.132), because z has a different meaning in those equations.

Third, we present the *polar form* of all conic sections

$$r_1(\delta) = \frac{\rho}{1 + e\cos\delta} \tag{2.103}$$

(see Figure 2.29), which will prove to be very useful in many derivations.

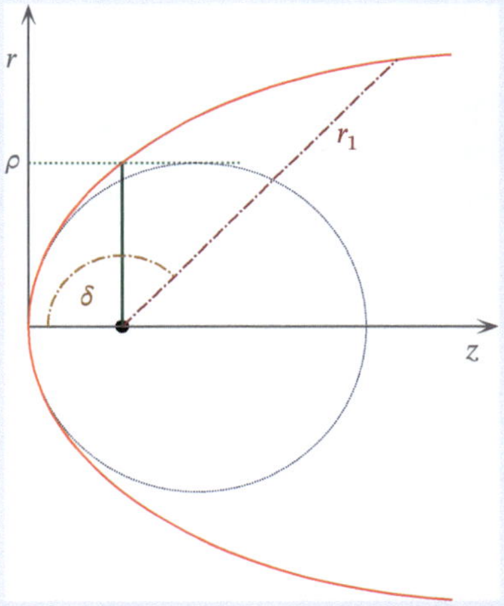

Figure 2.29 Polar form of a conic section (here, an ellipse with eccentricity $e = 0.8$). Again, z is the optical axis, r is the radial distance from the axis, r_1 is the distance from focus (black dot) to a point on the conic section, ρ is both the semilatus rectum and the radius of curvature at the vertex, and the curve is parametrized by angle δ, which is measured from negative z-axis.

Exercise 2.14: Vertex form and asphere representation

The vertex equation of a conic section introduced in Eq. 2.100 is a quadratic equation in z. Express the sag z as an explicit function of r (see Figure 2.29 for the geometric meaning of r, z and ρ).

Solution 2.14

We can simply divide Eq. 2.100 by $(1 + k)$ to obtain

$$z^2 - \frac{2\rho}{1+k}z + \frac{r^2}{1+k} = 0 \tag{2.104}$$

(Continued)

Solution 2.14 (Continued)

and solve by completing the square, with the solution

$$z(r) = \frac{\rho}{1+k}\left(1 \pm \sqrt{1-(1+k)\frac{r^2}{\rho^2}}\right) \tag{2.105}$$

Dr. Wisenheimer:

The result is correct, but the standard notation used in optics is different. With a little algebra (hint: multiply Eq. 2.100 by $r^2/z^2\rho^2$), we find

$$z(r) = \frac{r^2/\rho}{1+\sqrt{1-(1+k)\frac{r^2}{\rho^2}}} \tag{2.106}$$

which is the standard representation of the conic term in optics.

The minus sign in front of the root is ruled out by the requirement that for all k (also for $k = -1$, the parabola) we will not allow an infinite denominator.

Exercise 2.15: Intersection of a ray and a conic surface

In a meridional section, i.e. a z-r plane containing the optical axis, a ray may be given by

$$r(z) = r_0 + t(z - z_0) \tag{2.107}$$

with starting point (r_0, z_0) and slope $t = \tan\theta$.

Calculate the intersection point(s) of the ray with a conic surface (Eq. 2.100 or 2.105).

Solution 2.15

If $t = 0$ (ray parallel to optical axis), then $r(z) = r_0$, and the intersection point is simply $(r_0, z(r_0))$, using Eq. 2.105. Otherwise, without loss of generality, we move the ray starting point along the ray to the optical axis, by first setting $z_0 \leftarrow z_0 - r_0/t$ and then $r_0 \leftarrow 0$. Now, z_0 is the distance between the on-axis ray starting point and the surface vertex at $z = 0$: $r(z) = t(z - z_0)$.

We use the vertex equation, Eq. 2.100, and get

$$2\rho z - (1+k)z^2 = t^2(z - z_0)^2 \tag{2.108}$$

(remember: ρ represents the vertex curvature) with the solution

$$z_{1,2} = \frac{\rho + t^2 z_0 \pm \sqrt{\rho^2 + t^2(2\,\rho z_0 - z_0^2(k+1))}}{t^2 + 1 + k} \tag{2.109}$$

In general, there are two candidate intersection points, $(r(z_{1,2}), z_{1,2})$. Typically, you would take that point which is found first while traversing along the ray from its starting point (using the minus sign before the root).

Dr. Wisenheimer:
The easy analytic solution with little numerical effort for the intersection point of a ray with a conic section is the reason why conic surfaces parameterized by their conic constants k are available in virtually all lens design software packages.

We just looked at the general properties of conic sections. Now, we will provide specific parameters and formulas for parabola, ellipse, and hyperbola, for later use in illumination optics, and we will discuss their optical properties.

2.5.1 Parabola

Just like the shape of a circle, the shape of a parabola is defined by a single degree of freedom. Accordingly, the following useful parameters are not independent:

- focal length f (distance from focal point to vertex)
- vertex radius ρ
- semilatus rectum (equals vertex radius ρ, see Figure 2.31).

The conic constant is fixed by definition, $k = -1$, and focal length f and vertex curvature radius ρ are related by

$$\rho = 2f \tag{2.110}$$

Thus, the vertex equation Eq. 2.100 simply becomes

$$r^2 = 2\rho z = 4fz \tag{2.111}$$

allowing us to derive the key optical property of a parabola by calculating it slope,

$$2r\,\mathrm{d}r = 4f\,\mathrm{d}z \quad \Rightarrow \quad \frac{\mathrm{d}r}{\mathrm{d}z} = \frac{2f}{r} = \frac{\rho}{r} \tag{2.112}$$

Using (i) the definitions in Figure 2.30, (ii) the polar form relation Eq. 2.102 with eccentricity $e = 1$, and (iii) the half-angle tangent relation Eq. 2.6, we obtain

$$\tan\varphi = \frac{\mathrm{d}r}{\mathrm{d}z} = \frac{\rho}{r} = \frac{\rho}{r_1 \sin\delta} = \frac{\rho}{\rho\frac{\sin\delta}{1+\cos\delta}} = \frac{1}{\tan\frac{\delta}{2}} = \tan\frac{\pi-\delta}{2} = \tan\frac{\gamma_1}{2} \tag{2.113}$$

Thus, a ray from the focus with slope $\tan\gamma_1$ is reflected at a point with exactly half that slope. Consequently, the resulting reflected ray has slope *zero* for all γ_1: All rays from the focal point are collimated upon reflection at the parabola.

$$\theta(\delta) = \frac{g}{r_1}\sin\delta = \frac{g}{\rho}\sin\delta\cdot(1+\cos\delta) \tag{2.114}$$

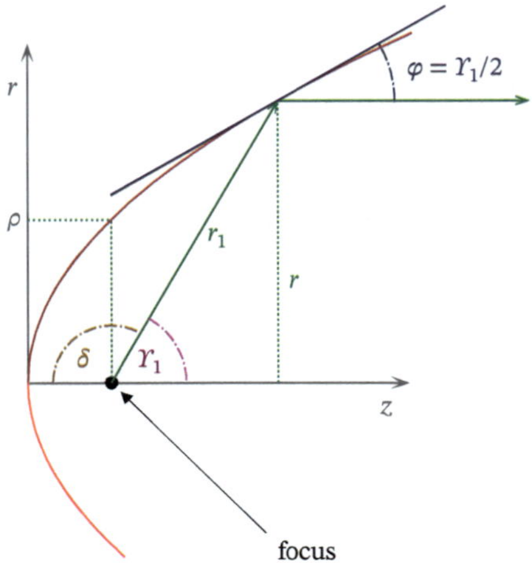

Figure 2.30 A parabola (red line) and its parameters, with a single ray (green arrow) starting at the focal point. The semilatus rectum ρ determines the parabola shape. The polar form angle δ selects the single ray (green arrows), whose angle with the axis is $\gamma_1 = \pi - \delta$. The polar form ray segment length is r_1, and thus the intersection point's radial distance to the axis is $r = r_1 \sin\theta = r_1 \sin\delta$. The ray's angle with the parabola tangent φ must equal $\gamma_1/2$ to make the reflected ray parallel to the optical axis, which we prove in Eq. 2.113.

Exercise 2.16: Off-axis parabola

A point source emits a $\pm 10°$ beam. Sketch a reflector that simultaneously

- collimates the beam to a diameter of $D = 40$ mm
- reflects it by 90°.

Where is the vertex located?

Solution 2.16

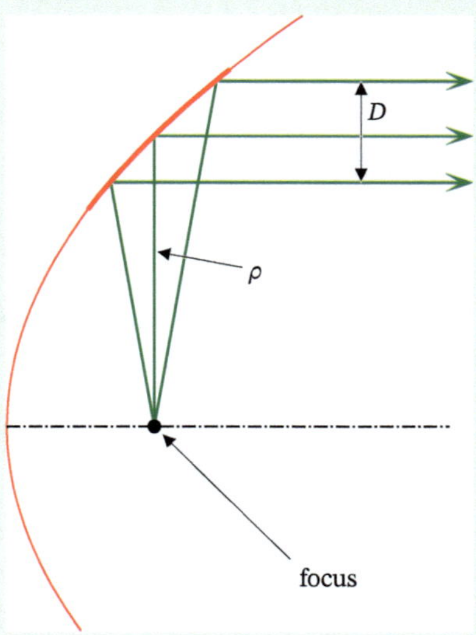

Figure 2.31 Off-axis parabola.

The requested optical function is delivered by an off-axis parabola (a segment of a parabolic reflector, see Figure 2.31). The $\pm 10°$ beam acquires $D = 40$ mm at a distance of

$$\rho = \frac{20\,\text{mm}}{\tan 10°} = 113.4\,\text{mm} \tag{2.115}$$

which defines the semilatus rectum ρ of the parabola (not its focal length!). Accordingly,

$$f = \frac{\rho}{2} = 56.7\,\text{mm} \tag{2.116}$$

and the vertex is at $f = 56.7$ mm to the left of the focus.

We refer to Exercise 4.3 for the discussion of a bulb in such a parabolic reflector.

Knowledge Corner Topic 2.16: Parabola magnification

Imagine a small source of length $\pm g$ in the focus of a parabola and oriented along the optical axis z [44] (see Figure 2.32). "Small" means that the source appears under a small angle when seen from any point on the reflector surface, in other words, $g \ll f$. We apply the small-angle approximation where appropriate HANDLE WITH CARE.

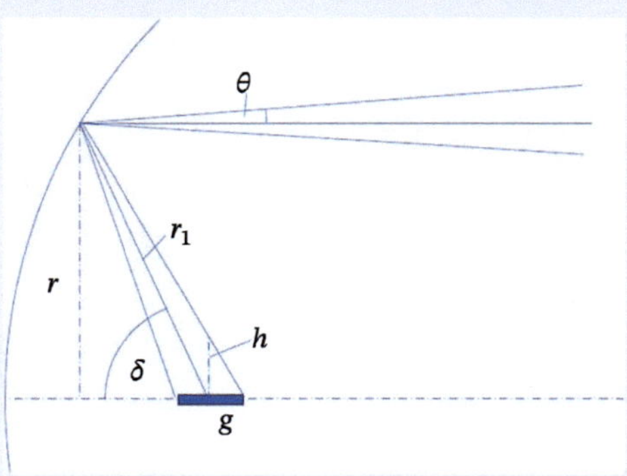

Figure 2.32 Geometry of a source in the focal point of a parabola (cf. Figure 4.1).

We consider an arbitrary surface point at height r, from which light from the source is reflected into an angle of $\pm\theta$, the same angle under which source appears from there. We apply Eq. 2.102 with $e = 1$ for the parabola and find

$$\theta(\delta) = -\frac{g}{r_1}\sin\delta = -\frac{g}{\rho}\cos\delta\,(1 + \cos\delta) \tag{2.117}$$

(Continued)

Knowledge Corner Topic 2.16: (Continued)

The *axial* magnification, i.e. exit angle divided by source size, is

$$M_g(\delta) = \frac{\theta}{g} = -\frac{1}{\rho} \sin\delta \cdot (1 + \cos\delta) \tag{2.118}$$

Now imagine a small source of length $\pm h$ in the focus but oriented perpendicularly to z (cf. Exercise 2.21, Figure 2.33, and [45]).

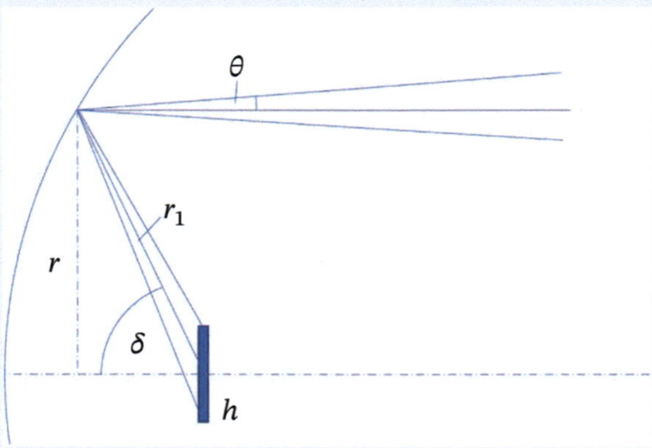

Figure 2.33 Geometry of a source in the focal point of a parabola (cf. Figure 2.32).

The angle $\theta(\delta)$ is now

$$\theta(\delta) = -\frac{h}{r_1} \cos\delta = -\frac{h}{\rho} \cos\delta\,(1 + \cos\delta) \tag{2.119}$$

and the radial or transverse magnification is

$$M_h(\delta) = \frac{\theta}{h} = -\frac{1}{\rho} \cos\delta\,(1 + \cos\delta) \tag{2.120}$$

In a similar way, we obtain the magnification of a small sphere with radius R centered at the focus (see Figure 4.1),

$$M_{\text{sph}}(\delta) = \frac{\theta}{R} = \pm\frac{1}{\rho}\,(1 + \cos\delta) \tag{2.121}$$

The corresponding magnification curves (scaled for $\rho = 1$) are shown in Figure 2.34.

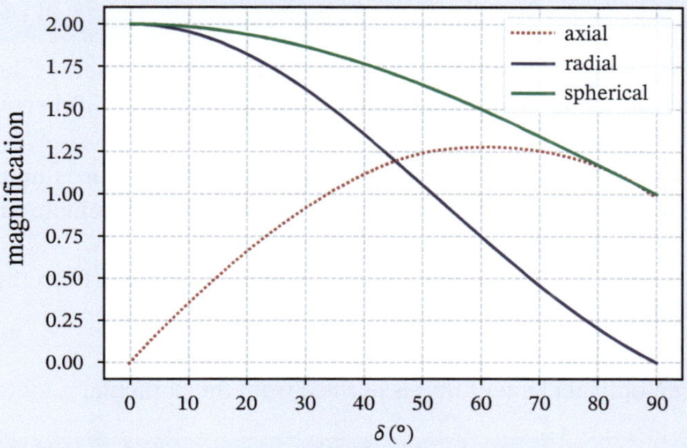

Figure 2.34 Magnification for a linear source oriented along (axial, Figure 2.32) and perpendicular to (radial, Figure 2.33) the optical axis, and for a spherical source (cf. Figure 4.1), as functions of δ. The angular cone generally decreases for the outer regions of the parabola, but the magnification depends on source shape and orientation.

Figure 2.34 shows that the magnification of the parabola depends both on the angle w.r.t. the focus, equivalent to position on the aperture, and on source shape and orientation. Below, we will see that this applies to all conic sections.

See [46] for a discussion on caustics near the focus, and how to deal with these caustics using additional optics.

Dr. Wisenheimer:

Consider a ray coming from the far field at an angle θ with the optical axis and hitting the reflector at a distance r from the axis. The reflector front plane acts as the aperture stop and we ask for the position $R(\theta, r)$ where the ray ends in the source region (see Figure 4.1).

We remember the polar form of a conic (Eq. 2.102)

$$r_1 = \frac{2f}{1 + \cos\delta} \tag{2.122}$$

Now, we eliminate δ using

$$r_1 = \frac{r}{\sin\delta} \tag{2.123}$$

and obtain

$$r_1 = f\left(1 + \frac{r^2}{4f^2}\right) \tag{2.124}$$

as well as

$$R(\theta, r) = \theta r_1 = \theta f\left(1 + \frac{r^2}{4f^2}\right) \tag{2.125}$$

(Continued)

Knowledge Corner Topic 2.16: (Continued)

Thus, the deviation R is a function of the field coordinate θ times the square of the aperture coordinate r, which is called *coma* in the framework of imaging aberrations.

As the illuminance distribution on the front of a parabolic reflector is non-uniform, too, an integrator is usually added to homogeneously illuminate an imager [47]. Fly's eye condensers are ubiquitous here.

Their lenslet size can be adapted to the different angular cone sizes from the parabolic reflector [48].

Now we introduce an optical element that is derived from the parabola.

Knowledge Corner Topic 2.17: Compound parabolic concentrator/collimator

The compound parabolic concentrator (CPC) is of extraordinary academic (theoretical and conceptual) importance – it literally shaped the field of non-imaging optics – but has not really found too many practical use cases. We therefore introduce the CPC, but only briefly, referring you to a comprehensive range of literature on the topic for further information [49–53].

The CPC was initially developed as a concentrator for collimated light from a distant source. When used in this manner, the CPC collects light from infinity (within a given angular range $\pm\theta$) to a disk target (radius r_T). Its shape is obtained by rotating a parabola

- that is inclined by the desired maximum angle of incidence θ
- with its focus at the opposite rim of the disk

around an axis through the center of the target disk (Figure 2.35).

The edge ray principle [15] ensures that meridional rays of smaller angle of incidence will hit the target disk while rays of a larger angle to the optical axis will be rejected. For skewed rays, that is not entirely the case, though.

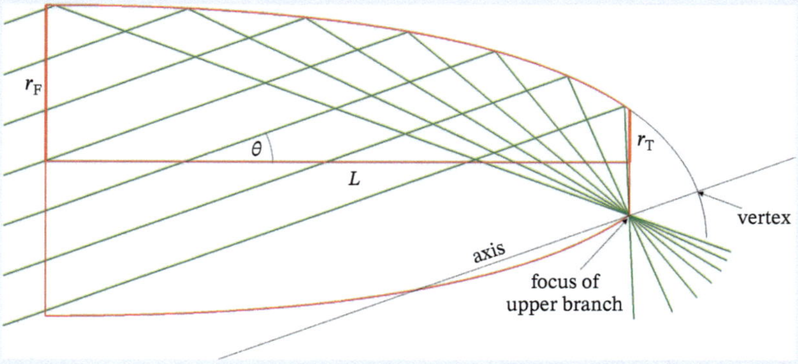

Figure 2.35 CPC layout in air. Collimated rays with angle θ (green) enter a hollow CPC (red) with large end radius r_F, small end radius r_T, collection angle $\pm\theta$ and length L; these rays are focused at the lower edge of the small face [54].

Acceptance angle and disk radius determine the size of the entrance aperture (radius r_F) via etendue conservation

$$r_\mathrm{F} \sin\theta = r_\mathrm{T} \sin 90° \tag{2.126}$$

and also the full shape of the CPC.

Concentration C is the ratio of the irradiance values at the entrance and at the exit of an optical system. In the absence of losses, and ignoring the slightly non-perfect performance for skewed rays, flux conservation implies

$$C = \left(\frac{r_\mathrm{F}}{r_\mathrm{T}}\right)^2 = \sin^2\theta \tag{2.127}$$

In illumination optics, the CPC is mostly used in reverse direction: As a collimator, with the source at the small end. Unfortunately, for optimal performance, the source would need to completely fill the small aperture up to its edge, which is not practical for single LEDs.

A sibling of the CPC is needed when we would like to use a dielectric (glass or plastic) with TIR. Then, the design angles are those *within* the dielectric, and the resulting structure is known as the θ_i–θ_o concentrator/collimator. For the same ratios of face diameters and thus acceptance/collimation angles, these structures are similar but longer, and they have a cone section at the small face (see Figure 2.36). For details of the construction, see [52, Appendix H]. Both the CPC and the θ_i–θ_o concentrator are available as design elements in some optical design software packages.

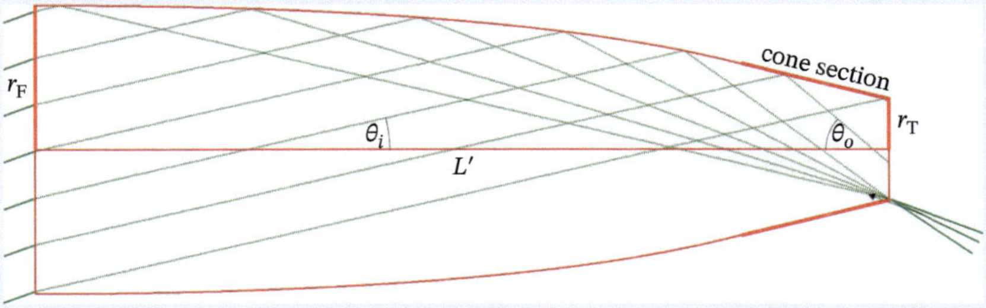

Figure 2.36 The θ_i–θ_o concentrator/collimator, which converts radiation within $r_\mathrm{T} = 1$ mm to radiation within θ_o, here shown as the dielectric variant of the CPC. The CPC itself is a special case, with $\theta_\mathrm{o} = 90°$ and vanishing cone section.

Exercise 2.17: Compound parabolic concentrator

Consider a receiver at $z = 0$ with $r_\mathrm{T} = 1$ mm, upon which light of an angular range of $\theta = \pm 20°$ is incident from infinity.

Sketch (we really mean sketch, with pencil and paper) a CPC [49] (hollow, i.e. with reflecting surfaces and air inside) and calculate its parameters.

Solution 2.17

Before any actual work, we need a little preparation: We use the polar form Eq. 2.102 to calculate the focal length of the parabola. We know that

$$2r_T = 2\,\text{mm} = \frac{\rho}{1 + \cos(90^\circ - \theta)} \tag{2.128}$$

and obtain the semilatus rectum $\rho = 2.68$ mm, which defines the vertex location together with the 10° inclination of the parabola's axis. The focal length is then

$$f = \frac{\rho}{2} = 1.34\,\text{mm} \tag{2.129}$$

The lower receiver rim is the focus of the upper parabola and vice versa.

Etendue conservation tells us that

$$\pi \cdot r_T = \pi \cdot r_F \sin 20^\circ \tag{2.130}$$

which yields a front radius of $r_F = 2.92$ mm. Our resulting sketch is displayed in Figure 2.37.

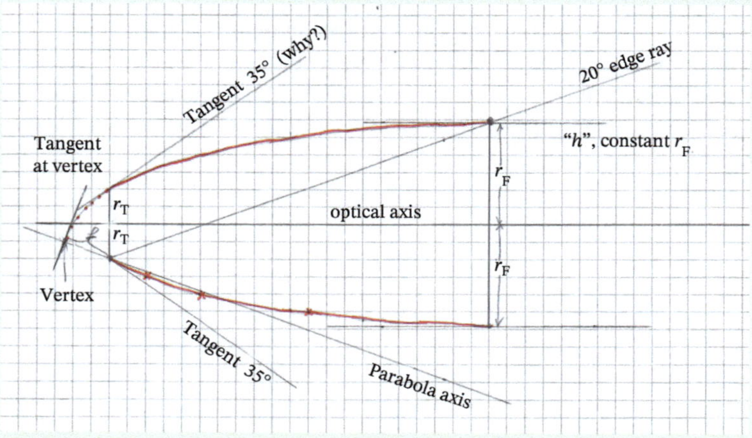

Figure 2.37 2D CPC design sketch for collection: We draw (in this order, and with a 10:1 magnification, i.e. 1 mm → 1 cm) the optical axis, the receiver, the axis of the upper parabola (−20° and crossing the lower end of the receiver), the focal point (1.34 cm back on that axis), the +20° edge ray from the focus and a horizontal line h at $r_F = 2.92$ cm above the optical axis. The large face is where the edge ray has height r_F, i.e. where it crosses h. Next, we draw two tangents: One at the vertex, perpendicular to the parabola axis, and one at the upper receiver edge, with 35° slope (why?). The third tangent is h itself at the large face edge. With these tangents, we can sketch the upper parabola curve by hand, just taking into account how we guess the curvature decreases from left to right. Finally, we sketch the lower parabola as a mirror image, just using the corresponding tangents.

You can do similar sketches in CAD software or even in Excel®. In Excel, we choose the lower focal point as the origin and plot the upper parabola with the help of the polar Eq. 2.102; the result is shown in Figure 2.38.

Provided file CPC.xlsx.

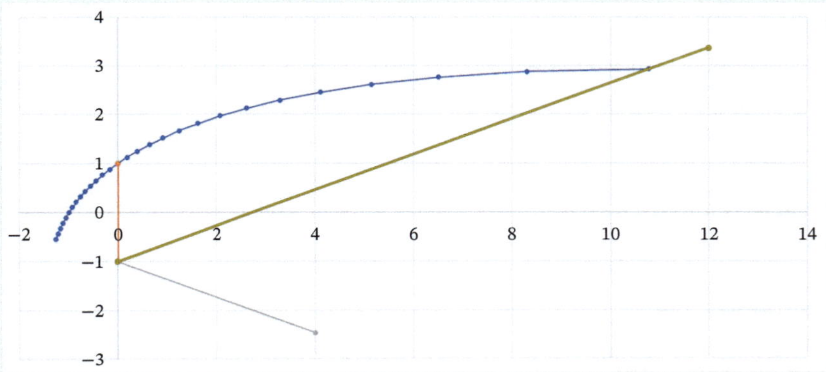

Figure 2.38 2D CPC parabola branch sketch for ±20° collection drawn in Excel.

Dr. Wisenheimer:

The use of a CPC for source collimation [55] is not easy (though widespread in design considerations) because if the entrance aperture is not perfectly filled, the homogeneity on the exit side will be rather poor. The situation can be improved by applying facets or incorporating homogenizer features such as a square or hexagonal CPC cross sectional area. But especially for small exit angles, CPCs get pretty long (cf. Exercise 4.22).

Taking into account alignment tolerances between an LED source and a CPC as well as the necessary gap, we conclude CPCs may be appropriate for large, circular LED sources such as chip-on-board (CoB) LED arrays, but rather not for small individual LEDs.

Exercise 2.18: CPC length

Calculate the nominal length L of the CPC from Exercise 2.17.

Solution 2.18

To calculate the length of the CPC, we use the diagonal ray (cf. Figure 3.6) of the (r_F, r_T, L) light tube (see Section 3.3)

$$\tan\theta = \frac{r_F + r_T}{L} \tag{2.131}$$

giving

$$L = \frac{1.0\text{ mm} + 2.92\text{ mm}}{\tan 20^\circ} = 10.8\text{ mm} \tag{2.132}$$

A CPC may be ideally collecting incident radiation, but for (angularly) small sources, Eq. 2.130 tells us that it will get rather long. This is the reason why a CPC would be combined with a lens or reflector for solar collection [53, Ch. 5] or even avoided altogether [56].

Dr. Wisenheimer:

An image web search would probably yield 99 sketches and one picture of a real CPC. For LED optics, a few lens elements are usually cheaper than molding a CPC. Still, the CPC remains an extremely useful academic exercise.

2.5.2 Ellipse

In contrast to a parabola, which extends to infinity (unless trimmed) and whose other focus is at infinity, an ellipse is a closed, quadrant symmetric curve which has two foci at a finite distance (Figure 2.39).

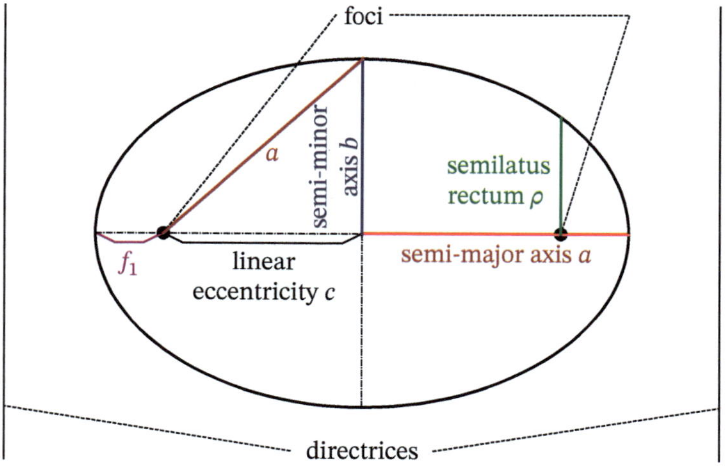

Figure 2.39 Commonly used parameters of an ellipse. Semilatus rectum ρ and focal length f_1, however, are applicable to all other conic sections as well.

Ellipse parameters: While shape and size of an ellipse have only two degrees of freedom, a host of parameters is commonly used, all or them useful in some contexts (see Figure 2.39):

- semi-major axis a
- semi-minor axis b
- linear eccentricity $c = \sqrt{a^2 - b^2}$
- numerical eccentricity $e = c/a$
- conic constant $k = -e^2$
- semilatus rectum $\rho = b^2/a$
- vertex radius of curvature ρ (equal to the semilatus rectum)
- small focal length $f_1 = a - c$
- large focal length $f_2 = a + c$
- focal ratio $m = f_1/f_2$

For a more thorough discussion, see [57].

The standard form or center equation (origin at $r = z = 0$) of an ellipse is

$$\frac{r^2}{(a+z)(a-z)} = \frac{b^2}{a^2} \tag{2.133}$$

which is equivalent to

$$\frac{z^2}{a^2} + \frac{r^2}{b^2} = 1 \tag{2.134}$$

The stigmatic property of an ellipse (all rays from one focus are reflected to the second one) is related to equal path length (Figure 2.40)

$$r_1 + r_2 = 2a \tag{2.135}$$

The corresponding angles of the rays with the optical axis are also interrelated [57]:

$$\tan\frac{\gamma_1}{2}\tan\frac{\gamma_2}{2} = m = \frac{1-e}{1+e} \tag{2.136}$$

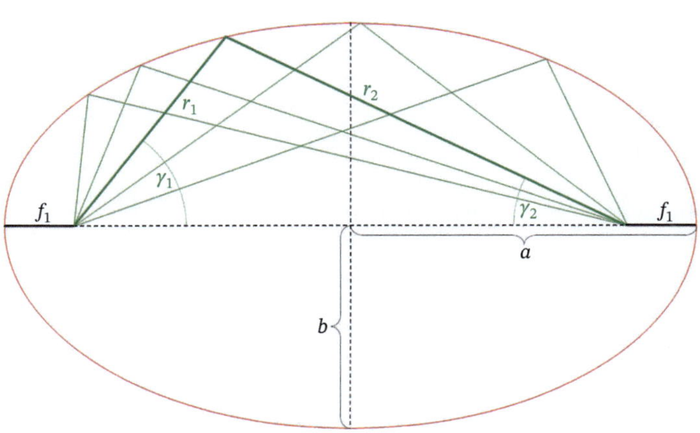

Figure 2.40 Stigmatic property of an ellipse. Foci, ray segment lengths and angles are shown.

The angle-dependent magnification of an ellipse is derived and discussed in [57], identifying again coma as the essential aberration, just like for the parabola in Knowledge Corner Topic 2.16.

Exercise 2.19: Numerical eccentricity of an ellipse (1)

In the 2D plane, a point source is located at (0,1), and the target point at (1,0). There is a family of ellipses which share just those two focal points (Figure 2.41). One of them touches the origin. Calculate its numerical eccentricity.

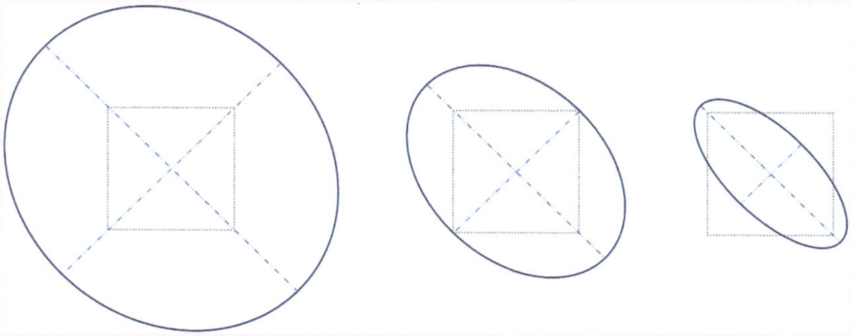

Figure 2.41 Ellipses with same focal points (the top left and bottom right corners of the square) but different numerical eccentricity.

Solution 2.19

Obviously, the distance between the foci is

$$2c = \sqrt{2} \tag{2.137}$$

An ellipse that also crosses the origin needs a semi-minor axis of

$$b = \frac{1}{\sqrt{2}} \tag{2.138}$$

Hence,

$$a = \sqrt{c^2 + b^2} = 1 \tag{2.139}$$

and the required numerical eccentricity is therefore

$$e = \frac{c}{a} = \frac{1}{\sqrt{2}} \tag{2.140}$$

Exercise 2.20: Numerical eccentricity of an ellipse (2)

A pencil of parallel rays is to be focused by an ellipsoid made of glass (index n, Figure 2.42). Is an ellipsoid even suitable for this purpose? If yes, which numerical eccentricity e is needed to do the job properly?

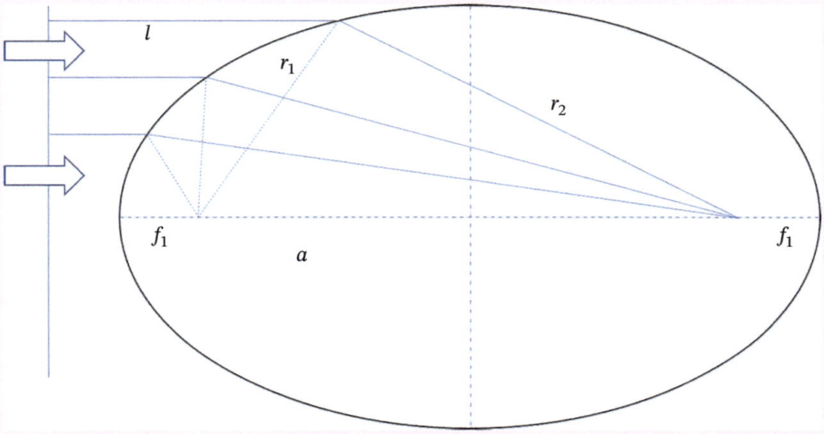

Figure 2.42 Focusing by an ellipsoid made of glass.

Solution 2.20

The criterion for a perfect image is an equal optical path length (OPL) for all rays. Parallel rays are perpendicular to the directrix, introduced in Knowledge Corner Topic 2.15. When continued to the distant focus, they must obey (see Figure 2.39):

$$\mathrm{OPL} = l + nr_2 = l + n\,(2a - le) = 2an + l\,(1 - ne) = \mathrm{const} \tag{2.141}$$

To determine e, we use the basic ellipse properties from Eqs. 2.133 and 2.99. This is fulfilled for all values of l if the term in parentheses vanishes. Consequently,

$$e = \frac{1}{n} \tag{2.142}$$

ensures perfect imaging by the glass ellipsoid. To answer the first question last: Yes, an ellipsoid is suitable.

Another way to see this: We know from the reflective stigmatic properties of the ellipse that $nr_1 + nr_2 = \mathrm{const}$. We need $l + nr_2 = \mathrm{const}$, i.e. we need $l = nr_1$. In other words, l must be a fixed multiple of r_1 for our constant OPL requirement, but that is exactly what the directric property, Eq. 2.99, tells us!

Simulation 2.3: Focusing by an ellipse

We set up an ellipsoid with foci distance $2c = 110$ mm made from N-SF6 ($n =$ 1.818 @ 530 nm, Figure 2.43). Thus we need

$$e = \frac{1}{n} = 0.55 \tag{2.143}$$

Hence, we need to set up an ellipsoid with

$$a = \frac{c}{e} = \frac{55\text{ mm}}{0.55} = 100\text{ mm} \tag{2.144}$$

and

$$b = \sqrt{a^2 - c^2} = 83.5\text{ mm} \tag{2.145}$$

Depending on which software you use, you may need the radius of curvature, i.e. the semilatus rectum ρ and the conic constant k instead of a and b. We calculate

$$\rho = \frac{b^2}{a} = a(1 - e^2) = 69.3\text{ mm} \tag{2.146}$$

and

$$k = -e^2 = -0.3025 \tag{2.147}$$

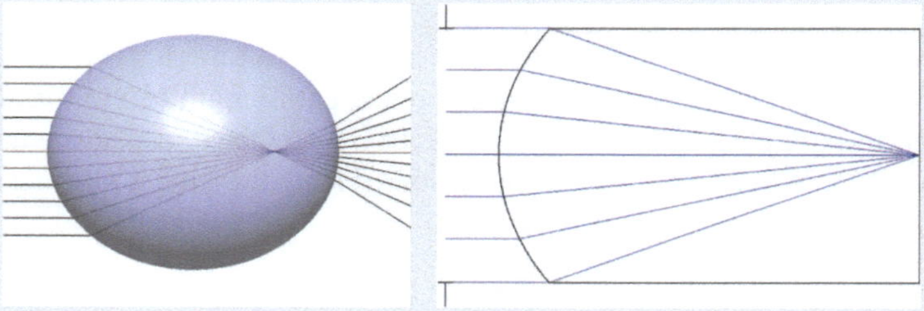

Figure 2.43 Ellipsoid made of N-SF6 focusing a parallel beam. Left: Ellipsoid modeled non-sequentially by the calculated semi axes. Right: Elliptical surface modeled in a sequential way from vertex curvature and conic constant.

The simulation shows that our calculation is correct, because the parallel beam is perfectly focused.

Provided LightTools file.

Exercise 2.21: Numerical eccentricity of an ellipse (3)

A small LED, modeled as a surface that emits light into one hemisphere, is placed in the focal point of a large ellipsoid, facing the vertex. We want the reflected beam to fill an angular cone of $\pm\theta = \pm 30°$, but not beyond that.

Calculate the numerical eccentricity of the ellipse.

Solution 2.21

From Figure 2.44, we see that the "vertical" ray that is tangent to the LED source is the one that passes the second focus with the desired maximal angle θ. The distance to this ray's intersection point with the ellipse is the *semilatus rectum* ρ, which is equal to the vertex radius of curvature, as we know from Figure 2.29, and is given by

$$R = \rho = \frac{b^2}{a} \tag{2.148}$$

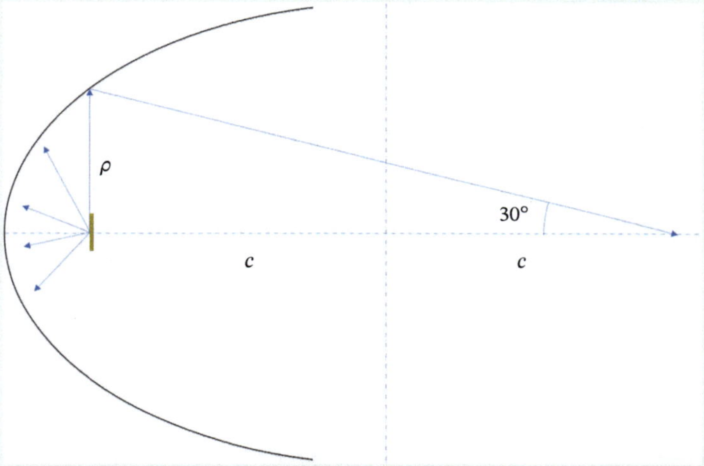

Figure 2.44 The maximum angle of incidence on the secondary focal plane for a hemispheric source inside an elliptical reflector results from the "vertical" angle.

Using the ellipse parameters given at the start of Section 2.5.2, we find that

$$\tan\theta = \frac{\rho}{2c} = \frac{b^2}{2ac} = \frac{a^2 - c^2}{2ac} = \frac{1 - e^2}{2e} \tag{2.149}$$

which, for $\theta = 30° \Rightarrow \tan\theta = 1/\sqrt{3}$ results in a numerical eccentricity of

$$e = \frac{1}{\sqrt{3}} \tag{2.150}$$

Alternatively, we could have used [57, Eq. 6, $\tan\frac{\gamma}{2}\tan\frac{\gamma'}{2} = m$]. With a small source at the focus and facing the near vertex, any elliptical reflector of exactly this numerical eccentricity would deliver just this angle, independent of its size. Of course, any finite source extension will blur the angular boundary.

Exercise 2.22: Cone angle from elliptical reflector

The semi-major and -minor axes of an elliptical lamp reflector are $a = 50$ mm and $b = 27$ mm (Figure 2.45). Assume that the front surface of the reflector is located in the center of the ellipse (the maximum possible size).

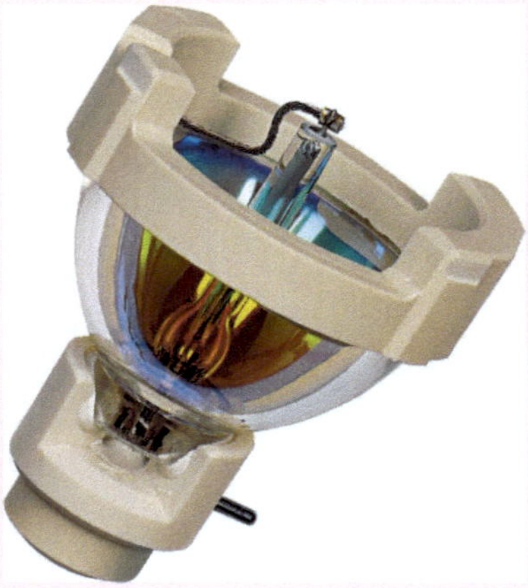

Figure 2.45 Short-arc mercury lamp (HBO) with an elliptical reflector. Courtesy of ams-OSRAM.

Calculate the angular spread of the beam from the reflector (neglecting source size).

Solution 2.22

The linear eccentricity of the ellipse is

$$c = \sqrt{a^2 - b^2} = 42.1 \text{ mm} \tag{2.151}$$

The maximum angle from the reflector with respect to the axis is then given by

$$\tan \gamma_2 = \frac{b}{c} = 0.64 \tag{2.152}$$

with $\gamma_2 = 32.7°$.

This may be too simple as the edge of the light distribution from the source may not exactly coincide with the reflector rim due to lamp electrodes blocking higher angles. The angular size of the lamp's beam may therefore be smaller as a result of the discharge vessel's intensity distribution, especially for DC lamps.

For example, let us assume a lamp emission angle of $\pm 45°$ w.r.t. the optical axis. We apply Eq. 2.134 with a numerical eccentricity of

$$e = \frac{c}{a} = \frac{42.1 \text{ mm}}{50 \text{ mm}} = 0.84 \tag{2.153}$$

which yields a focal length ratio of

$$m = \frac{1 - e}{1 + e} = 0.086 \tag{2.154}$$

and with

$$\tan\frac{45^\circ}{2}\tan\frac{\gamma_2}{2} = 0.086 \tag{2.155}$$

we obtain an angle $\gamma_2 = 23.5^\circ$, smaller than what is supported by the reflector.

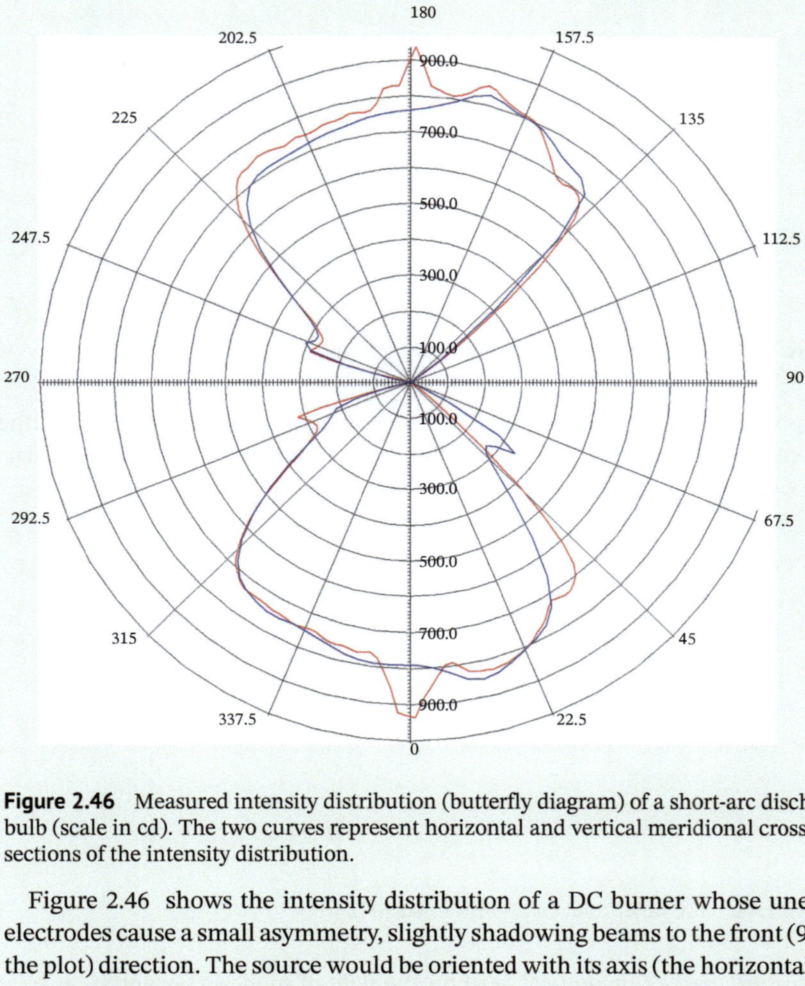

Figure 2.46 Measured intensity distribution (butterfly diagram) of a short-arc discharge bulb (scale in cd). The two curves represent horizontal and vertical meridional cross sections of the intensity distribution.

Figure 2.46 shows the intensity distribution of a DC burner whose unequal electrodes cause a small asymmetry, slightly shadowing beams to the front (90° in the plot) direction. The source would be oriented with its axis (the horizontal line in the diagram) along the optical axis of a reflector (its front opening located on the right, cf. Figure 2.40). The asymmetry suggests a DC "burner", which because of the large anode at the font side does not provide light in directions closer than 45° to the axis.

This is a light distribution curve (LDC) *of a source.* In Knowledge Corner Topics 4.2 and 4.10, we will introduce and use LDCs of *luminaires.*

The ellipse does not only enable stigmatic optics such as lens elements and reflectors, but also non-imaging devices such as mirror chambers and the compound elliptical concentrator (CEC), a close relative of the CPC.

Knowledge Corner Topic 2.18: Elliptical mirror chamber and CEC

In 1967, Martin Ploke published a paper on an *elliptical mirror chamber* [58] and related devices (Figure 2.47), driven by considerations to transport energy to a detector.

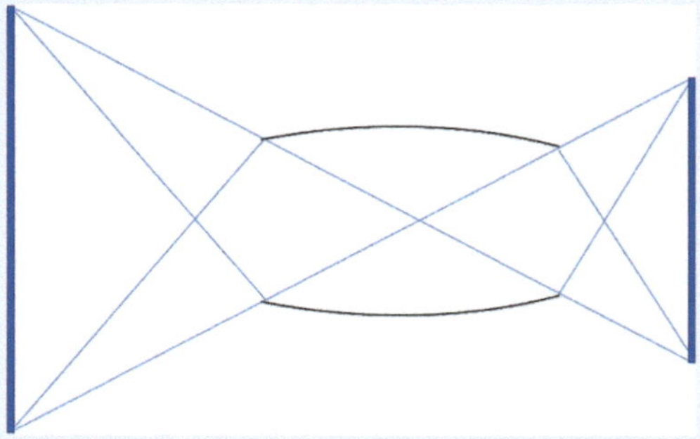

Figure 2.47 Elliptical Mirror Chamber. *Source:* Adapted from [58]/Public Domain.

Approximately ten years later, Roland Winston introduced the compound elliptical concentrator (CEC) as a relative to the ubiquitous CPC for solar collection (Figure 2.48), properly citing Ploke and earlier work [59].

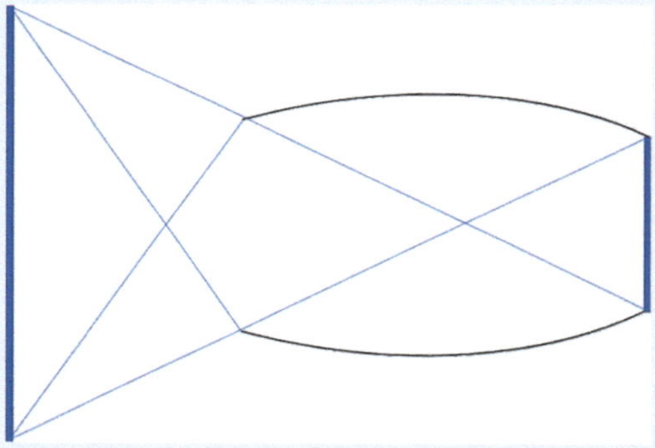

Figure 2.48 CEC (after Rabl and Winston [60, 61]).

This was part of theoretical work in the field of highest-concentration solar optics. A target (e.g. a high-temperature ceramic receiver or a multi-junction solar cell) receives light from a hemisphere and not just from an exit pupil such as in Figure 2.47.

There are many features that are identical to both designs:

- the entrance side (left), forming a light tube (Section 3.3) of source and aperture, defining the system etendue,
- the lower points of source and target act as the foci for the opposite elliptical curve,

- the coma rays (Figure 3.6) of the source and target light tubes coincide and define the length of the device by intersection with the ellipse.

The difference is at the exit side. In Ploke's version, there is a finite exit pupil, whereas in Winston's design, rays arrive from the whole hemisphere onto a receiver.

Dr. Wisenheimer:
While both Ploke's chamber and Winston's CEC are ideal as drawn, the ellipses may be shortened or extended. Then, there is not necessarily the same etendue on the entrance and exit sides.

Exercise 2.23: Elliptical mirror chamber

A source is shining into a pupil, and the beam needs to be transferred to a target (Figure 2.49).

Quantity	Value
Source area	$z = -5,\ r = [-1, 1]$
Pupil area	$z = -3,\ r_p = [-1.5, 1.5]$
Target area	$z = 5,\ R = [-1, 1]$

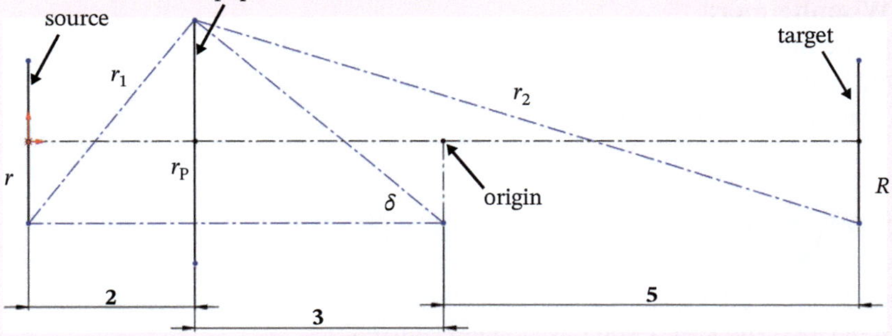

Figure 2.49 Design sketch for a mirror chamber (1). The angle δ is the polar angle of the ellipse (cf. Eq. 2.102).

Devise and sketch a mirror chamber for this purpose and calculate its parameters (ellipse and exit pupil).

Solution 2.23

The lower end points of source and receiver make the focal points of the upper ellipse. Hence the ellipse center is at ($z = 0$, $r = -1$), and obviously $c = 5$ mm (and in our case, the major axis of the ellipse is parallel to the centerline of source and target). We draw a radius r_1 to the upper end of the entrance pupil and similarly r_2.

We read from the sketch (with some assistance by Pythagoras) $r_1 = 3.2$ mm and $r_2 = 8.4$ mm, thus (cf. Eq. 2.133) $a = 5.8$ mm and $e = c/a = 0.86$. Indeed, such an ellipse has the required properties (Figure 2.50).

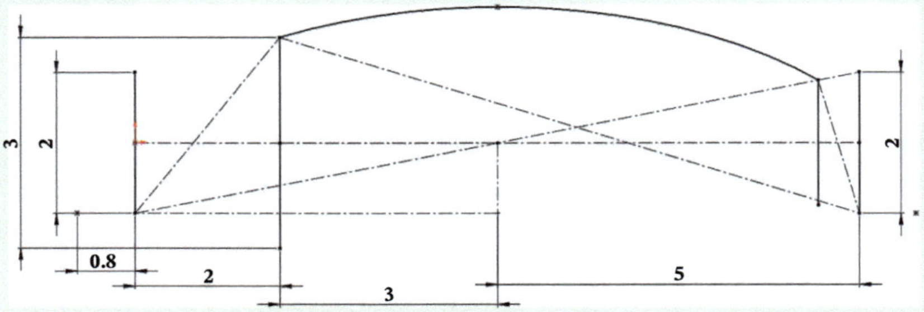

Figure 2.50 Design sketch for a mirror chamber. [2]/Princeton University Press.

The end of the chamber on the target side is found by requiring it to lie on the diagonal line that connects the lower end of the source and the upper end of the target (cf. Figure 2.50). This finalizes the design.

Dr. Wisenheimer:
There are not really many real-world appearances of CECs, but mostly beautiful sketches and theory and some designs [62–64]. You may be the one to change this.

Simulation 2.4: CEC in illumination software

Tools to set up a CPC come with some illumination software, but for the simulation of CEC (and CHC), you have to help yourself.

One way is to calculate surface points and shape some kind of tube (in LightTools: "skinned sheet") based on splines (Figure 2.51). Another method would be to create an ellipsoidal tool to carve out a hollow CEC from a bulk volume.

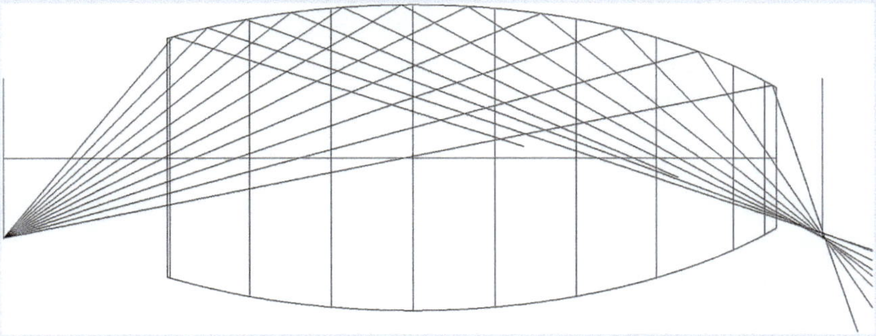

Figure 2.51 Mirror chamber set up using a calculated profile.

Provided files: Hottel_CEC.xlsx
CEC.lts.

In the simulation, we have a transfer efficiency of more than 95% (limited by geometric imperfections). The illuminance profile is inhomogeneous (Figure 2.52).

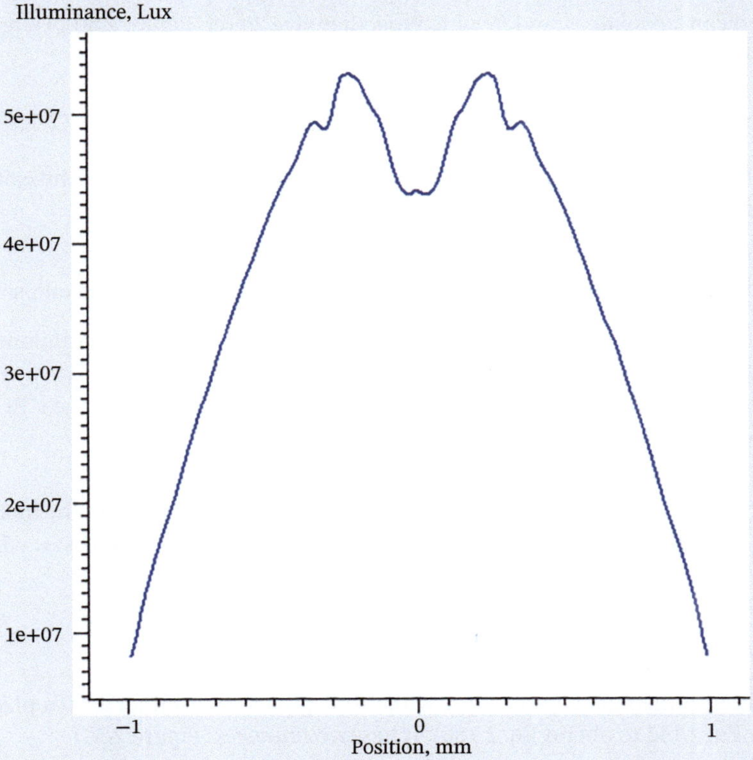

Figure 2.52 Illuminance distribution in the target area of the CEC. In contrast to the original papers, we have a situation with a source smaller than the entrance pupil.

(Continued)

Simulation 2.4: (Continued)

We finish with the comment that the etendue is not necessarily the same on the source and target sides.

Exercise 2.24: CEC length

Calculate the length of the CEC in Exercise 2.23.

Solution 2.24

In our particular case, the solution $L = 6$ mm seems straightforward because of symmetry.

In general, we would proceed as follows: The distance of the two focal points (lower rim of source and target) yields the linear eccentricity of the ellipse, in our case $2c = 10$. We have one point on the periphery (the upper point of the entrance pupil) and can calculate r_1 and r_2 which together give $2a$ (cf. Figure 2.40). Here, we have $r_1 = 3.2$ mm, $r_2 = 8.4$ mm, hence $2a = 11.6$ mm, $e = 0.86$ and

$$\rho = \frac{b^2}{a} = a(1 - e^2) = 1.5\,\text{mm} \tag{2.156}$$

The position of the rear end of the CEC will depend on how we choose the exit pupil. It could be either

- based on symmetry, or
- based on the intersection of the source–target diagonal ray with the ellipse.

These options may be different. The effect of various light tube configurations has yet to be investigated.

In any case, we have two points with their z coordinates which give the CEC length. See also [65].

There are not many applications of *oblate ellipses* in illumination optics. In the imaging world, there may be some aspheric lenses or perhaps cylindrical surfaces with $k > 0$ [66].

2.5.3 Hyperbola

A hyperbola is mathematically an ellipse with an imaginary semi-minor axis (replace b by $i \cdot b$ in Eq. 2.133 to obtain Eq. 2.156). It has two *branches* (Figure 2.53).

Its parameters are:

- semi-major axis a
- semi-minor axis b

- linear eccentricity $c = \sqrt{a^2 + b^2}$ (cf. Figure 2.53)
- numerical eccentricity $e = c/a$
- conic constant $k = -e^2$
- semilatus rectum $\rho = b^2/a$
- vertex radius of curvature ρ (equal to semilatus rectum)
- asymptotes $\tan\phi = \pm(b/a)z$
- small focal length $f_1 = c - a$
- large focal length $f_2 = c + a$
- focal ratio $m = f_1/f_2$

Its standard form or center equation (origin at $r = z = 0$) is

$$\frac{r^2}{(z+a)(z-a)} = \frac{b^2}{a^2} \tag{2.157}$$

(just a tiny difference to Eq. 2.132), which is equivalent to

$$\frac{z^2}{a^2} - \frac{r^2}{b^2} = 1 \tag{2.158}$$

For the hyperbola, we have a constant *difference* of the distances (Figure 2.54)

$$r_1 - r_2 = 2a \tag{2.159}$$

and the angles are transformed according to

$$\tan\frac{\gamma_1}{2}\cot\frac{\gamma_2}{2} = -m^{\pm 1} \tag{2.160}$$

which may be important for the design of a hyperbolic reflector as described in [16].

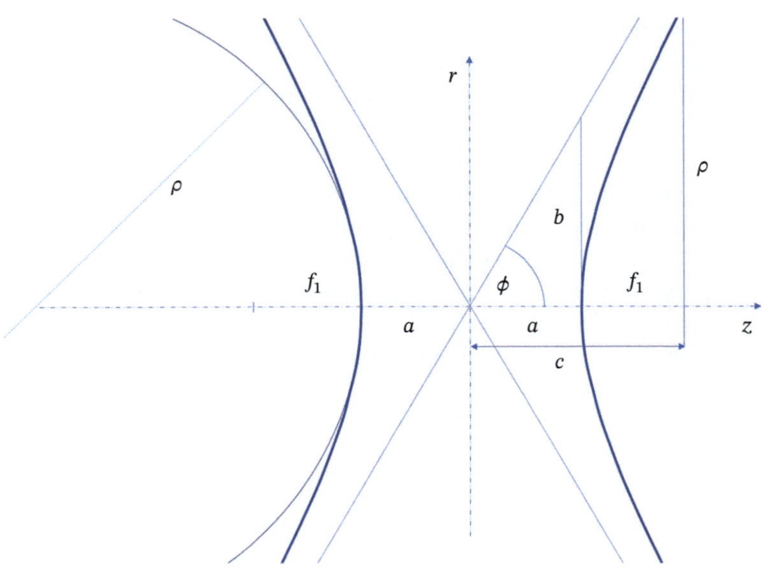

Figure 2.53 The hyperbola and its parameters.

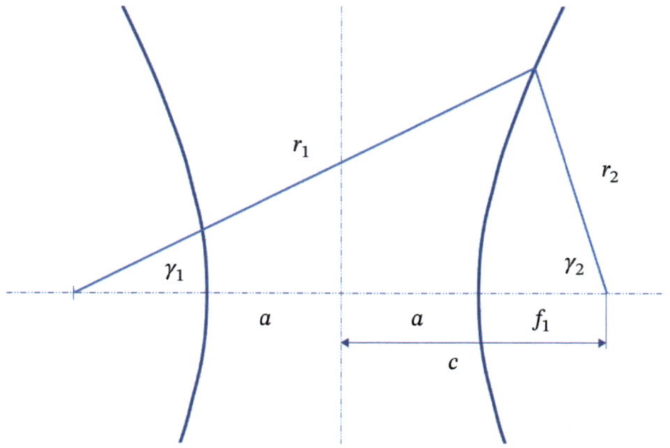

Figure 2.54 Hyperbola radii and angles.

Knowledge Corner Topic 2.19: CHC and trumpet

There are two non-imaging devices featuring a hyperbolic shape: the "trumpet" and the compound hyperbolic concentrator (CHC).

Both elements are hyperbolic and compound, but their setup is slightly different. Both connect a low-concentration region and a high-concentration region [67] (we will use the terms "LCR" and "HCR" for brevity).

When acting as a *concentrator*, both devices will "non-image" a beam that is incident on the LCR to the HCR.

When working as a *collimator*, a source in the HCR will be transformed into a virtual source of larger area (LCR) and smaller angular distribution. The LCR is always virtual.

Etendue conservation between HCR and LCR is ensured because the area increase is accompanied by an angular decrease.

We discuss both the trumpet and CHC just in a 2D cross section and rotate the design.

Trumpet

A trumpet is generated by rotating the branches of a hyperbola around an axis through the center but perpendicular to the hyperbola axis (Figure 2.55). Its function is to perfectly concentrate a beam that is directed to the LCR onto the HCR. The LCR connects the focal points of the hyperbola. Therefore, rays that are directed to one focal point are reflected on the other. Depending on the actual ray, *multiple reflections* may occur, rendering the principle less efficient for a reflectivity smaller than 100 %.

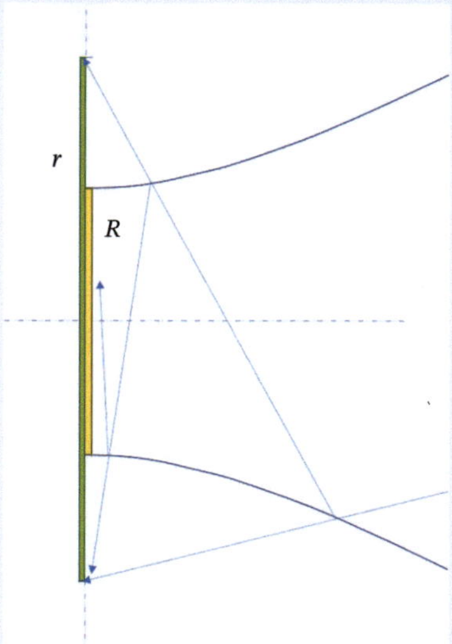

Figure 2.55 Trumpet. High concentration region (HCR) in yellow, low concentration region (LCR) in green.

The design parameters of a trumpet are the diameters of the HCR ($2R = 2a$) and the LCR ($2r = 2c$), which completely define the hyperbola, its focal points and branches. Thus numerical eccentricity is

$$e = \frac{c}{a} = \frac{r}{R} \tag{2.161}$$

CHC

Compared to the ubiquitous CPC, the CHC is not really abundant in the literature [68], but some math regarding CEC and CHC is given in [65, 69].

In contrast to the trumpet, a CHC works via *inner reflection* from the hyperbola. The design parameters are the size of the LCR ($2r$) and the HCR ($2R$) and their distance D. This arrangement seems to form kind of a light tube, not in the sense that it defines an etendue, but we still may benefit from using the terms.

(Continued)

Knowledge Corner Topic 2.19: (Continued)

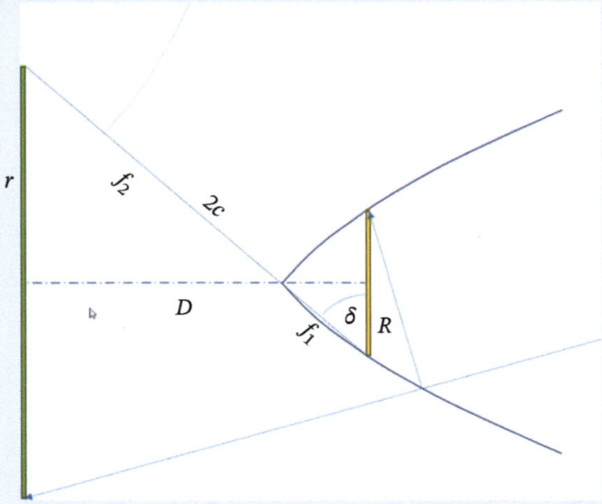

Figure 2.56 CHC design. HCR in yellow, LCR in green. The upper rim of the LCR and the lower rim of the HCR are the focal points of the top hyperbola (only right branch shown). The actual CHC consists of the thick hyperbola sections right of the HCR.

We extract from Figure 2.56 that

$$\tan\delta = \frac{D}{r+R} \tag{2.162}$$

and the distance of the foci is

$$2c = \frac{D}{\sin\delta} \tag{2.163}$$

The diagonal angle θ_D of the light tube (to be introduced in Figure 3.6) is related to

$$\delta = 90° - \theta_D \tag{2.164}$$

which is identical to the polar angle of the hyperbola (cf. Figure 2.56), and r_2 of Figure 2.54 is equal to $2R$ of the light tube. So Eq. 2.102 states

$$2R = \frac{\rho}{1 + e\cos\delta} \tag{2.165}$$

and for $\delta = 0$ we have

$$f_1 = \frac{\rho}{1+e} = c\,\frac{e-1}{e} \tag{2.166}$$

which together deliver the unknown e and ρ.

Dr. Wisenheimer:

What happens for a decreasing distance D? Does the CHC possibly converge to a trumpet? Well, almost.

Rays that are shot against an upper hyperbola focus (upper rim of the LCR in Figure 2.57) are reflected.

- toward the lower rim of the HCR (CHC)
- toward the lower rim of the LCR (trumpet).

The CHC variant appears to be superior because it needs only one reflection to reach the HCR target.

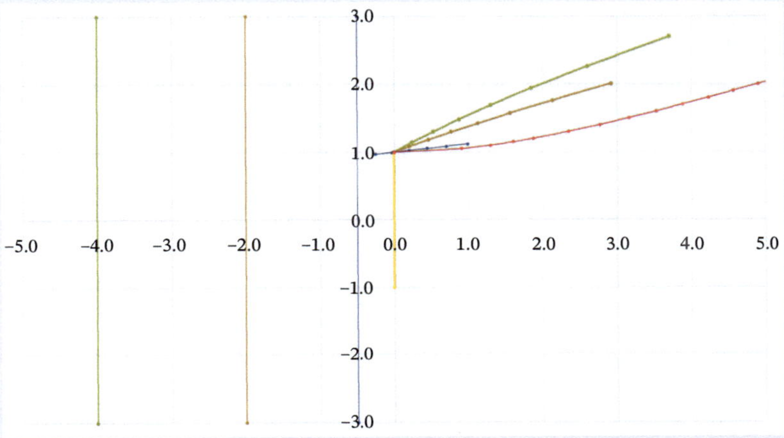

Figure 2.57 CHC almost approaching a trumpet: Three light tubes of different center distance but same LCR size (green, orange, blue) and identical HCR (yellow) are defining three CHCs (only upper branches shown). In addition, a trumpet is drawn (red) with the definition similar to the blue CHC.

Provided file hyperbola.xlsx with some calculus on CHC and trumpet.

Exercise 2.25: Trumpet

A beam is incident on the target $z = 0, r = [-2, 2]$. Your task is to concentrate it to the new target $z = 0, r = [-1, 1]$. Calculate the parameters of a trumpet that will concentrate the beam to the new target.

Solution 2.25

From the task we take the vertex distance (HCR)

$$2a = 2 \tag{2.167}$$

and the distance of the foci (LCR) is

$$2c = 4 \tag{2.168}$$

(Continued)

Solution 2.25 (Continued)

and

$$b = \sqrt{3} \tag{2.169}$$

Hence,

$$e = \frac{c}{a} = 2 \tag{2.170}$$

so we can calculate the branches on the basis of Eq. 2.156 or 2.102 (Figure 2.58).

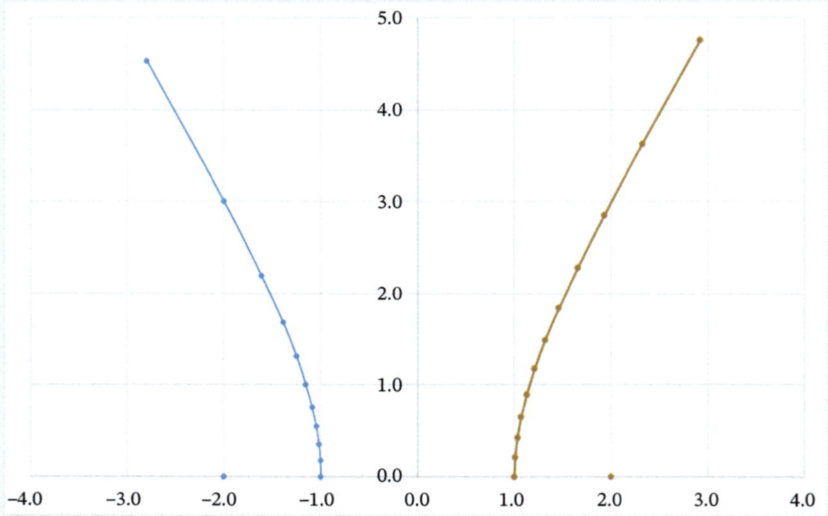

Figure 2.58 Trumpet solution.

Exercise 2.26: CHC

A source is located at $z = 4$, $r = [-1, 1]$. Sketch a CHC to convert it into a virtual source that is radiating from $z = 0$, $r = [-2, 2]$. Calculate the parameters of a CHC.

Solution 2.26

Eq. 2.161 delivers a focal distance of $c = 2.5$. We solve for numerical eccentricity and get $e = 2.4$. Hence the semi-major axis is $a = 1.1$, $\rho = 4.8$ and we can plot using Eq. 2.102. The result is shown in Figure 2.59.

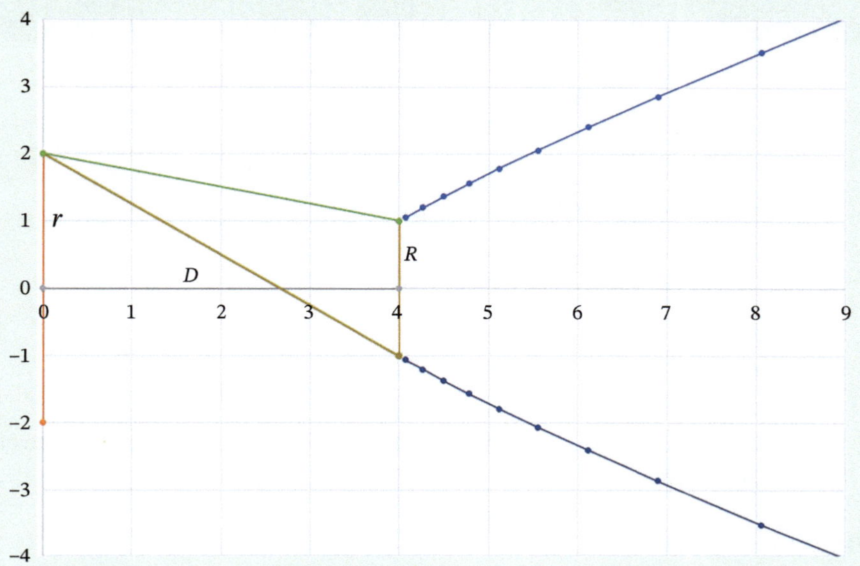

Figure 2.59 CHC design for given parameters.

So far we have only sketched the branches, but a CHC also has a well-defined length L for proper function.

True light tubes are formed by the front area (radius R_F) of the CHC and the LCR (R_F, r, $L + D$) (defining the acceptance of the CHC) and the front and HCR (R_F, R, L) areas, with equal etendue.

Exercise 2.27: CHC length

Calculate the length and front diameter of a CHC based on the above HCR and LCR in a distance D.

Solution 2.27

Their diagonal rays from the rim of the front and the coma ray (slope θ_C, Figure 3.6) of the pseudo light tube (r, R, D) coincide [65].

Together with the polar form from Eq. 2.102 we obtain (Figure 2.60)

$$L = \frac{\rho \cos \theta_C}{1 - e \sin(\delta - \theta_C)} \tag{2.171}$$

Again, we used a combination of the polar equation of a conic with some angular term to calculate the length.

(Continued)

Solution 2.27 (Continued)

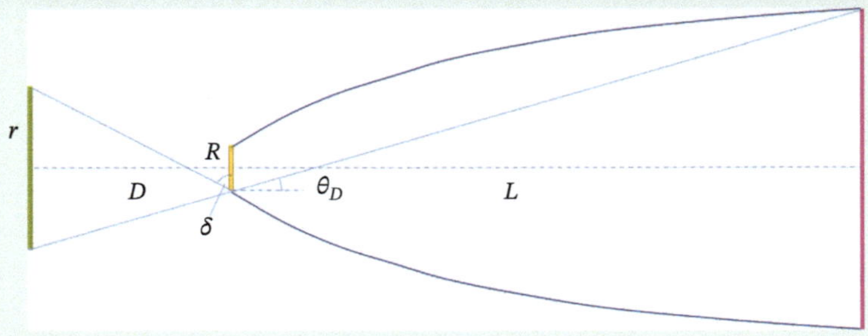

Figure 2.60 CHC length calculation.

Have a look at Knowledge Corner Topic 3.8, where we investigate the etendue of such arrangements.

Dr. Wisenheimer:
There are a few hardware realizations of CHCs [68] and trumpets [70, 71], but many more numerical investigations [72] and "proposed systems" [72, 73].

References

1 Prando, G. (2020). A spectral unit. *Nature Physics* 16 (8): 888. https://doi.org/10.1038/s41567-020-0997-3.

2 Needham, T. (2021). *Visual Differential Geometry and Forms: A Mathematical Drama in Five Acts*. Princeton, NJ: Princeton University Press.

3 Euler, L. (1778). De mensura angulorum solidorum. *Opera Omnia* 26. https://scholarlycommons.pacific.edu/euler-works/514/

4 Hopf, H. (1940). Selected chapters of geometry, https://pi.math.cornell.edu/~hatcher/Other/hopf-samelson.pdf.

5 Van Oosterom, A. and Strackee, J. (1983). The solid angle of a plane triangle. *IEEE Transactions on Biomedical Engineering* 30 (2): 125–126. https://doi.org/10.1109/TBME.1983.325207.

6 Eriksson, F. (1990). On the measure of solid angles. *Mathematics Magazine* 63 (3): 184–187. https://doi.org/10.1080/0025570X.1990.11977515.

7 Khadjavi, A. (1968). Calculation of solid angle subtended by rectangular apertures. *Journal of the Optical Society of America* 58 (10): 1417. https://doi.org/10.1364/JOSA.58.001417.

8 Mathar, R.J. (2020). Solid angle of a rectangular plate, https://rxiv.org/pdf/2001.0603v1.pdf

9 Illuminating Engineering Society of North America (2012). *LM-75-01 Goniophotometer Types and Photometric Coordinates*. Lighting Measurements Testing & Calculation Guides. Illuminating Engineering Society of North America.

10 Wolf, B. (2004). *Geometric Optics on Phase Space*. Springer.
11 Born, M. and Wolf, E. (1999). *Principles of Optics*, 7e. Cambridge University Press.
12 Feynman, R.P. (1985). *QED: The Strange Theory of Light and Matter*. Alix G. Mautner Memorial Lectures. Princeton, NJ: Princeton University Press.
13 Yoder, P.R. (2005). *Opto-Mechanical Systems Design*, 3e. CRC Press.
14 Roeder, J.L. (2007). Why light won't refract through adjacent faces of a cube. *The Physics Teacher* 45 (3): 182. https://doi.org/10.1119/1.2709680.
15 Ries, H. and Rabl, A. (1994). Edge-ray principle of nonimaging optics. *Journal of the Optical Society of America A* 11 (10): 2627. https://doi.org/10.1364/JOSAA.11.002627.
16 Muschaweck, J. and Rehn, H. (2022). *Designing Illumination Optics*. SPIE.
17 Arecchi, A.V., Koshel, R.J., and Messadi, T. (2007). *Field Guide to Illumination*. Bellingham WA: SPIE.
18 Simoff, D.A., Stolov, A.A., Wu, H. et al. (2017). Evolution of fluoropolymer clad fibers. Proceedings of the 66th IWCS Conference. https://iwcs.conferencespot.org/event-data/pdf/a379/fl00353.
19 Greivenkamp, J.E. (2004). *Field Guide to Geometrical Optics. Number 1 in SPIE Field Guides*. Bellingham, WA: SPIE Press.
20 Riedl, M.J. (2009). *Optical Design*. Bellingham, WA: SPIE. https://doi.org/10.1117/3.835815.
21 Herman, E., Aikens, D., and Youngworth, R.N. (2022). *Modern Optics Drawings: The ISO 10110 Companion*. SPIE Press.
22 Geary, J.M. (2002). *Introduction to Lens Design: with Practical ZEMAX Examples*. Richmond, VA: Willmann-Bell.
23 Marshall, G.F. and Stutz, G.E. (2012). *Handbook of Optical and Laser Scanning*, 2e. Boca Raton: CRC Press
24 Kogelnik, H. and Li, T. (1966). Laser beams and resonators. *Proceedings of the IEEE* 54 (10): 1312–1329. https://doi.org/10.1364/AO.5.001550.
25 Brooker, G. (2002). *Modern Classical Optics*, 1e. Oxford: Oxford University Press, Optical and Laser Physics.
26 Zangwill, A. (2013). *Modern Electrodynamics*. New York, NY: Cambridge University Press. https://doi.org/10.1364/AO.5.001550.
27 Marshall, G.F. (2017). *Laser Beam Scanning: Opto-Mechanical Devices, Systems, and Data Storage Optics*, 1e. Routledge.
28 Paschotta, R. (2008). *Field Guide to Lasers. Number 12 in SPIE Field Guides*. Bellingham, WA: SPIE Press.
29 Self, S.A. (1983). Focusing of spherical Gaussian beams. *Applied Optics* 22 (5): 658. https://doi.org/10.1364/AO.22.000658.
30 Dandelin, G.P. (1822). Mémoire sur quelques propriétés remarquables de la focale parabolique. *Nouveaux mémoires de l'Académie royale des sciences et belles-lettres de Bruxelles*.
31 Heath, T.L. (1896). *Appolonius of Perga – Treatise of Conic Sections*. Cambridge University Press.
32 Taylor, C. (1881). *An Introduction to the Ancient and Modern Geometry of Conics*. Cambridge: Deighton Bell. https://www.abebooks.com/9781523930067/Introduction-Ancient-Modern-Geometry-Conics-1523930063/plp.

33 Ayers, S. (2010). String theory on medieval masons and carpenters drawing an ellipse, http://www.sbebuilders.com/tools/geometry/treatise/String-Theroy-Ellipse.html.

34 Van Maanen, J. (1992). Seventeenth century instruments for drawing conic sections. *The Mathematical Gazette* 76 (476): 222–230. https://doi.org/10.2307/3619131.

35 Blake, E.M. (1900). The ellipsograph of proclus. *American Journal of Mathematics* 22 (2): 146–153. https://doi.org/10.2307/2369752.

36 Strubecker, K. (1967). *Vorlesungen Über Darstellende Geometrie.* Göttingen: Vandenhoeck & Ruprecht. https://people.math.harvard.edu/~knill/history/darstellend/Strubecker.pdf#page=37\&zoom=110,-142,564.

37 Kleinhans, W.A. (1976). Aberrations of a sequence of conic mirrors. *Applied Optics* 15 (9): 2283–2287. https://doi.org/10.1364/AO.15.002283.

38 Ostroumov, N., Gordon, J.M., and Feuermann, D. (2009). Full landscape of aplanatic optics for concentration near the thermodynamic limit. *Proceedings of SPIE* 7423. https://doi.org/10.1117/12.824189.

39 Korsch, D. (2012). *Reflective Optics.* Burlington: Elsevier Science.

40 Gordon, J.M. (2010). Aplanatic optics for solar concentration. *Optics Express* 18 (101): A41–A52. https://doi.org/10.1364/OE.18.000A41.

41 True, T. (1984). Constant magnification light collection system. US4642740, https://patents.google.com/patent/US4642740A/en.

42 Sekiguchi, A., Sasagawa, T., and Goto, Y. (2004). Étendue-density homogenization lens optimized for high-pressure mercury lamps. *Journal of the Society for Information Display* 12 (1): 105. https://doi.org/10.1889/1.1824246.

43 Centina, A.D. (2016). On Kepler's system of conics in Astronomiae pars optica. *Archive for History of Exact Sciences* 70 (6): 567–589. https://doi.org/10.1007/s00407-016-0175-2.

44 Brennesholtz, M.S. (1996). Light collection efficiency for light valve projection systems. *Projection Displays SPIE* 2650: 71–79. https://doi.org/10.1117/12.237020.

45 Rehn, H. (2014). Luminaire for illuminating a target area by means of retroreflection of light from a light-emitting diode module on a reflector. US8801230, https://patents.google.com/patent/US8801230B2/en.

46 Ries, H. and Spirkl, W. (1996). Nonimaging secondary concentrators for large rim angle parabolic troughs with tubular absorbers. *Applied Optics* 35 (13): 2242–2245. https://doi.org/10.1364/AO.35.002242.

47 Muschaweck, J. and Rehn, H. (2019) Illumination design patterns for homogenization and color mixing. *Advanced Optical Technologies* 8 (1): 13–32. https://doi.org/10.1515/aot-2018-0051.

48 Brennesholtz, M.S. and Stupp, E.H. (2008). *Projection Displays.* New York: Wiley. https://www.wiley.com/en-us/Projection+Displays%2C+2nd+Edition-p-9780470770917.

49 Winston, R. (1970). Light collection within the framework of geometrical optics. *Journal of the Optical Society of America* 60 (2): 245. https://doi.org/10.1364/JOSA.60.000245.

50 Rabl, A. (1976). Comparison of solar concentrators. *Solar Energy* 18 (2): 93–111. https://doi.org/10.1016/0038-092X(76)90043-8.

51 Gordon, J.M. and Rabl, A. (1992). Nonimaging compound parabolic concentrator-type reflectors with variable extreme direction. *Applied Optics* 31 (34): 7332–7338. https://doi.org/10.1364/AO.31.007332.

52 Winston, R., Miñano, J.C., and Benitez, P. (2005). *Nonimaging Optics*. Academic Press.

53 Chaves, J. (2016). *Introduction to Nonimaging Optics*, 2e. CRC Press.

54 Haring, J. and Pastuszka, N. (2018). Design considerations for compound parabolic concentrators. *Optiforms White Paper,* 8.

55 Bierhuizen, S.J. and Harbers, G. (2009). Illumination system with optical concentrator and wavelength converting element. https://patents.google.com/patent/US7543959B2/en?oq=US7543959

56 Benítez, P., Miñano, J.C., Zamora, P. et al. (2010). High performance Fresnel-based photovoltaic concentrator. *Optics Express* 18 (S1): A25. https://doi.org/10.1364/OE.18.000A25.

57 Rehn, H. (2004). Optical properties of elliptical reflectors. *Optical Engineering* 43 (7): 1480. https://doi.org/10.1117/1.1755231.

58 Ploke, M. (1967). Lichtführungseinrichtungen mit starker Konzentrationswirkung. *Optik* 25: 31–43.

59 Baranov, V.K. and Melnikov, G.K. (1966). Study of the illumination characteristics of hollow focons. *Soviet Journal of Optical Technology* 33 (5): 408–411.

60 Rabl, A. and Winston, R(1976). Ideal concentrators for finite sources and restricted exit angles. *Applied Optics* 15 (11): 2880. https://doi.org/10.1364/AO.15.002880.

61 Winston, R. (1978). Cone collectors for finite sources. *Applied Optics* 17 (5): 1539–4522. https://doi.org/10.1364/AO.17.000688.

62 Lu, Y., Zhang, X., and Wu, R. (2024). Utilizing an aspheric lens and compound ellipsoidal cavity for a laser uniform illumination system. *Applied Optics*, 63 (27): 7225–7231. https://doi.org/10.1364/AO.530431.

63 Canavarro, D., Chaves, J., and Collares-Pereira, M. (2017). New dual asymmetric CEC linear Fresnel concentrator for evacuated tubular receivers. *AIP Conference Proceedings* 1850: 040001.

64 Garcia-Botella, A., Fernandez-Balbuena, A.A., and Bernabeu, E. (2006). Elliptical concentrators. *Applied Optics* 45 (29): 7622. https://doi.org/10.1364/AO.45.007622.

65 Eichhorn, W.L. (1982). Generalized conic concentrators. *Applied Optics* 21 (21): 3887. https://doi.org/10.1364/AO.21.003887.

66 Wang, W., Lin, L., Yang, X. et al. (2008). Design of oblate cylindrical perfect lens using coordinate transformation. *Optics Express* 16 (11): 8094–8105. https://doi.org/10.1364/OE.16.008094.

67 Winston, R. and Gallagher, J. (1980). Fundamentals and techniques of nonlmaging optics for solar energy concentration, https://www.osti.gov/servlets/purl/5415106

68 Kong, C.R., Barman, I., Dingari, N.C. et al. (2011). A novel non-imaging optics based Raman spectroscopy device for transdermal blood analyte measurement. *AIP Advances* 1 (3): 032175. https://doi.org/10.1063/1.3646524.

69 Jenkins, D. and Winston, R. (1996). Integral design method for nonimaging concentrators. *Journal of the Optical Society of America A* 13 (10): 2106. https://doi.org/10.1364/JOSAA.13.002106.

70 O'Gallagher, J. and Winston, R. (1986). Test of a "trumpet" secondary concentrator with a paraboloidal dish primary. *Solar Energy* 36 (1): 37–44. https://doi.org/10.1016/0038-092X(86)90058-7.

71 Spencer, C., Xie, Z., and Geralds, T. (2020). Hyperbolic ceiling reflector for directional light sources. US10670227B2, https://patents.google.com/patent/US10670227B2/en.

72 Ustaoglu, A., Ozbey, U., and Torlaklı, H. (2020). Numerical investigation of concentrating photovoltaic/thermal (CPV/T) system using compound hyperbolic –trumpet, V-trough and compound parabolic concentrators. *Renewable Energy* 152: 1192–1208. https://doi.org/10.1016/j.renene.2020.01.094.

73 Georlette, O. and Gordon, J.M. (1994). Generalized nonimaging compound elliptical and compound hyperbolic luminaire designs for pair-overlap illumination applications. *Applied Optics* 33 (19): 4171. https://doi.org/10.1364/AO.33.004171,

3

Theory

This chapter introduces the knowledge required for radiometric calculations, which are ultimately the basis for solving illumination design tasks. You will learn how to evaluate light distributions both from a source and toward a target, modify them as needed for the task, and understand fundamental limitations.

3.1 Rays

In mathematics, a ray is a one-dimensional straight line, specified by six numbers: the coordinates of its origin position and its direction. When we try to apply the ray concept in optics, we note that a single ray can hardly be properly prepared experimentally, but the whole concept still remains useful as "pencils behave as they were composed of single rays" [1, Ch 1]. Hence, light beams and light distributions in an optical system can be successfully described by a continuous multitude of rays. And the *tracing* of rays through an optical system has been introduced to computers a long time ago.

The use of rays in illumination optics is based on the introduction of *screens* and the definition of a ray at the selected screen.

Knowledge Corner Topic 3.1: Screens

A screen is a virtual surface introduced into an optical system (Figure 3.1). It is not identical to any actual optical surface but may be as close to it as necessary. Moreover, it does not need to be planar and may be situated inside or outside a medium.

(Continued)

Engineering Illumination Optics: Case Studies and Exercises, First Edition. Henning Rehn and Julius Muschaweck.
Published 2026 by Wiley-VCH GmbH.

Knowledge Corner Topic 3.1: (Continued)

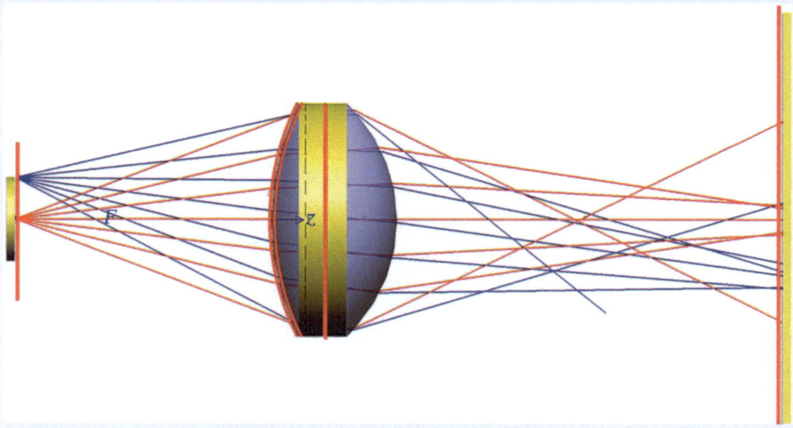

Figure 3.1 Screens (shown as thick red lines) in an optical system.

A smart screen setup can minimize the mathematical effort necessary to calculate the etendue. A smart choice, for example, is at a position where the beam is collimated, i.e. where the same angular distribution is found at all spatial positions. If this is the case, one can separate angles and areas as in Eq. 3.8. Other common choices are directly outside the source, directly before a target surface, or at the opening of a baffle.

We define two coordinate systems on the screen, one for spatial coordinates and one for ray directions. Consequently, we can describe a ray by only *four* numbers, in contrast to the six numbers required in 3D space. We save one spatial coordinate by describing the known position on the 2D screen in real space by two coordinates. One directional coordinate is eliminated by describing the direction in terms of a *momentum* or *ray vector* [2] $\mathbf{k}$ with the requirement that the length of the ray vector be equal to the (local) refractive index,

$$|\mathbf{k}| = \sqrt{k_x^2 + k_y^2 + k_z^2} = n \tag{3.1}$$

With this normalization, the components k_x, k_y, which are tangent to the screen, suffice to determine the direction vector. This is the natural choice for phase-space coordinates. However, it is sometimes more convenient, or intuitive, to use the latitude θ (the angle of a ray with respect to the optical axis),

$$k_z = n \cos \theta \tag{3.2}$$

and the longitude φ determined by

$$\tan\varphi = \frac{k_y}{k_x} \tag{3.3}$$

(an actual implementation would use some sort of ATAN2 function).

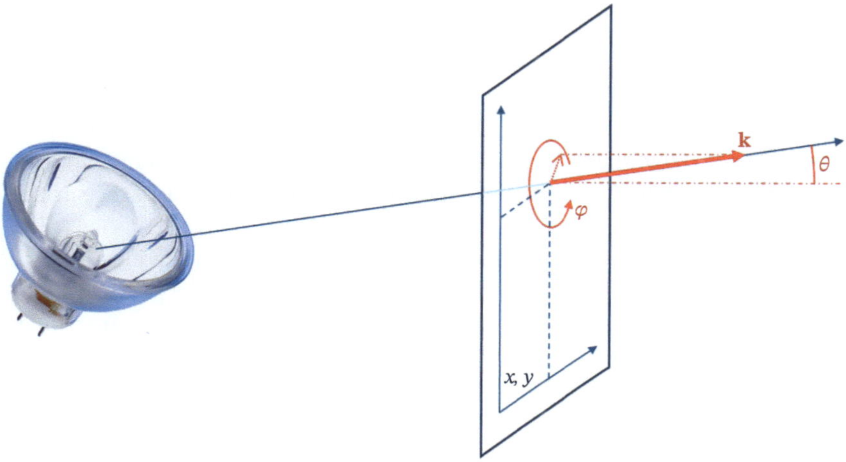

Figure 3.2 A ray crossing a screen. Spatial screen coordinates x, y (in blue) result from defining a spatial coordinate system. Angular coordinates (in red) are specified as polar angle θ and azimuth angle φ, or as the in-plane tangential components (small dotted red arrow) of the optical momentum vector $\mathbf{k}$ (thick red arrow).

Thus, we describe any ray r at its intersection point with the screen by a quadruplet of numbers $r = (x, y, k_x, k_y)$ combining spatial Cartesian coordinates x, y on the screen with directional coordinates k_x, k_y (Figure 3.2). These four numbers span a four-dimensional space, the phase space of geometrical optics [3, 4]. An extended ray bundle crossing a screen fills a volume in this phase space, very much like the points of which an object consists fill a volume in ordinary 3D space. In phase space, the volume of a ray bundle is called its *étendue* (commonly – and in this book as well – also written *etendue*, without the accent on the e). The term has been used in illumination optics for several decades now, but related concepts and definitions were known before [5–7], and other fields use different terms for very similar concepts as well (see Knowledge Corner Topic 3.5).

3.2 Etendue

Knowledge Corner Topic 3.2: Etendue

Etendue, i.e. volume in phase space, is a *purely geometrical measure* of the spatial and angular extent of a ray bundle that is determined whenever this ray bundle intersects a certain screen. The infinitesimal building block of etendue, dU, is obtained from an infinitesimally thin light tube (Section 3.3) around a ray.

(Continued)

Knowledge Corner Topic 3.2: (Continued)

We consider an infinitesimal area element $\mathrm{d}A$, embedded in a material with index n, which sends radiation along that ray into an infinitesimal solid angle $\mathrm{d}\Omega$. The ray forms an angle θ with the surface normal to $\mathrm{d}A$. Then, $\mathrm{d}U$ is given by

$$\mathrm{d}U = n^2 \cos\theta \, \mathrm{d}\Omega \, \mathrm{d}A \tag{3.4}$$

Etendue is essentially the product of area times solid angle, but with two additional factors: The $\cos\theta$ factor accounts for the geometrical fact that when we look back along the ray toward $\mathrm{d}A$, the tilt angle θ makes the area element appear smaller. The n^2 factor accounts for the physical fact that refraction into an optically denser material decreases the angle of the ray with respect to the surface normal.

Since n is the local refractive index, and θ is the angle with the local surface normal, the procedure nicely generalizes to curved screens and/or gradient-index materials, but in most cases, materials are homogeneous, and screens are planar.

To calculate the etendue in a certain situation or for a certain element, we need to integrate within defined limits. The angular extent, introduced in Knowledge Corner Topic 2.9 as

$$X = \int n^2 \cos\theta \, \mathrm{d}\Omega = \int n^2 \, \mathrm{d}\Omega_\mathrm{p} \tag{3.5}$$

may actually depend on the location of the surface element $\mathrm{d}A$. The etendue U is therefore obtained by integrating X over the area,

$$U = \int X \, \mathrm{d}A \tag{3.6}$$

or, combined into one equation,

$$U = \iint n^2 \cos\theta \, \mathrm{d}\Omega \, \mathrm{d}A \tag{3.7}$$

Dr. Wisenheimer:

Etendue is a purely geometric property of an optical component or system, and it does not depend on the specific rays or the light distribution you may send into it (or on whether you shine some light at it at all).

Figure 3.3 illustrates the steps toward etendue.

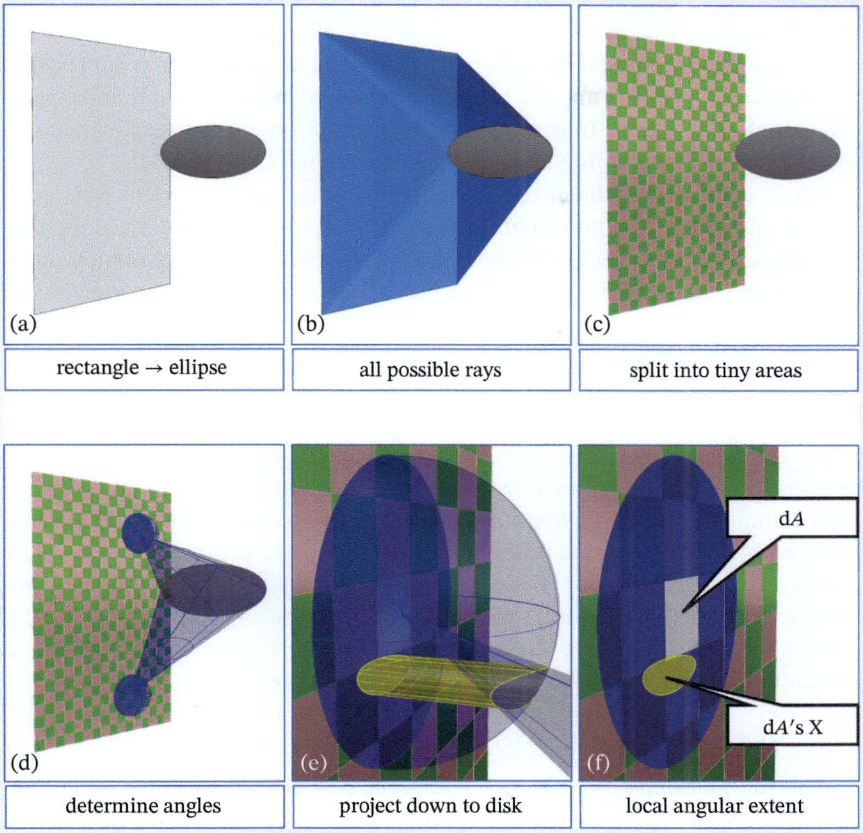

Figure 3.3 Assembling etendue. a) We consider a rectangular screen and an elliptical aperture. b) The goal is to compute the etendue of all rays that leave the screen and intersect the elliptical aperture. c) The first step is to split up the screen area into infinitely many, infinitesimally small area elements. d) For each area element, we determine which rays from that area element belong to the desired ray bundle. e) To each area element, we apply a tangent circular disk, with a hemisphere on top of it, with dimensionless radius n, and find the area on the hemisphere which is formed by the rays that hit the elliptical target from here. This area on the hemisphere is then projected down to the tangent disk. f) For each area element $\mathrm{d}A$, we have its associated angular extent X: the size of this down-projected shape, measured in "dimensionless units" of steradians (sr).

We use a simple rectangle-to-ellipse light tube as our ray bundle, but the procedure easily generalizes to more complicated scenarios. The screen may be curved, may have holes, may even consist of disconnected pieces. The angular extent is

(Continued)

Knowledge Corner Topic 3.2: (Continued)

actually a scalar function of position as, in general, it varies from point to point on the screen (see (d)), and the angular extent, too, may have holes and consist of disconnected pieces. Integration of this function "angular extent" of position on the screen area, finally, yields the desired etendue: $U = \int X\,\mathrm{d}A$.

While every area element may have its individual angular extent, it is quite common that the angular extent is the same everywhere. When this is the case and the screen is embedded in a homogeneous material, the integration reduces to a simple multiplication $U = XA$, because we can then separate the integration:

$$U = \underbrace{n^2 \int \cos\theta \,\mathrm{d}\Omega}_{X=n^2\Omega_\mathrm{p}} \underbrace{\int \mathrm{d}A}_{A} = n^2\,\Omega_\mathrm{p}\,A = XA \tag{3.8}$$

A good analogy is volume in 3D: to compute the volume of a complex shape, we would need to integrate over this shape, essentially adding up all infinitesimal volume elements dV. When the shape is simple, such as a cylinder, then the volume is simply base area times height. Our simplest, yet important, example is the task to compute the etendue of all possible rays intersecting a screen with area A embedded in a material with index n. Here, the angular extent is the full size of the tangent disk (see (f) above), $X = n^2\pi\mathrm{sr}$, everywhere on the screen, and the full-phase space etendue on this screen is therefore

$$U_\mathrm{full} = n^2\pi\mathrm{sr}\cdot A \tag{3.9}$$

This full-phase space volume is but a special case of a screen which sends light into a telecentric cone with a half-opening angle θ. Telecentric means that the rays through each area element form a cone whose central ray (the chief ray) is perpendicular to the screen.

Then, the angular extent becomes $X = (n\sin\theta)^2\pi\mathrm{sr}$, and we obtain an etendue of

$$U_\text{telecentric cone} = (n\sin\theta)^2\pi\mathrm{sr}\cdot A \tag{3.10}$$

This is probably the single most important etendue formula. We strongly advise you to commit it to memory. In illumination optics, this situation occurs, for example, at the exit of an integrator rod, at the entrance and exit of optical fibers (or fiber bundles), when a rotationally symmetric imaging subsystem is used which is approximately telecentric, or when the target is a far-field circle (e.g. when designing a stage light).

When the screen area is circular with radius *r*, the etendue when sending light into a telecentric cone is

$$U_{\text{circle} \to \text{cone}} = (n\,\pi\,r \sin\theta)^2 \text{ sr} \tag{3.11}$$

In case the circular cone is given in terms of the $F/\#$ of a lens,

$$F/\# = \frac{f}{d_{\text{ENP}}} \tag{3.12}$$

(d_{ENP} being the diameter of the entrance pupil [8]). In the paraxial case (target sufficiently far away from the entrance pupil, i.e. small θ) and for aplanatic systems, we have

$$F/\# = \frac{1}{2\sin\theta} \tag{3.13}$$

and we may express the etendue from circle to lens as

$$U_{\text{circle} \to \text{lens}} = \left(\frac{n\pi r}{2F/\#}\right)^2 \text{ sr} \tag{3.14}$$

Knowledge Corner Topic 3.3: 2D etendue

In a 2D disc world [9], where systems are reduced to meridional cross-section planes, screens are lines of length a, and angles are planar and not solid, the 2D etendue u is defined according to

$$u = \int_{-a}^{a} \int_{-\theta}^{\theta} n \cos\theta' \, d\theta' da' \tag{3.15}$$

(we use lowercase letters such as u and a for 2D quantities). The unit of the 2D etendue is mm rad.

If area and angle can be separated (equal angular distribution at every position), we can easily integrate:

$$u = n \cdot 2a \cdot 2\sin\theta \tag{3.16}$$

2D etendue is closely related to the Lagrange invariant Λ of a lens system [10],

$$u = 4\Lambda \tag{3.17}$$

The 2D etendue can be calculated as the difference of optical path lengths [2], leading to Hottel's formula (Eq. 3.36).

For rotationally symmetric systems, we can get the 4D etendue by rotating the 2D case via [11]

$$U = \left(\frac{\pi}{4}u\right)^2 \tag{3.18}$$

To summarize, the etendue is a geometrical quantity that captures the spatial and angular extent of a ray bundle as that ray bundle propagates through a screen. The angular extent of the rays that belong to the ray bundle may vary across the screen, the refractive index may vary across the screen, and the screen need not be planar. However, the procedure we illustrated to determine the etendue is applicable in all these cases. It is useful to keep the analogy with everyday 3D volume in mind. Just like the volume of a general 3D shape (e.g. a potato), the etendue of a general ray bundle may be complicated to determine, but if there is symmetry, then the etendue of a ray bundle may be as easy to compute as the volume of a 3D cylinder.

Why is etendue so useful? As we shall see in Knowledge Corner Topic 3.4, etendue is conserved as a ray bundle propagates from screen to screen, under certain conditions. This allows us to make very strong, very general, rock-solid statements about feasibility, and to gain insight into the nature of the problem at hand. Why, then, is etendue not as widely used as it should be, and instead is often thought of as a mystery? First, the definition of etendue must be understood in its admittedly abstract and not quite straightforward general form in order to apply it properly. Second, it is the "under certain conditions" part of the etendue conservation law that is often poorly understood and may therefore make it difficult to apply etendue conservation with confidence. Our own experience tells us, however, that understanding and applying etendue is supremely useful in our everyday work, and that such an understanding is required to become an expert illumination optics designer.

Studying a beam of light traveling through an optical system and various screens, we would find that the distribution of rays will change. This can be described as a *transformation of the phase space* itself [4]. Along with this transformation of phase space, the beam is also transformed.

We further notice that the etendue of the beam as a whole will remain unchanged during phase-space transformations, and consequently the *etendue* of the beam is conserved during its propagation through an optical system. The phase space is uncompressible; the mathematical basis is Liouville's theorem [12–14].

Knowledge Corner Topic 3.4: Etendue conservation

Any phase-space transformation, any optical system between source and target will transform the source phase space, but will keep the etendue of the source unchanged [4]. A comparison of the source etendue U_S with the target etendue U_T provides a first estimate for the maximum optical efficiency possible:

$$\eta_{\max} = \min\left(\frac{U_T}{U_S}, 1\right) \tag{3.19}$$

This defines only an upper limit of the optical efficiency (and there may be more such limits).

The actual efficiency achievable is usually smaller than this for a variety of reasons. Plus there is no guarantee that you will be able to find or design an optical system that is that perfect.

Knowledge Corner Topic 3.5: Other expressions of etendue

The laser optics community uses the *beam parameter product* (BPP) instead of a 2D etendue:

$$BPP = w \sin \theta = M^2 \frac{\lambda}{\pi} \tag{3.20}$$

with the unit mm. Hence (using Eq. 3.44),

$$U = (M^2 \lambda)^2 \mathrm{sr} = (BPP \cdot \pi)^2 \mathrm{sr} \tag{3.21}$$

which for $M^2 = 1$ yields the *etendue quantum* [15]

$$U_\mathrm{Q} = \lambda^2 \,\mathrm{sr} \tag{3.22}$$

The fiber optics community knows the *V number* [16] (comparing a Gaussian beam with step-index fiber modes)

$$V = \frac{2\pi}{\lambda} r_\mathrm{core} \cdot \mathrm{NA} = \frac{2}{\lambda}\sqrt{\frac{U}{\mathrm{sr}}} \tag{3.23}$$

(using Eq. 3.11). A value of $V < 2.4$ describes a single-mode fiber. In other words, a single-mode fiber is a fiber with an etendue of

$$U < 1.44\, U_\mathrm{Q} \tag{3.24}$$

(the factor is just obtained by calculating $2.4^2/4$).

In the field of Thermal Radiative Transfer, the term *configuration factor* [17] (also *shape factor* or *viewing factor*) is used instead of etendue. It is defined as

$$\eta = \frac{U_\mathrm{T}}{U_\mathrm{S}} \tag{3.25}$$

where U_T is a target etendue defined, e.g., by a light tube, and U_S is the etendue of a Lambertian source (Eq. 3.9) at the entrance area. Figure 3.9 illustrates this definition, with U_S being represented by a Lambertian source at the entrance pupil and U_T given by the following optical system.

Exercise 3.1: Etendue of area and half-plane

Consider two half-planes forming a corner (Figure 3.4). Calculate the etendue between an area $\mathrm{d}A'$ on one half-plane and the whole opposite half-plane.

(Continued)

Exercise 3.1: (Continued)

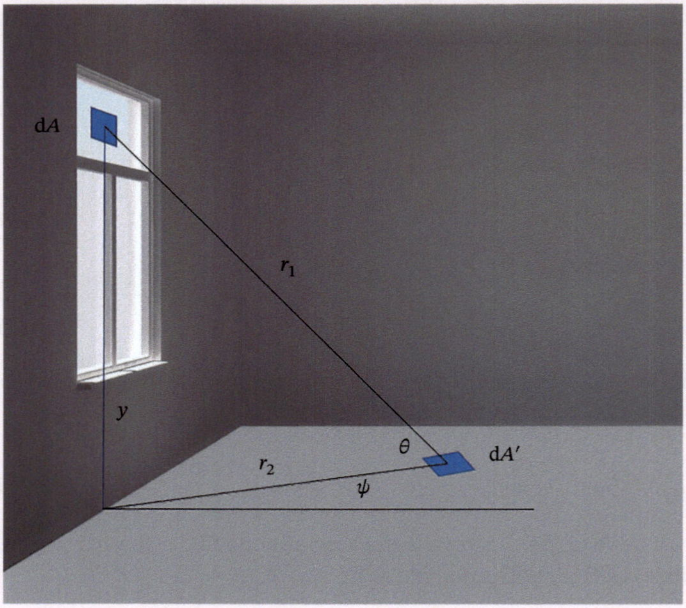

Figure 3.4 Two half-planes forming a corner and their respective area elements.

Solution 3.1

Our first approach is as follows: the projected solid angle defined by one half plane as seen from an area element on the other half plane is half that of a hemisphere (with value π),

$$\Omega_{\mathrm{p}} = \frac{\pi}{2}\mathrm{sr} \tag{3.26}$$

and accordingly the etendue will be

$$U = \frac{\pi}{2}\mathrm{sr}\,\mathrm{d}A' \tag{3.27}$$

A more detailed argument would be the following: the etendue between both area elements of Figure 3.4 (using Eq. 3.42 with consideration of the angles between the normal vectors to the area elements and their connecting line) is

$$\mathrm{d}U = \frac{\mathrm{d}A\cos\theta\cos\psi\cdot\mathrm{d}A'\sin\theta}{r_1^2}\mathrm{sr} \tag{3.28}$$

which describes a light tube (see Section 3.3) between dA and dA', with the distance r_1, the inclination $\cos\theta$ of dA, the inclination $\sin\theta$ of dA', and an out-of-plane factor $\cos\psi$.

The distance r_1 between the area elements is

$$r_1 = \frac{y}{\sin\theta} \tag{3.29}$$

and

$$r_2 = \frac{y}{\tan\theta} \tag{3.30}$$

with

$$\mathrm{d}r_2 = \frac{y}{\sin^2\theta}\mathrm{d}\theta \tag{3.31}$$

We use polar coordinates for the horizontal half-plane to write

$$\mathrm{d}A' = r_2\,\mathrm{d}r_2\,\mathrm{d}\psi = y^2\frac{\cos\theta}{\sin^3\theta}\mathrm{d}\theta\,\mathrm{d}\psi \tag{3.32}$$

Thus, the differential etendue (Eq. 3.28) becomes

$$\mathrm{d}U = \mathrm{d}A\ \cos^2\theta\,\mathrm{d}\theta\ \cos\psi\,\mathrm{d}\psi\,\mathrm{sr} \tag{3.33}$$

independent of y.

Integration over the floor ($\theta = 0 \ldots \pi/2$ and $\psi = -\pi/2 \ldots \pi/2$) yields

$$\mathrm{d}U = \mathrm{d}A\left[\frac{\theta}{2} + \frac{1}{4}\sin 2\theta\right]_0^{\pi/2}\left[\sin\psi\right]_{-\pi/2}^{\pi/2}\ \mathrm{sr} = \mathrm{d}A\cdot\frac{\pi}{2}\mathrm{sr} \tag{3.34}$$

identical to our above (first) computation.

We found that in many tasks, etendue is given between areas at finite distances, in contrast to the standard area/angle definition (the latter being equivalent to an infinitely distant exit pupil in the language of imaging optics). The following section provides the corresponding knowledge.

3.3 Light Tubes

A light tube, first introduced by Zimmer [18] in the seventies, is an arrangement of two areas in space at a certain distance (Figure 3.5) [19]. It has no physical side walls. The light tube can be in vacuum or embedded in a refractive-index distribution $n(\vec{r})$. Light tubes appear in many branches of optics [19] such as imaging optics (object and entrance pupil), the coupling of an LED to a light guide, or compound concentrators (Knowledge Corner Topic 2.18). The two surfaces play often an equal role (such as for etendue calculations), but in some situations we have to distinguish between source/target or object/entrance pupil sides.

We introduce several *special rays* in a light tube [10, 18], some of which are also known from imaging optics.

The following table introduces the special rays and their spatial and angular properties (see Figure 3.6):

Ray	from	to	angle
Marginal	0	$+R$	θ
Chief	$+r$	0	$\bar{\theta}$
Coma	$-r$	$-R$	θ_C
Diagonal	$-r$	$+R$	θ_D

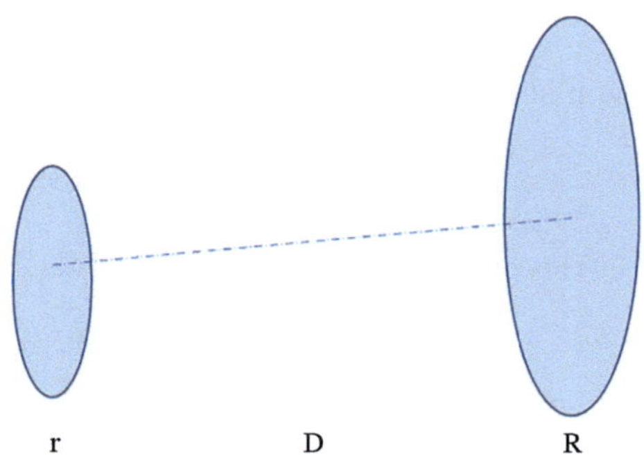

Figure 3.5 A light tube.

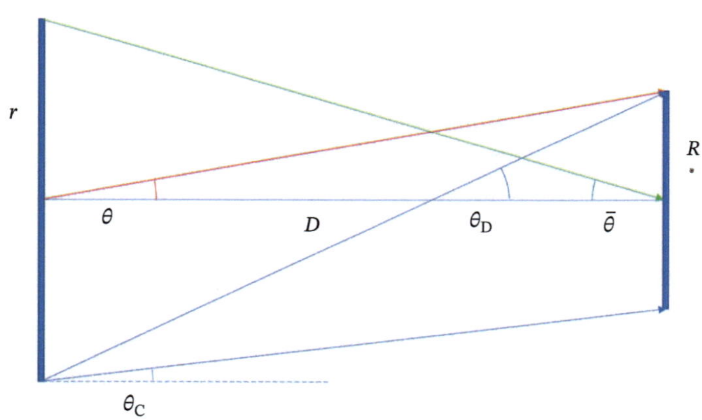

Figure 3.6 Special rays in a light tube: chief ray (green), marginal ray (red), coma and diagonal rays (blue).

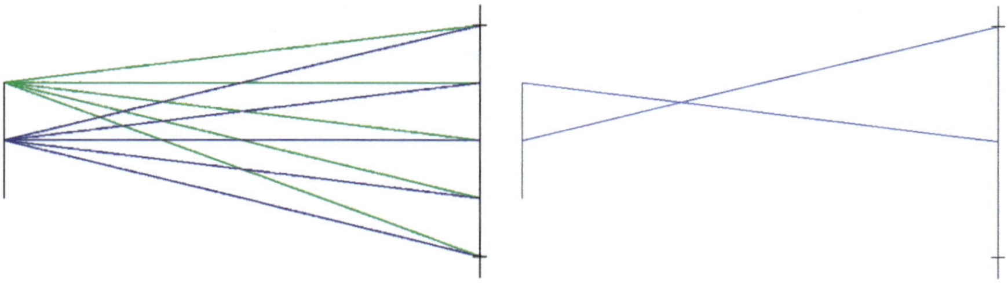

Figure 3.7 The object and the entrance pupil of an imaging optical system form a light tube (left). On the right, the marginal and chief rays are shown which support the calculation of the light tube's etendue [18].

The *Lagrange invariant* (or *Helmholtz invariant*) of an optical system is defined on the basis of the marginal and chief rays [20, Section 2.5]:

$$\Lambda = \theta \bar{y} - \bar{\theta} y \tag{3.35}$$

In this equation, y and $\bar{y}$ are the heights of the marginal and chief rays at an arbitrary screen within the light tube. This is a paraxial definition and only valid for light tubes with $D \gg R, r$.

Knowledge Corner Topic 3.6: Etendue of a light tube

A light tube defines an etendue (Figure 3.7) that can be calculated by means of Hottel's formula [21, 22]

$$U_{\mathrm{H}} = \frac{\pi^2}{4}\left(\sqrt{(r+R)^2 + D^2} - \sqrt{(r-R)^2 + D^2}\right)^2 \mathrm{sr} \tag{3.36}$$

instead of Eq. 3.8. Here, r is the semi-diameter of a source disk and R that of a target disk or pupil, and D is their center distance perpendicular to both disks. We have set $n = 1$ and will discuss the case of a dielectric in the light tube below (Knowledge Corner Topic 3.9).

There are different, but related forms present in the literature [23]. A derivation based on optical path length is given in [2].

Dr. Wisenheimer:

In the phase-space diagram [24], the standard etendue definition (Eq. 3.8) with area and angle forms a rectangle, whereas a light tube (the two areas defining a spatially dependent angular extent) would be represented by a parallelogram [11, 25].

Exercise 3.2: Etendue of a projection lens

In the 2014 IODC illumination design contest [26], a *F*/4.5 projection lens (a Cooke triplet) is defined as an accessory to set up a projector (Figure 3.8) using the following specifications:

Surface #	Radius	Thickness	Glass	Clear aperture	Mech. aperture
0	∞	30.372		12	12
1	11.69	1.575	SK16	4.9	5.3
2	−230.23	3.283		4.55	5.3
3	−15.4	0.875	F4	3.6	4.3
4	13.37	3.682		3.14	4.3
5	86.87	1.4	SK16	5	5.3
6	−15.04	5000		4.9	5.3

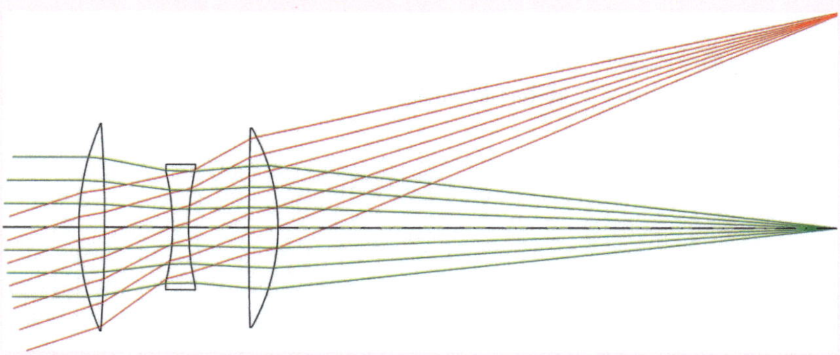

Figure 3.8 IODC 2014 contest projector Cooke triplet (film gate on the right).

The field (film gate) size is a 20.95 mm by 11.33 mm rectangle. Calculate the system etendue of the projector.

Solution 3.2

We choose the film gate as the screen on which we estimate the etendue, and approximate the angular extent by the *F*/# of the lens [HANDLE WITH CARE]. Using Eq. 3.14, we obtain

$$U_{\mathrm{T}} \approx A \cdot \frac{\pi}{4F/\#^2}\mathrm{sr} = \frac{\pi \cdot 237.4\,\mathrm{mm}^2}{4 \cdot 4.5^2}\mathrm{sr} = 9.2\,\mathrm{mm}^2\mathrm{sr} \tag{3.37}$$

which is probably a little too large, since this simple approach disregards $\cos^4$ (cf. Exercise 3.17) and vignetting issues.

Alternatively, we may ask the lens design software for the exit pupil (film side pupil) of the lens. The answer is a pupil with radius $R = 4.1$ mm at a distance of 6.5 mm from lens surface 1 (the surface closest to the film gate). A light tube is formed by the circumscribed circle of the film gate and the entrance pupil. We have

$$D = 6.5\,\text{mm} + 30.4\,\text{mm} = 36.9\,\text{mm} \tag{3.38}$$

The semi-diagonal of the gate is $r = 12$ mm and Hottel's formula (Eq. 3.36) yields the etendue of the light tube as 15.7 mm^2sr. This has to be corrected by the fraction the actual film area covers within its circumscribed circle. We obtain

$$U_\text{T} = \frac{237.4\,\text{mm}^2}{\pi(12\,\text{mm})^2} \cdot 15.7\,\text{mm}^2\text{sr} \approx 8.2\,\text{mm}^2\text{sr} \tag{3.39}$$

which is probably a little to small (why?).

When you need more precision, or when the situation becomes really complicated so that no quick approximations are applicable, we recommend computing the etendue by simulation [27]. We present this method right here where it is needed, although some basics will be introduced only later, in Section 3.4.

Knowledge Corner Topic 3.7: Etendue estimation by simulation

Consider an optical arrangement with an entrance gate, some arbitrary combination of apertures and optical elements, and an exit gate (Figure 3.9). Your task is to determine the system etendue. In principle, the etendue could be calculated by integration, but an analytic solution may well be impossible or at least extremely complicated. For example, consider a complete DLP projector, from the entrance face of the integrator rod to the projected image on the wall. How can you obtain a more precise estimate of the etendue other than just the estimate for the imager itself?

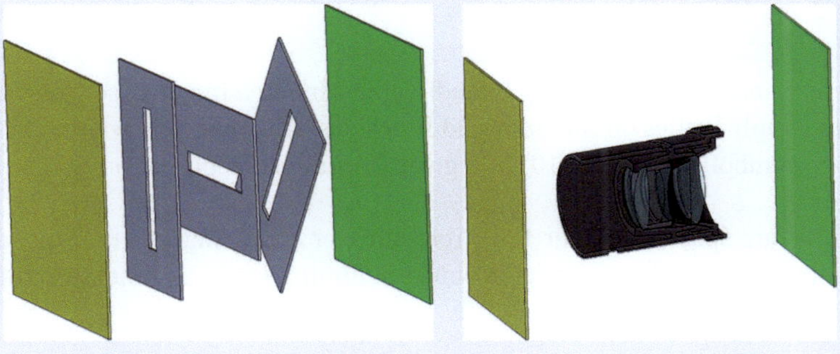

Figure 3.9 Optical systems of unknown etendue.

(Continued)

Knowledge Corner Topic 3.7: (Continued)

This seemingly daunting task becomes amazingly simple when you use simulation. Anticipating some results from Section 3.4, we use the facts that (i) the luminance remains conserved along a ray, if losses and scattering are absent, and (ii) luminance is density in phase space, $L = \mathrm{d}\Phi/\mathrm{d}U$ (Eq. 3.91), just like mass density ρ in 3D space is $\rho = \mathrm{d}m/\mathrm{d}V$. You can easily determine the volume of an intricately shaped vessel by filling it with water and then weighing the water: the volume has as many liters as the water has kilograms! So let's do the same with light, using a simulation.

The procedure is as follows [27]:

- Put a Lambertian source of luminance L_S in front of the entrance gate, with some overfill.
- Place a receiver beyond the exit gate, also with some overfill.
- Ensure that no absorption, no partial reflection, no ray splitting and no scattering occurs.
- Run a large number of rays and measure receiver flux Φ_T.

The system etendue is then (cf. Eq. 3.108)

$$U_\mathrm{T} = \frac{\Phi_\mathrm{T}}{L_\mathrm{S}} \tag{3.40}$$

Dr. Wisenheimer:

Note that this approach is purely geometric, and *does not* measure efficiency. Actual sources may be inhomogeneous, they may underfill the system, and real systems do always have losses, partial reflection and scattering. Be sure to modify your real system before measuring etendue, and be sure to undo the modifications before returning to photometric simulations – a price worth paying for being able to apply all these etendue calculations to systems with complicated geometries.

Simulation 3.1: Etendue estimation (1)

We set up the lens from Exercise 3.2 including the film area in a non-sequential software and apply baffles on the front and film sides. A Lambertian source (the hemispheric symbol in Figure 3.10) of a given luminance is placed outside the entrance baffle.

All surfaces are specular, either 100% refracting or absorbing. We then flood the arrangement with light of a specific constant luminance L_S and measure the flux Φ_T on the receiver.

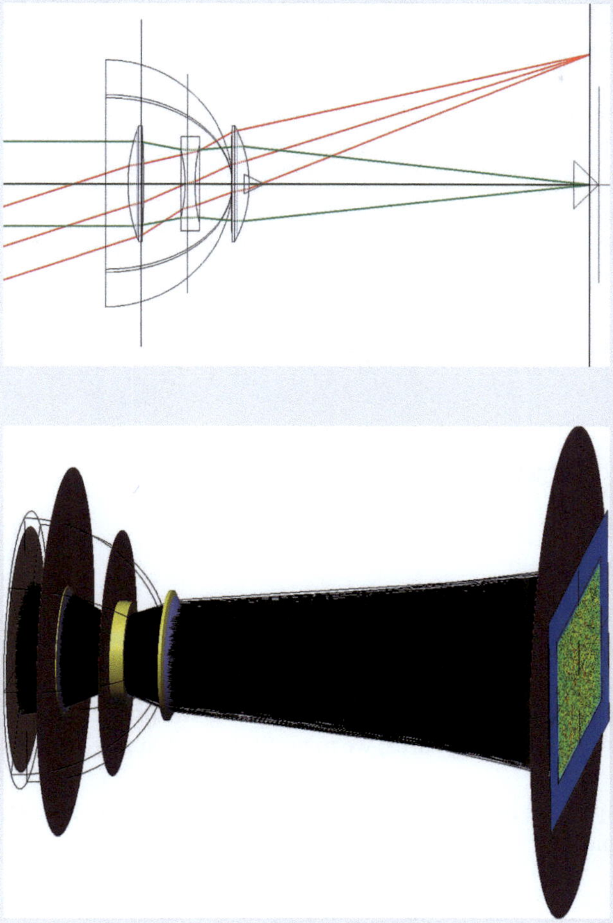

Figure 3.10 Lens system with baffles to block light outside the system etendue. A Lambertian source is positioned in front of the system, a receiver collects all light that passes the film area baffle. An extra baffle was needed to block light that would pass outside of the middle negative lens.

In our case, we have

$$U_T = \frac{\Phi_T}{L_S} = \frac{8.5\,\text{lm}}{1\,\text{Mnit}} = 8.5\,\text{mm}^2\text{sr} \tag{3.41}$$

which is slightly smaller than our first estimate in Exercise 3.2 because *vignetting*, i.e. rays blocked by lens aperture edges other than the stop surface, reduces the throughput. On the other hand, the result is slightly larger than our second estimate from Exercise 3.2.

The simulation files are provided here: cinderella.seq and cinderella.1.lts.

Knowledge Corner Topic 3.8: Calculus around Hottel's formula

We restrict our considerations to

- $n = 1$,
- rotationally symmetric systems, and
- source and target both centered on the axis.

Far-field approximation

This means that we use $D \to \infty$, i.e. r, $R \ll D$. With the series expansion introduced in Eq. 1.5, Eq. 3.36 becomes

$$U_H\Big|_{D\to\infty} \approx \frac{\pi^2}{4}\left(D + \frac{(r+R)^2}{2D} - D - \frac{(r-R)^2}{2D}\right)^2 \text{sr} = \left(\frac{\pi r R}{D}\right)^2 \text{sr} \tag{3.42}$$

(cf. [28, §126]).

The angular size of the target is

$$\tan\theta = \frac{R}{D} \approx \sin\theta \tag{3.43}$$

and we are back to Eq. 3.8,

$$U_H\Big|_{D\to\infty} = (\pi r \sin\theta)^2 \text{sr} = A_S\Omega_P \tag{3.44}$$

Here we see the origin of the unit sr in the formulas due to the projected solid angle. In the case of a small and a large area, we may still find an approximation (see Eq. 4.58).

The result can be expressed in terms of the Lagrange invariant (Eq. 3.35, cf. Knowledge Corner Topic 3.3)

$$U_H\Big|_{D\to\infty} = (\pi\Lambda)^2 \text{sr} \tag{3.45}$$

For the 2D etendue of a light tube, the same approximation yields

$$u_H = \frac{4rR}{D} \tag{3.46}$$

Equal areas

With $R = r$, Eq. 3.36 becomes

$$U_H = \frac{\pi^2}{4}\left(\sqrt{(2r)^2 + D^2} - D\right)^2 \text{sr} \tag{3.47}$$

Small gap

We now consider the case $D \ll r$, which is important for coupling LED light to a light guide with only a small air gap between them (Figure 3.11).

Figure 3.11 CoB LED shining into a cylindrical light guide [19].

For a light tube with unequal sides, we have to look for the sign of the second square root, and we obtain from Eq. 3.36

$$U_{\mathrm{H}}|_{D\to 0} = \begin{Bmatrix} (\pi R)^2\ \mathrm{sr} & \text{for } r > R \\ (\pi r)^2\ \mathrm{sr} & \text{for } R > r \end{Bmatrix} \tag{3.48}$$

In each case, the smaller of both disks of the light tube determines its etendue.

Small source

We now consider the case $r \to 0$. We use another version of Eq. 1.5,

$$\sqrt{(r+R)^2 + D^2} - \sqrt{(r-R)^2 + D^2} \approx \frac{2R}{\sqrt{R^2 + D^2}}\, r \tag{3.49}$$

so that Eq. 3.36 becomes

$$U_{\mathrm{H}}|_{r\to 0} = (\pi r)^2 \frac{R^2}{R^2 + D^2}\ \mathrm{sr} \tag{3.50}$$

which converges to Eq. 3.42 for $D \gg R$. And with $\tan\theta = R/D$ and Eq. 2.5, we have

$$\sin^2\theta = \frac{R^2}{R^2 + D^2} \tag{3.51}$$

and can now understand that Eq. 3.44 holds for $D \to \infty$ or $r \to 0$.

Finally, for $R \gg r$, D, Eq. 3.50 converges toward a Lambertian source.

$$U_{\mathrm{L}} = (\pi r)^2 \mathrm{sr} \tag{3.52}$$

Hottel's formula in terms of the configuration factor

We can divide Eq. 3.36 by Eq. 3.52 and get the *configuration factor* of the light tube

$$\eta_{\mathrm{H}} = \frac{U_{\mathrm{H}}}{\pi R^2} = \frac{1}{4}\left(\sqrt{(1+\rho_{\mathrm{H}})^2 + \delta_{\mathrm{H}}^2} - \sqrt{(1-\rho_{\mathrm{H}})^2 + \delta_{\mathrm{H}}^2}\right)^2 \tag{3.53}$$

(cf. Eq. 3.25) with $\rho_{\mathrm{H}} = R/r$ and $\delta_{\mathrm{H}} = D/r$. The use of those relative quantities will help us, for example, to concentrate on shape and not on size.

(Continued)

Knowledge Corner Topic 3.8: (Continued)

Constant etendue

Imagine you have to fill a light tube [29] with the light of a given source; both featuring equal etendue. Assume that the *r* side of the light tube is defined and fix, and you need to find combinations of length *D* and the semi-diameter of the other (*R*) side that will work (there is still one degree of freedom for the given etendue).

As the problem is scalable, we may use Eq. 3.53 instead and solve for $\rho_H(\delta_H)$ for a given η_H:

$$\rho_H(\delta_H) = \sqrt{\pi\eta_H\left(1 \pm \frac{\delta_H^2}{\pi - \eta_H}\right)} \tag{3.54}$$

This result is shown in Figure 3.12. Each curve shows the relative semi-diameter as a function of the relative length that yields the same relative etendue (three of them are shown in the diagram legend).

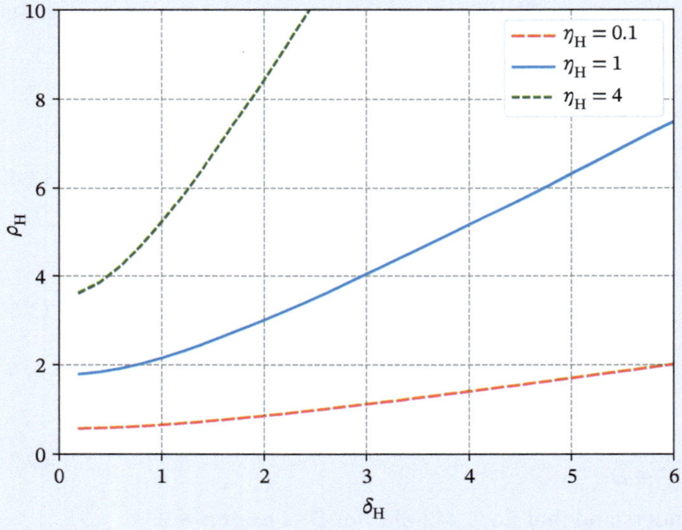

Figure 3.12 Relative size ρ_H of the other light tube disk as a function of relative distance δ_H for some configuration factors η_H.

Provided spreadsheet.

In other words: for a given system etendue and source size, this describes the corresponding pupil size as a function of the length of the light tube.

Can this constitute a means to analyze compound concentrators? Looking back at Figure 2.60, we modify the setup and look back into the CHC front R_F. In this case, we have

$$\eta_{\mathrm{H}} = \frac{U_{\mathrm{H}}}{\pi R_{\mathrm{F}}^2}$$

$$\rho_{\mathrm{H}} = \frac{R}{R_{\mathrm{F}}} \tag{3.55}$$

$$\delta_{\mathrm{H}} = \frac{D}{R_{\mathrm{F}}} \text{ or } \delta_{\mathrm{H}} = \frac{D+L}{R_{\mathrm{F}}}$$

The interpretation of Figure 3.12 is that valid ρ_{H} values related to R and r (cf. Figure 2.60) can be found by looking at a particular curve for any given L and D.

Generally speaking, different light tubes can still share the same etendue.

So far, we have discussed light tubes with $n = 1$ inside. Now let's have a look at light tubes with glass or another dielectric inside, either completely or partially filled. For example, consider the important case of an LED chip covered by a glass plate, defining a distance of the subsequent light guide that will collect the light (cf. Figure 3.11). This design is preferred by LED manufacturers (because otherwise customers would damage the bond wires and request compensation for the faulty article), but may constitute a disadvantage for some optical design approaches.

Knowledge Corner Topic 3.9: A dielectric in the light tube

We set up a light tube in an optical design software (with the object and the stop representing the areas) and insert a glass plate (Figure 3.13). The ray aiming capability of the software will find the proper marginal and chief rays.

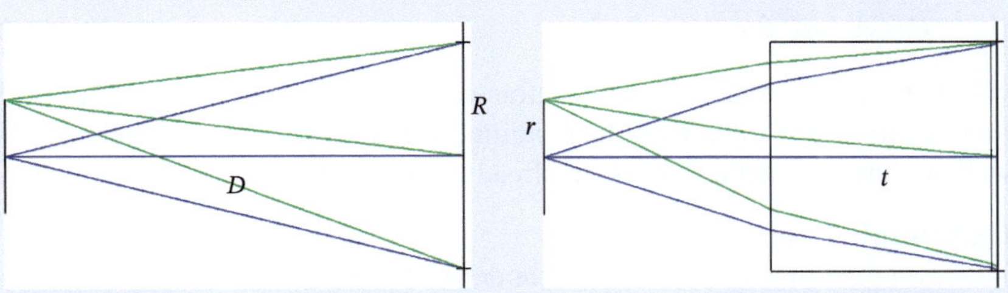

Figure 3.13 Light tubes of equal sizes with and without a glass plate (thickness t) inside.

The marginal and chief ray angles are increased by the glass and so is the etendue. The position of the plate is obviously irrelevant.

For a calculation of the etendue increase we first try the paraxial approximation, although the angles may be not really small in this case. The original marginal ray angle is

$$\tan\theta = \frac{R}{D} \tag{3.56}$$

(Continued)

Knowledge Corner Topic 3.9: (Continued)

With the glass plate, we have

$$R = \tan\theta_1(D - t) + \tan\theta_2 t \tag{3.57}$$

Using Snell's law and for small angles ($D \gg r, R$), we have

$$\tan\theta_1 \approx \frac{R}{D - t + t/n} \tag{3.58}$$

The etendue of the light tube is then (cf. Eq. 3.42)

$$U = (\pi r \sin\theta)^2 \mathrm{sr} \approx \left(\frac{\pi r R}{D - t + t/n}\right)^2 \mathrm{sr} \tag{3.59}$$

which is obviously true for $n = 1$.

For $t = D/2$, we obtain

$$U_{n,t=D/2} \approx \left(\frac{\pi r R}{D}\right)^2 \mathrm{sr} \cdot \left(\frac{2n}{n+1}\right)^2 = \left(\frac{2n}{n+1}\right)^2 U_{n=1} \tag{3.60}$$

For $t \to D$ (the tube is almost completely filled with glass, but the areas are still in air), we obtain

$$U_{n,t=D} = n^2\, U_{n=1} \tag{3.61}$$

This is a little surprising, as both source and receiver are still in air and not embedded in the medium with index n.

It turns out that embedding source and target into the higher-index medium is not required, and what we did here is to explain how the n^2 term appeared in Eq. 3.7. The etendue is given by the apparent size of the pupil with the semi-diameter y hit by the marginal ray. If the light tube is almost filled, we have with Eq. 2.44

$$U = A_S \sin^2\theta = A_S (n \sin\theta_n)^2 \tag{3.62}$$

with the marginal angle θ possibly refracted to θ_n inside the medium. If the light tube is completely filled with the medium, with source and target embedded as well, we have to use Eq. 3.8, which already includes the n^2 term.

Dr. Wisenheimer:

The cover glass on an LED chip is not as detrimental as it seems at first glance. For coupling light into a light guide, it may not be as good as zero distance, but it is still better than an air gap of the same thickness.

A related phenomenon occurs in optical designs for collimating a light source, where not only a high-index lens, but also a high-index parallel plate in front of an LED is of advantage [30].

The rod lens systems of rigid endoscopes work much in the same way, too. They feature an increased system etendue by filling *most of the space* between the intermediate images with glass [31].

Exercise 3.3: Real marginal and chief rays of a light tube

Calculate the *real* marginal ray angle of a half-filled light tube and compare it to the paraxial solution.

Solution 3.3

We may combine the ray heights

$$\tan\theta_1 + \tan\theta_2 = \frac{2R}{D} \tag{3.63}$$

and Snell's law (Knowledge Corner Topic 2.7),

$$\sin\theta_1 = n\sin\theta_2 \tag{3.64}$$

An analytic solution exists but does not look very practical.

Instead, we remember Fermat's principle and search numerically for a minimum of the optical path length (Figure 3.14).

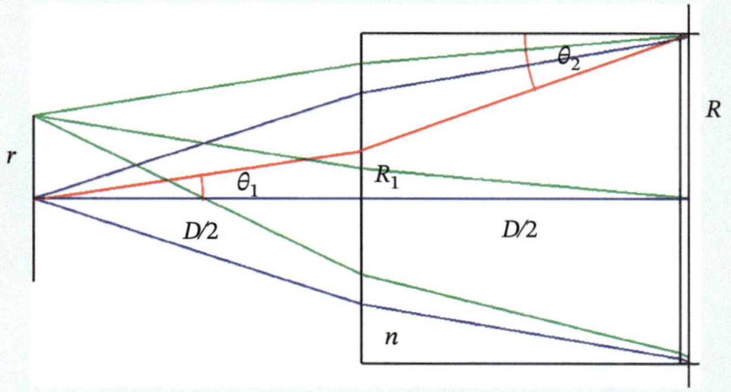

Figure 3.14 Light tube with a dielectric and a "candidate Fermat ray path" (in red). Fermat's principle selects the true (blue) path among all candidates.

A file dielectric_in_tube.xlsx is provided where the minimum path length and the chief ray angle are calculated with Excel's solver for a series of refractive index values (see Figure 3.15).

We now compare the etendue values calculated using those marginal ray angles with the paraxial approximations above, Eqs. 3.60 and 3.61 (Figures 3.16 and 3.17).

(Continued)

Solution 3.3 (Continued)

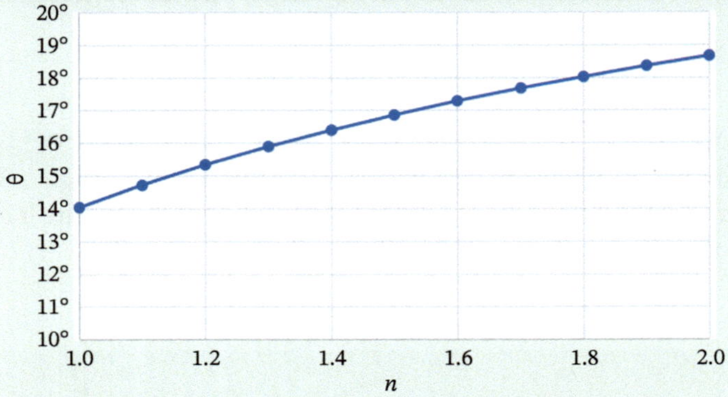

Figure 3.15 Marginal ray angle θ at the source as a function of the refractive index n of the dielectric.

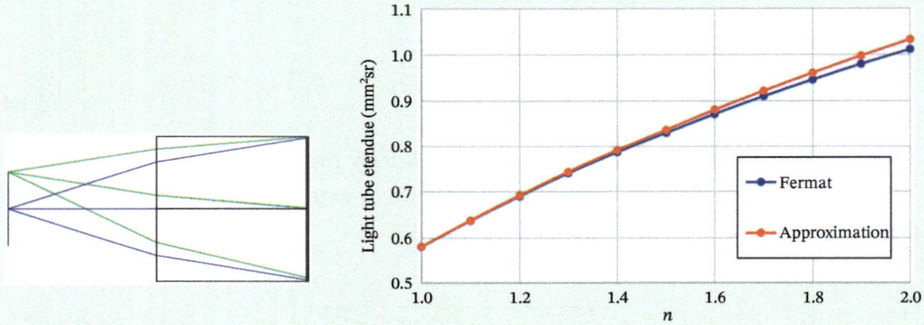

Figure 3.16 Left: Light tube semi-filled with a glass plate (thickness $t = D/2$). Right: Light tube etendue as a function of refractive index (according to Fermat's principle and Eq. 3.60).

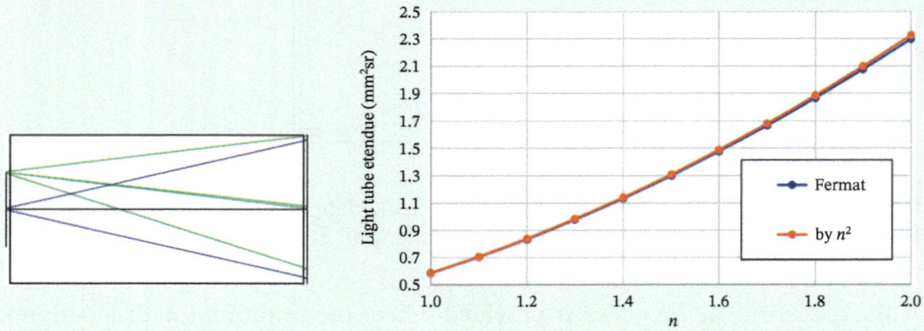

Figure 3.17 Left: Light tube almost completely filled with a dielectric medium. Right: Again, the Fermat results and the approximation based on Eq. 3.61 are in good agreement.

The paraxial approximation seems to work well even in this case.

From now on, we will leave the paraxial region and roam through wider fields.

Exercise 3.4: LED to light guide coupling and air gap

The light of a 1 mm^2 thin-film LED (a square Lambertian emitter) needs to be coupled into a glass light guide (with a square entrance area of 1 mm^2). On the LED is a N-BK7 cover glass with a thickness of 0.5 mm (no immersion, both sides in air). Calculate the geometrical loss due to the gap with and without cover glass.

Solution 3.4

First we calculate the etendue of two *disks* with $r = R = 0.5$ mm at a distance of 0.5 mm from each other in air (Eq. 3.47):

$$U_{\mathrm{H}} = \frac{\pi^2}{4}\left(\sqrt{1\,\mathrm{mm}^2 + 0.25\,\mathrm{mm}^2} - 0.5\,\mathrm{mm}\right)^2 \mathrm{sr} \approx 0.94\,\mathrm{mm}^2\mathrm{sr} \tag{3.65}$$

Using a "dirty" approximation (we use rotational symmetry and account for the square area by including a factor of $4/\pi$), we estimate the etendue of two squares (cf. Figure 3.18) to be

$$U_{\square\square} \approx \left(\frac{4}{\pi}\right)^2 U_{\mathrm{H}} = 1.6\,\mathrm{mm}^2\mathrm{sr} \tag{3.66}$$

We remember the etendue of a 1 mm^2 LED is

$$U_{\square} = \pi\ \mathrm{mm}^2\mathrm{sr} \tag{3.67}$$

so there is a $\eta = 50\%$ coupling efficiency due to the gap between source and light guide.

Now for a glass plate between LED and light guide. We try and go for the paraxial approximation: if the previously hollow light tube is now completely filled with the glass plate, we may want to apply the n^2 factor (Eq. 3.61). For N-BK7 ($n = 1.518$), we had

$$U_{\square\square n} \approx n^2\left(\frac{4}{\pi}\right)^2 U_{\mathrm{H}} = 3.5\,\mathrm{mm}^2\mathrm{sr} \tag{3.68}$$

which would in theory allow for complete light collection.

We are afflicted by doubts, however, and decide to double-check by means of a simulation.

Simulation 3.2: Etendue measurement (2)

We apply the etendue measurement method again to double-check the values calculated above (Figure 3.18).

(Continued)

Simulation 3.2: (Continued)

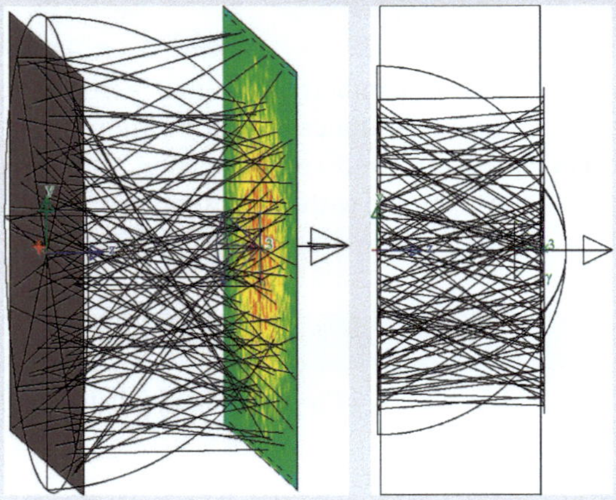

Figure 3.18 Simulation of the LED-to-light-guide coupling with a gap in between. Both pictures show the source exit area (left) and the light guide entrance area (right, both 1 mm^2, distance $D = 0.5$ mm). The luminous flux of the source is always set to π lm (thus $L_S = 1$ Mnit). Left picture: system with air gap, the receiver collects 1.3 lm. Right picture: system filled with a N-BK7 disk, the receiver shows $\Phi_T = 2.1$ lm now .

The result in air corresponds to a system etendue of

$$U_{\text{T, air}} = \frac{\Phi_T}{L_S} = 1.3\,\text{mm}^2\text{sr} \tag{3.69}$$

which is slightly smaller than the result from Eq. 3.66.

With the glass included, we obtain an increased etendue of

$$U_T = 2.1\,\text{mm}^2\text{sr} \tag{3.70}$$

but this is by far smaller than our estimate from Eq. 3.68. We conclude that this is not a paraxial situation and the approximation Eq. 3.61 fails.

Provided file: squares_etendue.lts.

Exercise 3.5: LED collimation (2)

Sketch optical systems for a Chip-on-Board (CoB) LED ($r_S = 11$ mm, Figure 3.19) that

(a) produce a telecentric image, and

(b) collimate its light.

For both tasks, the far-field angular spread should be ±10°. Assume that you can collect rays of up to 60° from the LED. What is the size of the exit pupil?

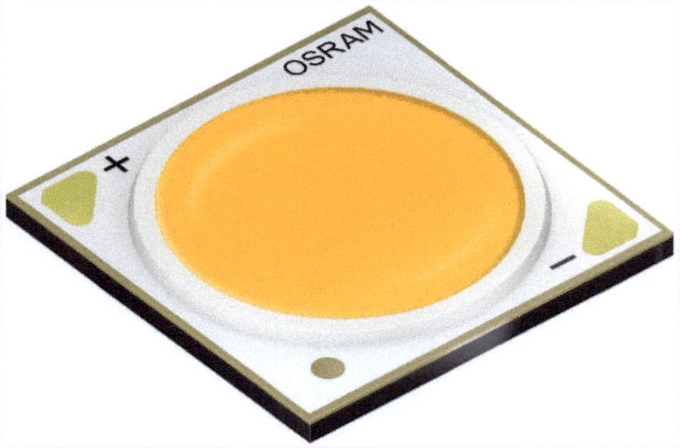

Figure 3.19 CoB LED, a sea of blue chips encapsulated in yellow phosphor providing partial conversion. Courtesy of ams-OSRAM.

Solution 3.5

The etendue on both sides of the optics must be equal:

$$U_\mathrm{S} = U_\mathrm{T} \tag{3.71}$$

which serves as a basis to lay out the design via

$$(\pi\, r_\mathrm{S} \sin 60°)^2 = (\pi\, r_\mathrm{T} \sin \theta)^2 \tag{3.72}$$

Using a lens system, we may collect rays with a maximum angle of 60° with respect to the optical axis. Steeper rays are not easily collected. We have to use a smaller source etendue in the calculations (which means an immediate loss of 25%) and thus obtain an exit pupil semi-diameter of

$$r_\mathrm{T} = \frac{11\,\mathrm{mm} \cdot \sin 60°}{\sin 10°} = 55\,\mathrm{mm} \tag{3.73}$$

independent of the actual optical design (cf. Figure 3.20). This is the value of the image size and of the exit pupil of the collimator.

Source imaging

If we *image* the source somewhere with a magnification of $\beta = 55/11$, we would automatically obtain a $\pm 10°$ cone. The optics will have to be approximately telecentric on both sides.

(Continued)

Solution 3.5 (Continued)

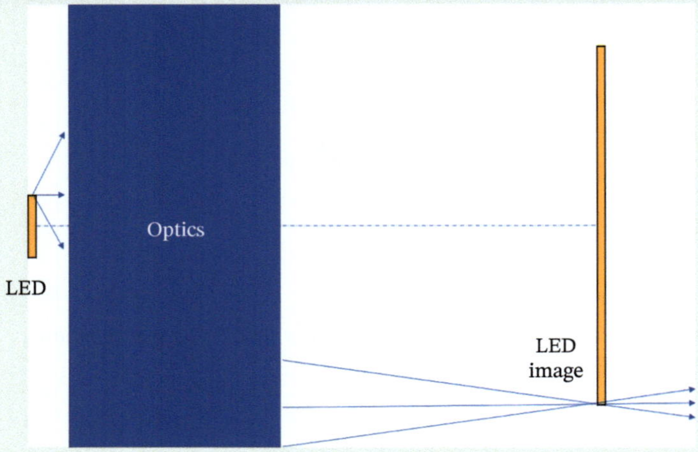

Figure 3.20 Imaging a CoB LED with a magnification of 5.

An actual design could look like the one shown in Figure 3.21.

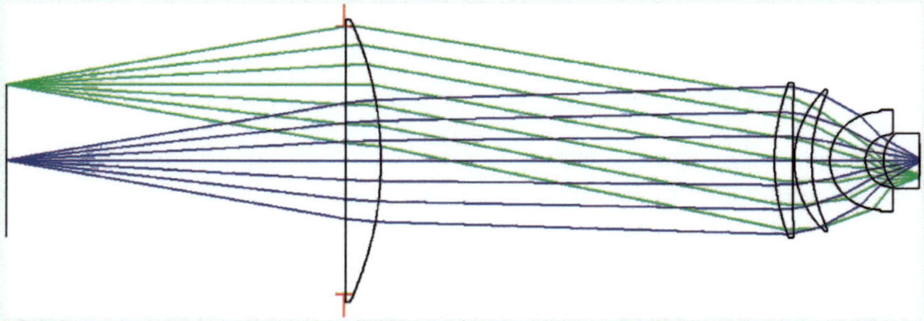

Figure 3.21 Optical design to image a CoB LED with a magnification of 5. The design starts on the left (the LED image, with the specified field of $r_T = 55$ mm and cone angle of $\pm 10°$), and the LED position is on the right. The maximum angles of incidence are slightly larger than 60° as required.

Collimation

On the other hand, we may *collimate* the LED light to $\pm 10°$ (Figure 3.22), and in this case the optics is expected to have an exit pupil semi-diameter of $r_T = 55$ mm (cf. Figure 3.23).

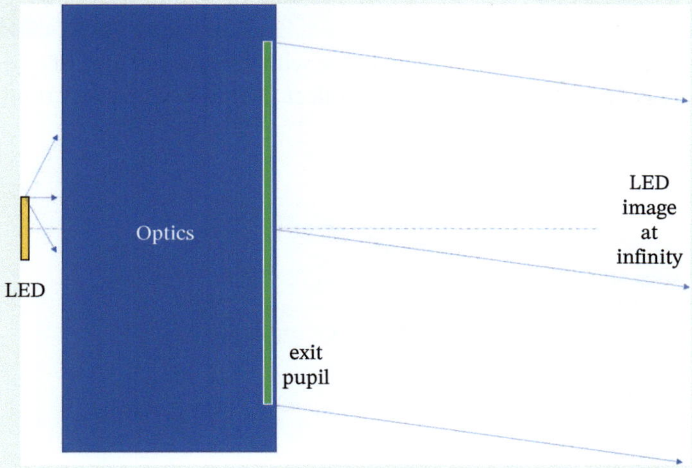

Figure 3.22 Collimating a CoB LED.

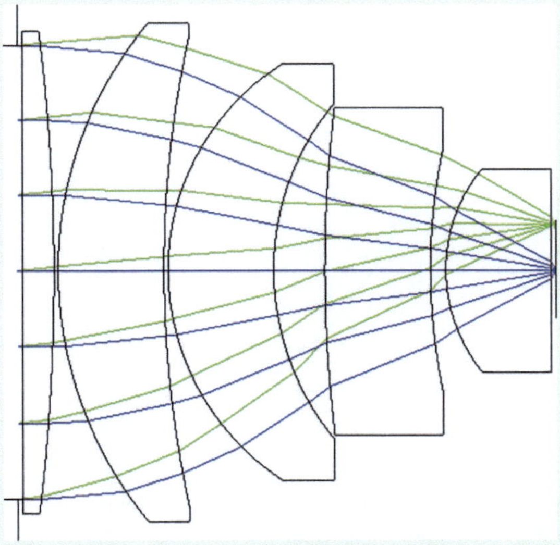

Figure 3.23 Collimator design: exit pupil on the left (field ±10°, aperture $r_T = 55$ mm $r_T = 63$ mm), LED plane on the right.

This is also the setup of Köhler illumination for a telecentric projection lens [32].

But we can also fulfill the task by non-imaging means as shown in Simulation 3.3.

Simulation 3.3: Tapered cylinder

For the sake of simplicity, we try a cylindrical cone with an entrance area of $r_S =$ 11 mm, an exit area of $r_T = 63$ mm (we want to collect LED rays of up to 90°) and a length of 500 mm (Figure 3.24).

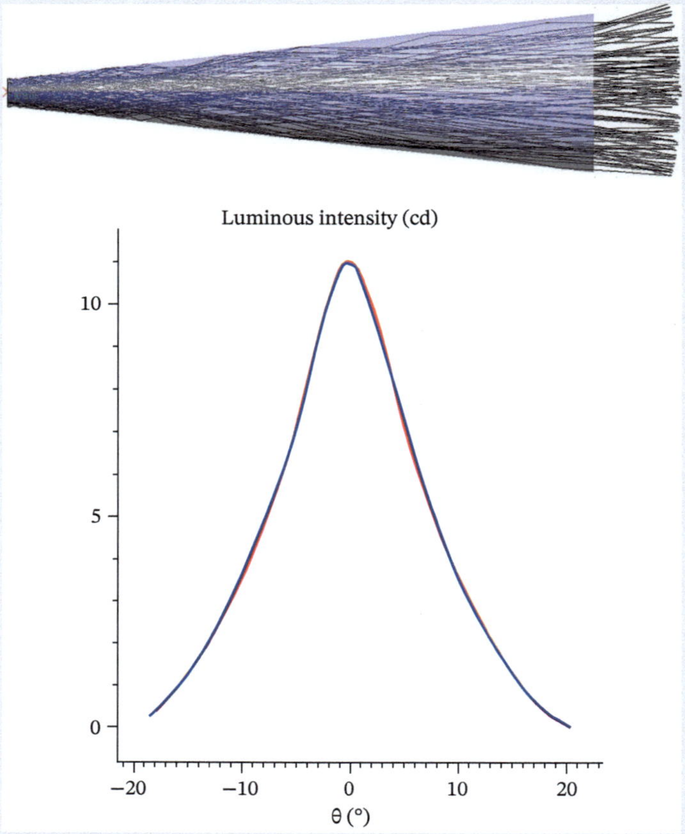

Figure 3.24 LED collimated by a tapered glass cylinder. The intended collimation to ±10° is not perfect and often has a trigonal shape, but it can be improved by an even longer device. A lens system (imaging collimator) may be more efficient and more practical.

What happened here – and how can we get rid of the triangular distribution? To analyze the situation, we place an *angular luminance meter* (a receiver accessory that plots luminance vs. angle) at various positions on the exit face of the cone (Figure 3.25).

Obviously, the beam is not telecentric, and we need to bend the beam angles at outer positions inward somehow. According to Eq. 2.61, this can be achieved by using a simple positive lens (Figure 3.26).

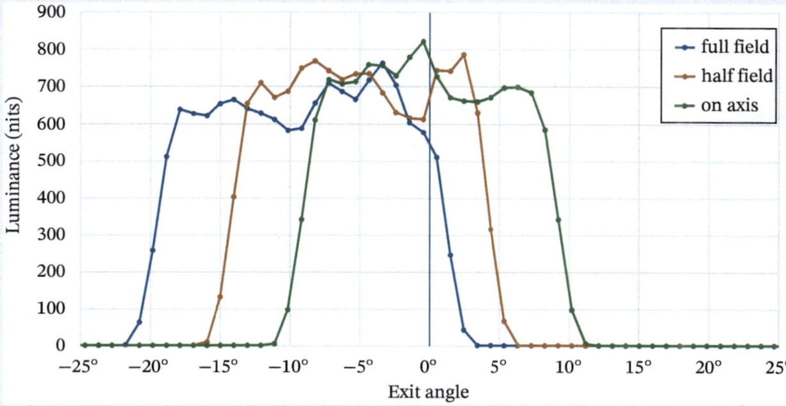

Figure 3.25 Angular luminance for various positions on the cone's exit surface.

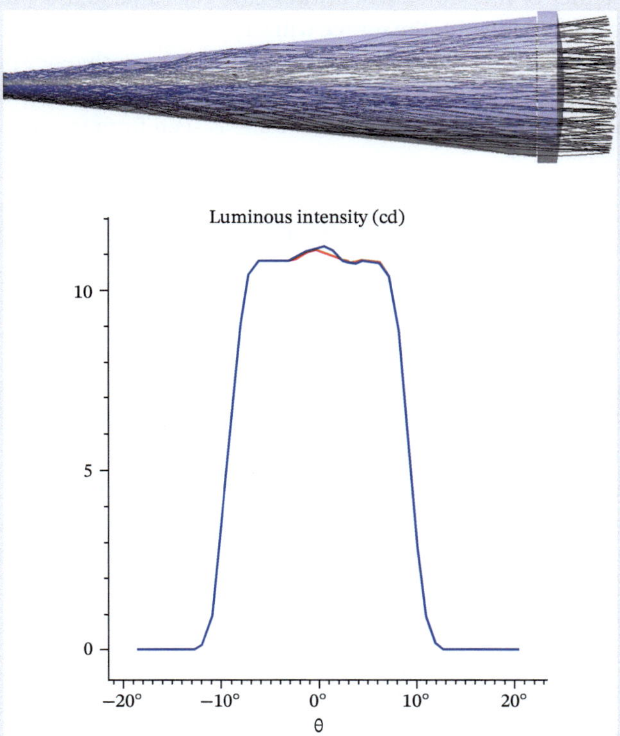

Figure 3.26 LED collimated by a tapered glass cylinder combined with a lens on the exit side (with a focal length roughly equal to the cylinder length). The light distribution on the exit side becomes telecentric with ±10°.

Provided file: CoB_LED_Colli.lts.

Now we want to use an extended source to illuminate Young's double slit setup (Figure 3.27). We will discuss the effect of spatial and angular dimensions of the source.

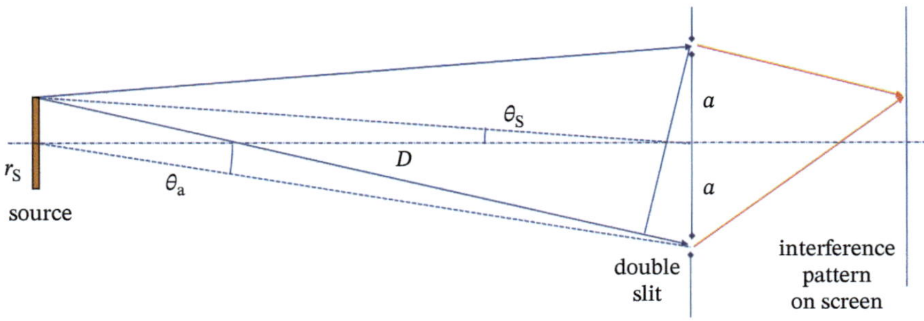

Figure 3.27 An extended source (radius r_S) illuminating a double slit (distance $2a$) from a distance D.

While rays from the center of the source will always have the same phase when arriving at the slits, rays from the edge of the source will show an optical path length difference. If this optical path length difference is small enough, an interference pattern with high contrast can be observed [33]. This condition will be fulfilled when both the source size and the angular size of the double slit as seen from the source are small enough.

Exercise 3.6: Coherence area

What is the maximum source etendue that allows for interference fringes to be visible?

Solution 3.6

In the following, we are obviously in the paraxial domain again, so r_S, $a \ll D$.

The interference pattern due to an on-axis source point is obviously symmetric. The pattern caused by an off-axis source point is shifted by the phase difference between a coma ray and a diagonal ray (the source and the slits form a kind of light tube, cf. Figure 3.6).

The length of a coma ray is (using Eq. 1.5)

$$l_C = \sqrt{D^2 + (r_S - a)^2} \approx D + \frac{(r_S - a)^2}{2D} \tag{3.74}$$

and that of a diagonal ray is

$$l_D = \sqrt{D^2 + (r_S + a)^2} \approx D + \frac{(r_S + a)^2}{2D} \tag{3.75}$$

resulting in an optical path difference (OPD) of

$$\text{OPD} = l_D - l_C = \frac{2ar_S}{D} \tag{3.76}$$

For an optical path difference of OPD = $\lambda/2$ [33, Eqs. 10.2–10.14], the interference pattern related to this source point will be shifted by a half period and as a consequence, contrast will be lost. Consequently, spatial coherence means that

$$\frac{ar_S}{D} < \frac{\lambda}{4} \tag{3.77}$$

and using Eq. 3.42, we obtain

$$U \le \left(\pi \frac{\lambda}{4}\right)^2 = \left(\frac{\pi}{4}\right)^2 U_Q \tag{3.78}$$

for the system etendue that allows for a clearly visible fringe pattern. The coherence of an extended source is related to the etendue quantum U_Q (Eq. 3.22).

And do you remember the factor $(\pi/4)^2$ from Eq. 3.18, connecting 2D and 4D etendue?

Let's invent something!

Quantum? Did you say quantum? I am thrilled by quantum stuff!

Heisenberg's *uncertainty principle* states that [34, Eq. 6]

$$\Delta x \Delta p \ge \frac{h}{4} \tag{3.79}$$

and we can relate this to spatial coherence.

To do so, we identify the source size with the lateral position uncertainty [35]

$$\Delta x = r_S \tag{3.80}$$

and from the photon momentum $p = h/\lambda$ (the de Broglie relation) we deduce that

$$\Delta p = \frac{h}{\lambda}\frac{a}{D} \tag{3.81}$$

Thus, Eq. 3.79 becomes

$$r_S \frac{h}{\lambda}\frac{a}{D} \ge \frac{h}{4} \tag{3.82}$$

and (interpreting uncertainty as incoherence)

$$U \le \left(\frac{\pi}{4}\lambda\right)^2 \tag{3.83}$$

is a coherence condition, perfectly equivalent to Eq. 3.78. This is magic – isn't it?

Dr. Wisenheimer:

Basically yes, as long as we do not care too much about little details such as a factor of π. And Heisenberg's early and rather symbolic formulation was actually

$$\Delta x \Delta p \ge h \tag{3.84}$$

(Continued)

Solution 3.6 (Continued)

Only shortly thereafter it was shown [36, 37] that the standard deviations of position and momentum are actually connected by

$$\sigma_x \sigma_p \geq \frac{\hbar}{2} = \frac{h}{4\pi} \tag{3.85}$$

So: yes, but somewhat HANDLE WITH CARE.

3.4 Radiometry

In the previous section, we introduced the phase space, defined the etendue and showed ways to calculate it. Now it is about time to fill it with light.

Knowledge Corner Topic 3.10: Radiometric basics

The fundamental radiometric relation connecting flux and radiance or luminance ([38, Ch. 3] and [39]) is presented here as an axiom

$$\Phi = \iint n^2 L(\vartheta, A) \cos\vartheta \, \mathrm{d}\Omega \, \mathrm{d}A \tag{3.86}$$

and

$$\mathrm{d}\Phi = n^2 L(\vartheta, A) \cos\vartheta \, \mathrm{d}\Omega \, \mathrm{d}A \tag{3.87}$$

A note on notation: We use the symbol Φ for both radiant and luminous flux (Φ_v) because the essential conclusions which will be discussed in the following are independent of the choice of units. Likewise, we freely use E, I, L, and not E_v, I_v and L_v, for illuminance, luminous intensity, and luminance. Moreover, most of these conclusions apply to their spectral siblings E_λ, I_λ and L_λ as well. To denote specifically the radiometric flux (in W), we use Φ_e. In most cases, however, we use the symbols without subscript and let them implicitly denote all the others as well.

We will use the differentiation between Φ_e and Φ_v only if necessary. Radiant and luminous flux are related to each other by the standardized eye sensitivity curve $V(\lambda)$ (see Knowledge Corner Topic 3.23). The units of flux are W and lm, respectively, for radiance W/m^{2}sr, and for luminance lm/m^{2}sr. We use the symbol E for both illuminance/irradiance and exitance by introducing exitance via a screen to the right on the source surface.

The term $L(\vartheta, A)$ shows that luminance or radiance is generally a function of position and direction.

Eq. 3.87 yields the *intensity*

$$I(\vartheta) = \frac{\mathrm{d}\Phi}{\mathrm{d}\Omega} = n^2 \cos\vartheta \int L(\vartheta, A) \, \mathrm{d}A \tag{3.88}$$

(units W/sr or lm/sr = cd) and the *irradiance/illuminance*

$$E(A) = \frac{d\Phi}{dA} = n^2 \int L(\vartheta, A) \cos\vartheta \, d\Omega \tag{3.89}$$

(units W/m^2 or lm/m^2 = lx).

When we consider both the angular and spatial flux distribution, i.e. when we ask "how strong is the emission from a certain location on the source aperture into a certain direction", we ask for the *luminance* of a source. Equation 3.87 yields

$$L = \frac{d\Phi}{dU} = \frac{d\Phi}{dA\, dX} = \frac{d\Phi}{dA\, n^2 \cos\vartheta \, d\Omega} \tag{3.90}$$

for the luminance radiating from an area element dA embedded in a material with index n into an infinitesimal solid angle $d\Omega$ around a direction that forms an angle ϑ with the local surface normal.

Officially, "our" luminance/radiance, which includes the n^2 factor, is termed "basic luminance/radiance" [40]. Historically, radiance was considered only in air or in a vacuum, where the inclusion of the n^2 factor made a negligible difference. Today, however, we consider radiance also in refractive media such as inside an LED chip, inside the silicone dome within an LED, or within liquids (e.g. in the human eye or bladder). Importantly, conservation of basic luminance holds across refraction, since both flux and etendue are conserved across refraction as well. "Standard" luminance is not conserved across refraction, however; clearly, basic luminance is more fundamental, more important and more general, and in air, there is no difference anyway. However, over the years, we got really tired of saying or writing "basic" a gazillion times. We strongly believe that in science and engineering, short names should be reserved for important quantities for which a conservation law holds, with a qualifying adjective for derived quantities. Accordingly, we take the liberty of omitting "basic" in this book. We will sprinkle a few reminders, though.

We note the appearance of $\cos\vartheta \, d\Omega$, which is known as $d\Omega_p$, the differential *projected solid angle* element. The units of radiance and luminance are W/sr and cd/m^2 = nit, respectively. In the world of specialty lighting, we often face the form Mnit = Mcd/m^2 = lm/mm^{2}sr.

We can equally define *radiance/luminance* using the differential etendue (Eq. 3.4)

$$L = \frac{d\Phi}{dU} \tag{3.91}$$

The *effective luminance* describes the average luminance within a given etendue,

$$L_{\text{eff}} = \frac{\Phi}{U} \tag{3.92}$$

Most radiometry textbooks present the connections between these four quantities in a diamond-shaped diagram such as the one shown in Figure 3.28.

(Continued)

Knowledge Corner Topic 3.10: (Continued)

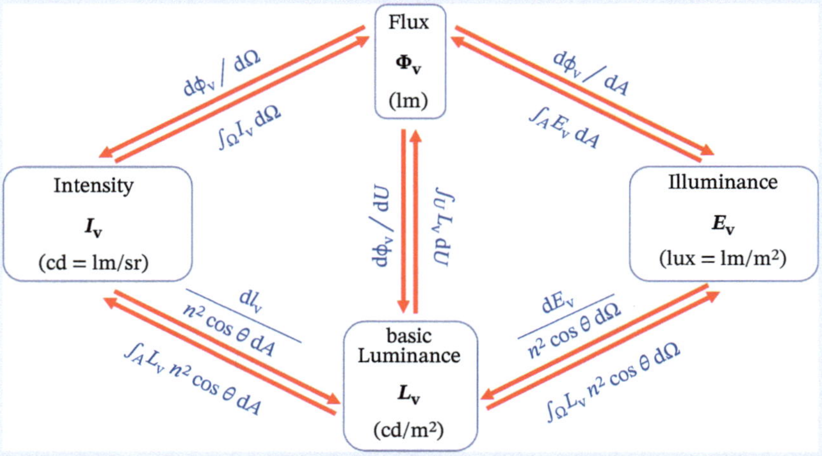

Figure 3.28 The four fundamental photometric quantities flux, intensity, illuminance and luminance, have differential and integral connections with each other.

Dr. Wisenheimer:

The way we introduce radiance and luminance (with the etendue including the refractive index) is pointing toward the fundamentally simple conservation law (Eq. 3.94) that will be introduced below.

"Our" radiance is called *basic radiance* in some references [40–42].

On the basis of these definitions, we can formulate some fundamental conservation laws that are discussed in Knowledge Corner Topic 3.11.

Knowledge Corner Topic 3.11: Conservation laws

The First Law of thermodynamics states that energy cannot be generated or eliminated, only transformed. In illumination optics, this law manifests itself in the form of flux conservation,

$$\Phi = \text{const} \tag{3.93}$$

There is no such thing as a perpetual motion machine of the first kind.

The Second Law of thermodynamics is expressed in different formulations, e.g. "there is no perpetual motion machine of the second kind", "without energy intake, entropy can only increase", and many more. In illumination optics, we know that the luminance cannot be increased by any passive optical system [43, 44], as will be illustrated in Exercise 3.19.

In the absence of absorption, scattering, or partial reflection, we have

$$L = \text{const} \tag{3.94}$$

This statement means constant luminance during phase-space transformations or inside an optical system. Make sure you avoid confusion with the constant (uniform) luminance of a source in the spatial or angular domain.

What about the common statement "the luminance of a ray is conserved" (as there is basically no such thing as the luminance of a ray)? Still, a ray can be assigned a flux value. If we split the source etendue into bins, all rays of a bin combined give the flux of that bin. The flux divided by the etendue of the bin yields the luminance of the bin, which may be formally assigned to each ray of the bin. A phase-space transformation will assign the bin to a bin from the target phase space, with the luminance conserved (and etendue and flux as well, of course).

Upper collection limit

The radiometric expression for the geometric statement from Eq. 3.19 is

$$\Phi_T \leq \min\,(L_S U_T, \Phi_S) \tag{3.95}$$

which is equal to (dividing both sides by Φ_S)

$$\eta_{\max} = \min\left(\frac{L_S}{\Phi_S} U_T, 1\right) = \min\left(\frac{U_T}{U_S}, 1\right) \tag{3.96}$$

which is a combination of both fundamental laws.

One implication is that light collection from an LED by a lens is limited by the acceptance angle θ or the *F*/# of the lens (via Eq. 3.14)

$$\eta_{\max} = \sin^2\theta = \frac{1}{4F/\#^2} \tag{3.97}$$

The initial radiometric equations (Eq. 3.86 and following) form the basis for all the calculations throughout this book. In many important situations, however, they can be simplified.

Knowledge Corner Topic 3.12: Constant luminance

We discuss a first special case.

Constant angular luminance

Luminance can vary with position, but at any position, it is constant over all angles.

$$L(\vartheta, A) = L(A) \tag{3.98}$$

This is true for laser phosphor sources, for example. Here, the Gaussian pump laser spot on the phosphor causes such a spatially inhomogeneous luminance distribution [45, Figure 5]. Another example is an LED surface with absorbing electrode structures.

(Continued)

Knowledge Corner Topic 3.12: (Continued)

In the initial equation Eq. 3.86, we can separate the integration over angle and position because all area elements share the same emission cone, using Eq. 2.14,

$$\Phi = n^2 \int \cos\vartheta \, d\Omega \int L(A) \, dA = \Omega_p I_0 \tag{3.99}$$

and introducing the on-axis intensity

$$I_0 = n^2 \int L(A) \, dA \tag{3.100}$$

sometimes also called the *center-beam candle power*, CBCP [46].

For off-axis directions ($\vartheta > 0$), the intensity is obtained as

$$I(\vartheta) = \frac{d\Phi}{d\Omega} = \cos\vartheta \, I_0 \tag{3.101}$$

featuring the characteristic cosine dependence.

Radiant or luminous exitance (here a function of position) is

$$E(A) = \frac{d\Phi}{dA} = n^2 L(A) \Omega_p \tag{3.102}$$

depending on luminance and projected solid angle (Eq. 2.12).

Dr. Wisenheimer:

The equation also applies to the task of concentrating a beam of luminance L onto a target surface and suggests using a large projected solid angle of incidence. This is the basis for all concentration ratio discussions in solar collection.

Next, we discuss a second special case.

Constant spatial luminance

In this case, the luminance varies with direction, but for any direction, it is constant over all positions:

$$L(\vartheta, A) = L(\vartheta) \tag{3.103}$$

A prominent example is the $\cos^n$ beam (Knowledge Corner Topic 3.15).

Again, we can separate Eq. 3.86 to obtain

$$\Phi = n^2 \int L(\vartheta) \cos\vartheta \, d\Omega \int dA = E_0 A \tag{3.104}$$

and the constant exitance

$$E_0 = n^2 \int L(\vartheta) \cos\vartheta \, d\Omega \tag{3.105}$$

(or irradiance, if the beam is incident on a surface). The intensity is then

$$I(\vartheta) = \frac{d\Phi}{d\Omega} = n^2 L(\vartheta) \cos\vartheta \cdot A \tag{3.106}$$

and on-axis, we have

$$I_0 = n^2 L(0) A \tag{3.107}$$

Finally, and most important, the luminance may be *constant for all positions and directions:* $L = \text{const}$. Such a source, when planar, is called a Lambertian emitter.

Lambertian emitter

This is not only an important academic example, but also highly relevant as LEDs are close to ideal Lambertian emitters.

Equation 3.91 then transforms into

$$L = \frac{\Phi}{U} \tag{3.108}$$

and from Eq. 3.86 we obtain

$$\Phi = n^2 L \int \cos\vartheta \, d\Omega \int dA = n^2 L \Omega_p A = n^2 \pi I_0 \text{ sr} \tag{3.109}$$

with

$$I_0 = LA \tag{3.110}$$

the same intensity distribution as above

$$I(\vartheta) = \frac{d\Phi}{d\Omega} = I_0 \cos\vartheta \tag{3.111}$$

and a constant exitance

$$E(A) = \frac{d\Phi}{dA} = E_0 = n^2 L\pi \text{ sr} \tag{3.112}$$

One important consequence of Eq. 3.109 is the following: if you have to image a source to a target of smaller etendue, magnification is irrelevant. In this case, the target etendue selects a portion $A\Omega_P$ from the source and because of $L = \text{const}$, any choice of angle and area will deliver the same flux.

Exercise 3.7: Source luminance

Two sources with $A = 1 \text{ mm}^2$ are embedded (1) in air and (2) in a material with $n = 1.5$ (Figure 3.29).

(a) What are the etendues of both sources? (More precisely, use a screen directly to the right of the source, and compute the etendue of the ray bundle consisting of all the rays that leave the source).

A luminance meter (outside the material in case (2)) measures $L = 1 \text{ cd/mm}^2$ in both cases.

(Continued)

Exercise 3.7: (Continued)

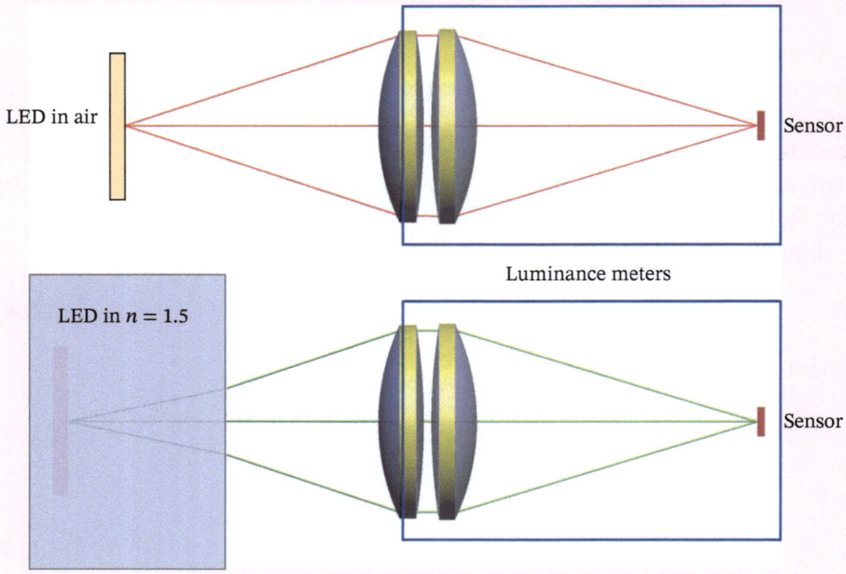

Figure 3.29 Two identical luminance meters measuring otherwise identical sources in air and inside a medium.

(b) Assuming that both sources have constant luminance (i.e. they are Lambertian sources), what is the luminous flux of both?

(c) From the ray tracing image above, you can see that the sensor side of the luminance meter sees the same ray bundle in both cases. On the source side, things are different. Notably, the angle of the ray cone leaving the sources is different, depending on the embedding. Discuss this difference in view of your previous results.

Solution 3.7

(a) The etendue is calculated from the definition in Eq. 3.7 with a separation of area and angle (Eq. 3.8),

$$\begin{aligned} U_1 &= \pi \cdot 1\,\text{mm}^2 \cdot \sin^2 90°\text{sr} = 3.1\,\text{mm}^2\text{sr} \\ U_2 &= 1.5^2\pi \cdot 1\,\text{mm}^2 \cdot \sin^2 90°\text{sr} = 7.1\,\text{mm}^2\text{sr} \end{aligned} \tag{3.113}$$

(b) Eq. 3.108 yields

$$\Phi_1 = L \cdot U_1 = 1\,\frac{\text{cd}}{\text{mm}^2} \cdot 3.1\ \text{mm}^2\text{sr} = 3.1\ \text{lm}$$
$$\Phi_2 = L \cdot U_2 = 1\,\frac{\text{cd}}{\text{mm}^2} \cdot 7.1\ \text{mm}^2\text{sr} = 7.1\ \text{lm} \tag{3.114}$$

(c) Using Snell's law in the form of a conservation law (Eq. 2.44), it is obvious that the angular cone of the pencil in the $n = 1.5$ medium has to be smaller than in air.

Another way to look at this is that all rays in air correspond to rays within the TIR cone in the material. The rays outside the TIR cone cannot exist in air, but the LED must still fill these ray paths with light. Therefore, the LED must emit a larger flux if the luminance is to be the same.

Knowledge Corner Topic 3.13: Solid angle or projected solid angle?

We start with Eq. 3.86, use $n = 1$, and carry out area integration:

$$\mathrm{d}\Phi = L(\vartheta, A)\cos\vartheta\,\mathrm{d}\Omega \cdot A \tag{3.115}$$

(a) Flux divided by *solid angle* (differentials) yields the (angle-dependent) intensity

$$\left.\frac{\mathrm{d}\Phi}{\mathrm{d}\Omega}\right|_{\vartheta} = L(\vartheta, A)\cos\vartheta \cdot A = I(\vartheta) \tag{3.116}$$

In the Lambertian case, $L = \text{const}$, so we use Eq. 3.110 to get

$$\left.\frac{\mathrm{d}\Phi}{\mathrm{d}\Omega}\right|_{\vartheta} = I_0 \cos\vartheta \tag{3.117}$$

which is plotted on almost every bare-chip LED datasheet.

(b) Flux divided by *projected solid angle* (differentials) yields

$$\left.\frac{\mathrm{d}\Phi}{\mathrm{d}\Omega_\mathrm{p}}\right|_{\vartheta} = L(\vartheta, A)A = \frac{I(\vartheta)}{\cos\vartheta} \tag{3.118}$$

For constant luminance, this reduces to

$$\left.\frac{\mathrm{d}\Phi}{\mathrm{d}\Omega_\mathrm{p}}\right|_{\vartheta} = I_0 \tag{3.119}$$

Dr. Wisenheimer:
The *solid angle* is used only in relation to intensity. For calculations related to etendue and luminance you always need to use the *projected solid angle*.

Exercise 3.8: Lambertian source shining into a light guide

A Lambertian source is located in front of a light guide, separated by a tiny air gap. Calculate the intensity distribution inside the light guide. Is it still Lambertian?

Solution 3.8

Flux is conserved at the interface (we neglect Fresnel losses). And luminance conservation is the link that connects the radiation inside and outside the medium.

We have

$$L(\beta) = L(\alpha) = L \tag{3.120}$$

which means that a Lambertian distribution remains unaffected by refraction. This is closely related to the law of Malus and Dupin [47, 48] (a congruence [49] remains unaffected by refractions and reflections [50, Ch. 3.3.3]). Figure 3.30 (right) shows the typical circular intensity distribution of the $\cos\beta$ dependence.

Equation 3.100 tells us that the intensity I_0 on-axis is increased by a factor of n^2 after refraction.

Simulation 3.4: Lambertian source shining into a light guide

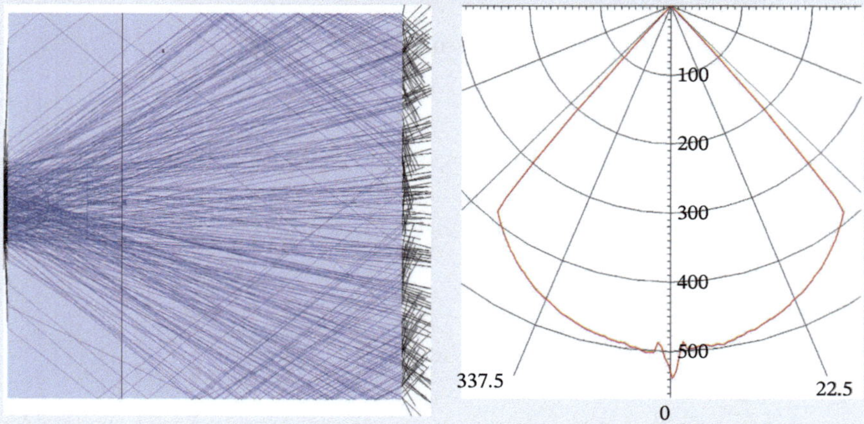

Figure 3.30 A Lambertian emitter shining into a light guide: simulation.

The noisy variations around the optical axis are the result of the triangular shape and size of the angular bins there. Each cross-section containing the optical axis may show such peaks or dips.

Exercise 3.9: Isotropic source inside a medium

Is an LED perfectly Lambertian? Let us assume that light generation by charge carrier recombination in the active region of a thin-film LED is characterized by an *isotropic* intensity distribution inside the medium (modeled as individual isotropic recombination sites in a transparent crystal). Assume furthermore that the crystal is thin and wide (disregard any edge effects), has a near-perfect specular mirror below the junction with tiny but nonzero absorption and a polished surface toward the outside air. The radiation from such an isotropic source is then refracted to the outside.

Why can you neglect Fresnel losses at the crystal–air interface? Calculate the intensity and luminance distributions over polar angle.

Solution 3.9

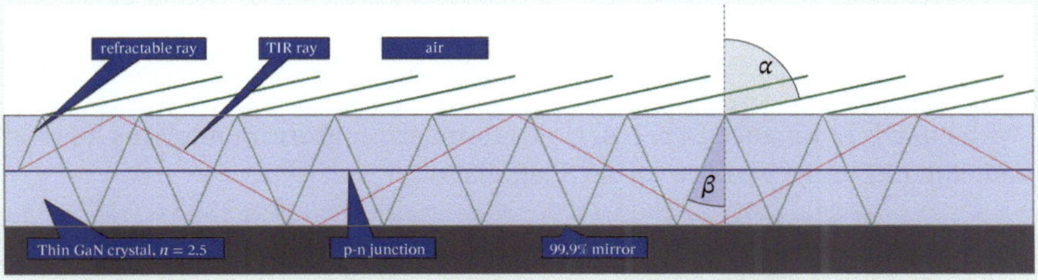

Figure 3.31 Our idealized thin-film LED model shows a thin GaN crystal with the transparent pn junction inside. At the bottom is a 99.9% reflective specular mirror. "Refractable" rays (green) with incidence angle α that are within the TIR cone (bonus question: what is the TIR angle?) are partially Fresnel-reflected, partially refracted outside with an exit angle β. TIR rays outside the TIR cone (red) are guided.

In our idealized model with a gallium nitride LED (see Figure 3.31), we find "refractable" rays within the TIR cone that are partially reflected at each GaN–air interface, but a large fraction of the light is refracted out, again and again, after returning from the nearly perfect mirror. Since losses at the mirror are negligible, almost all light will eventually be refracted out, all at the same angle. This is why Fresnel reflection at the GaN–air interface can be neglected in this case.

Rays outside the TIR cone will be guided and have a much longer "lifetime", but unlike the refractable rays, they will eventually be absorbed by the mirror. Essentially, the model behaves as if there were a perfect antireflective coating on the crystal, completely transmitting all refractable rays, and the mirror would be totally black. We can therefore argue on the basis of luminance and etendue conservation along the refractable rays.

(Continued)

Solution 3.9 (Continued)

If the far-field emission is Lambertian, with constant luminance L, then Eq. 3.101 yields

$$I(\alpha) = LA \cos \alpha \tag{3.121}$$

from the outside, and

$$I(\beta) = n^2 L \cos \beta \cdot A \tag{3.122}$$

inside. (It is a bit strange to talk about the intensity emitted by the junction inside because the crystal is way too thin for a "far field". However, this intensity inside is perfectly well-defined: We consider rays emitted by the junction that go into dΩ) around β. We remind you that "our" luminance, i.e. the one that is constant across refraction, is officially known as "basic luminance". The n^2 factor makes sure that everything is correct.)

A constant intensity I inside (isotropic emission) means that a ray in the direction of β must have a luminance of

$$L(\beta) = \frac{I}{A \cos \beta} = \frac{L_0}{\cos \beta} \tag{3.123}$$

with on-axis luminance $L_0 = I/A$. Now luminance conservation tells us that outside the LED chip, we have (using Snell's law)

$$L(\alpha) = \frac{L_0}{\sqrt{1 - \frac{\sin^2 \alpha}{n^2}}} \tag{3.124}$$

and therefore

$$I(\alpha) = L(\alpha) A \cos \alpha = \frac{L_0 A \cos \alpha}{\sqrt{1 - \frac{\sin^2 \alpha}{n^2}}} = I \frac{\cos \alpha}{\sqrt{1 - \frac{\sin^2 \alpha}{n^2}}} \tag{3.125}$$

which represents a Lambertian distribution with a minor deviation, but becomes perfectly Lambertian for large n and isotropic for $n = 1$ (see Figure 3.32).

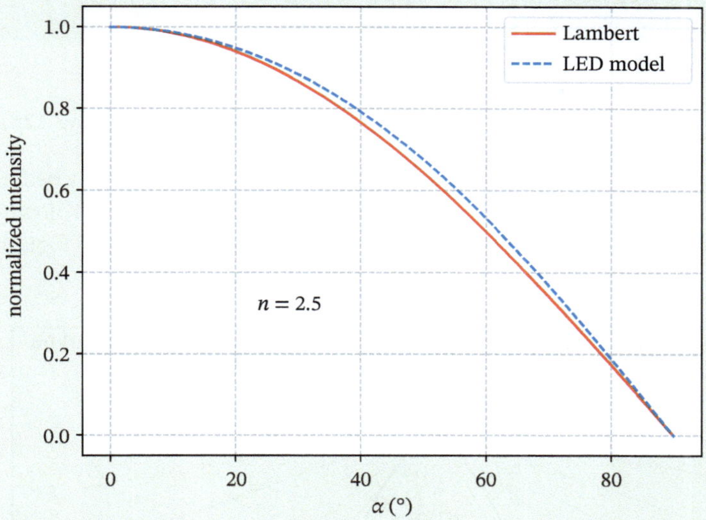

Figure 3.32 Comparing the intensity distribution of our LED model (Eq. 3.125) with an ideal Lambertian cosine distribution.

Dr. Wisenheimer:

Real InGaN thin-film LEDs do not behave like this model. Both their mirrors and their top surfaces are very rough. This is actually intentional and intended to cause diffuse scattering instead of the guided modes visible in Figure 3.31. This greatly increases the overall extraction efficiency and mitigates the ripple interference pattern in the intensity distribution that is actually seen with specular top surfaces. Anyway, the overall result is still very much Lambertian. For details about these processes from geometric-optical, thermodynamical and wave optical viewpoints, see "Extracting Light out of LEDs" [51].

There are many situations where the illumination designer needs to specify or describe a collimated beam with soft edges that is shining to the far field.

Knowledge Corner Topic 3.14: Collimation strength

The ratio

$$\kappa = \frac{I_0}{\Phi} \tag{3.126}$$

(peak intensity of a beam related to total flux) is a measure of the *shape* of a beam with the unit 1/sr or cd/klm. Light distribution curves of luminaires (Knowledge Corner Topic 4.2) are often scaled using their collimation strength (Figure 3.33).

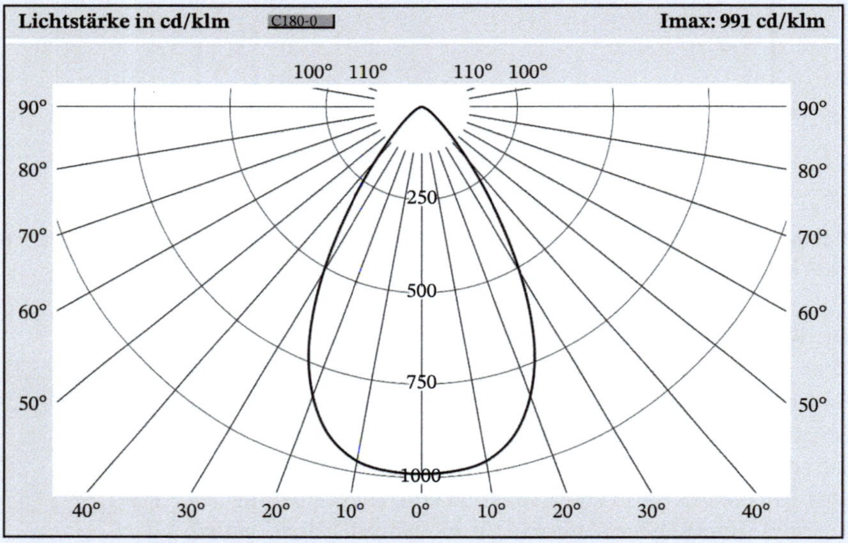

Figure 3.33 Light distribution curve of a downlight. *Source:* Image courtesy Siteco.

A Lambertian emitter has a collimation strength of

$$\kappa = \frac{LA}{LU} = \frac{1}{\Omega_p} = \frac{1}{\pi\,\text{sr}} \tag{3.127}$$

There are various ways to describe the shape of intensity distributions of collimated beams mathematically, for example, the Gaussian beam model from Eq. 2.76. The crucial point is here that a Gaussian distribution can be integrated analytically over an area (thanks to the $2\pi\, r\, \mathrm{d}r$ differential), but not in the angular domain. Therefore, we want to introduce the following angular beam model that we find to be extremely helpful in photometric estimations.

Knowledge Corner Topic 3.15: cos^n beams

A beam with an intensity distribution of

$$I(\vartheta) = I_0 \cos^n \vartheta \tag{3.128}$$

can be integrated. The total flux of such a beam is

$$\Phi = \int_0^{90°} I(\vartheta)\, d\Omega = 2\pi \int_0^{90°} I_0 \cos^n(\vartheta) \sin\vartheta\, d\vartheta \tag{3.129}$$

$$\Phi = 2\pi \mathrm{sr} I_0 \left[-\frac{\cos^{n+1}\vartheta}{n+1} \right]_0^{90°} = \frac{2\pi\ \mathrm{sr}}{n+1} I_0 \tag{3.130}$$

which means that

$$n = 2\pi\ \mathrm{sr} \frac{I_0}{\Phi} - 1 \tag{3.131}$$

Its collimation strength is

$$\kappa(n) = \frac{n+1}{2\pi\ \mathrm{sr}} \tag{3.132}$$

We cannot reasonably define a solid angle of the cos^n beam other than 2π sr, but the quantity $1/\kappa(n)$ may be considered as a kind of "effective solid angle" and the etendue of the beam will be

$$U = \frac{A_S}{\kappa} = \frac{A_S \cdot 2\pi sr}{n+1} \tag{3.133}$$

with the source area A_S.

The flux contained in a cone of angle θ is

$$\Phi(\theta) = \Phi\left[1 - \cos^{n+1}\theta\right] \tag{3.134}$$

The luminance of a source that emits a cos^n beam in the direction ϑ with a constant luminous exitance E_0 is

$$L(\vartheta) = \frac{I(\vartheta)}{A} = \frac{\Phi}{A}\kappa(n) \cos^{n-1}\vartheta = E_0 \kappa(n) \cos^{n-1}\vartheta \tag{3.135}$$

using Eqs. 3.88 and 3.130. Hence, the on-axis luminance becomes

$$L(0) = \frac{I_0}{A} = E_0 \kappa(n) \tag{3.136}$$

We can approximate a Gaussian beam reasonably well by a cos^n beam (Figure 3.34).

(Continued)

Knowledge Corner Topic 3.15: (Continued)

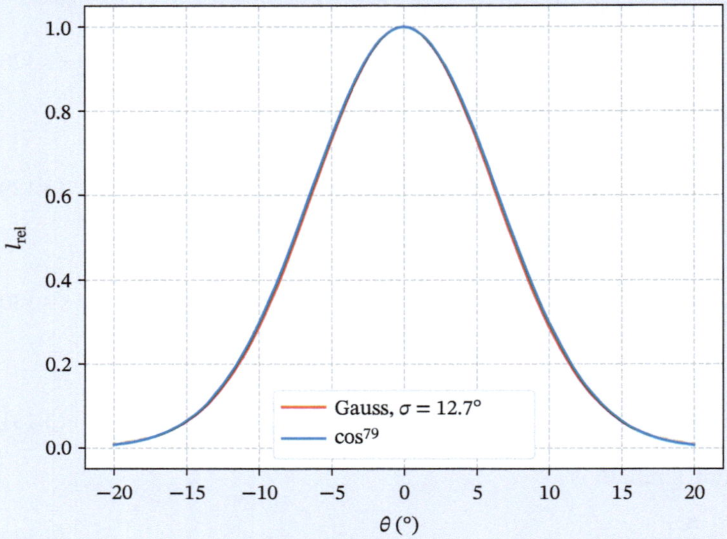

Figure 3.34 Gaussian beam of $\theta = 12.7°$ approximated by a $\cos^{79}$ beam.

We provide the $\cos^n$ calculator file.

Dr. Wisenheimer:

In some fields such as stage lighting, the *beam angle* (identical to the term FWHM) refers to the full angular width for a relative intensity of 50%,

$$\cos^n \theta_{0.5} = 0.5 \tag{3.137}$$

As Figure 3.35 illustrates, this corresponds to a cosine exponent of

$$n = \frac{\ln 0.5}{\ln \cos \theta_{0.5}} \tag{3.138}$$

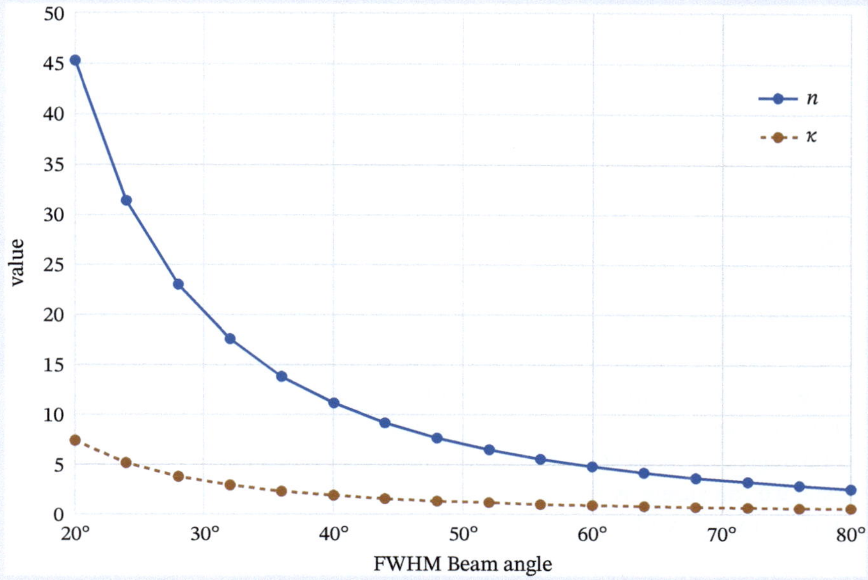

Figure 3.35 Cosine exponent n and collimation strength $\varkappa$ as a function of the beam angle (FWHM).

Provided file.

Exercise 3.10: $\cos^n$ beam

What luminous flux is contained in a $\cos^{100}$ beam with an on-axis luminous intensity of 100 kcd?

Solution 3.10

$$\Phi = \int_0^{90^\circ} I(\vartheta)\,\mathrm{d}\Omega = 2\pi\,\mathrm{sr} I_0 \int_0^{90^\circ} \cos^{100}\vartheta \sin\vartheta\,\mathrm{d}\vartheta \tag{3.139}$$

$$\Phi = -\frac{2\pi\,\mathrm{sr} I_0}{101}\left[\cos^{101}\vartheta\right]_0^{90^\circ} = \frac{2\pi\,\mathrm{sr}\cdot 100\,\mathrm{kcd}}{101} \approx 6200\,\mathrm{lm} \tag{3.140}$$

Exercise 3.11: Total included angle

In an LED datasheet, a quantity "total included angle" appears, implying that 90% of the total flux is contained within this angle. Calculate the total included angle of a Lambertian source.

Solution 3.11

We use Eq. 3.133 with $n = 1$

$$0.9 = 1 - \cos^2\theta \tag{3.141}$$

and obtain $2\theta = 143°$, which is slightly larger than the value given in the referenced datasheet.

We conclude the radiometry section with a derivation of further fundamental laws in the following Knowledge Corner Topics.

Knowledge Corner Topic 3.16: Inverse square law (1)

Imagine a point source illuminating a tiny circle with area A_T on a target at a large distance D. The circle is oriented toward the source. It therefore subtends a solid angle $\Omega = A_T/D^2$ (Eq. 2.8) as seen from the point source. Within this small cone, we can surely assume constant intensity, $I = I_0 = \text{const}$. The flux incident on A_T is then given by

$$\Phi_T = I_0\Omega = I_0\frac{A_T}{D^2}\,\text{sr} \tag{3.142}$$

Division by A_T yields the celebrated *inverse square law* that relates the source intensity I_0 to the target illuminance E_T:

$$E_T = \frac{\Phi_T}{A_T} = \frac{I_0}{D^2}\text{sr} \tag{3.143}$$

Its interpretation is straightforward: A beam with $I_0 = 1$ cd produces an illuminance of $E_T = 1$ lx at a distance of $D = 1$ m; the illuminance increases proportional to intensity and decreases proportional to $1/D^2$. A solid angle of 1 sr is, by definition, subtended by an area of D^2 on a sphere with radius D, as seen from the sphere's center. Thus, the D^2 factor really means "area per solid angle at that distance D", to be measured in m^2/sr instead of m^2, which nicely explains the appearance of sr.

In automotive lighting, for example, the ECE (Economic Commission for Europe, automotive regulations) test values are evaluated on a screen at a distance of $D = 25$ m: A beam of $I_0 = 100$ kcd gives rise to $E_T = 160$ lx at that distance.

The inverse square law is universally valid for the *far field* of a source. It is an indispensable tool to compute the irradiance on a distant target in lighting simulations. However, several related problems appear. First, it is sometimes hard to identify where the *near field* of the source ends and the *far field* begins (a distance also known as the *limiting photometric distance*, LPD [52–54]). Second, for more complicated sources, it is sometimes hard to tell from which point on the source the distance D to the target should be measured – from the center of the aperture, from an intermediate focus outside the lamp, from the light source within the lamp, or from somewhere else?

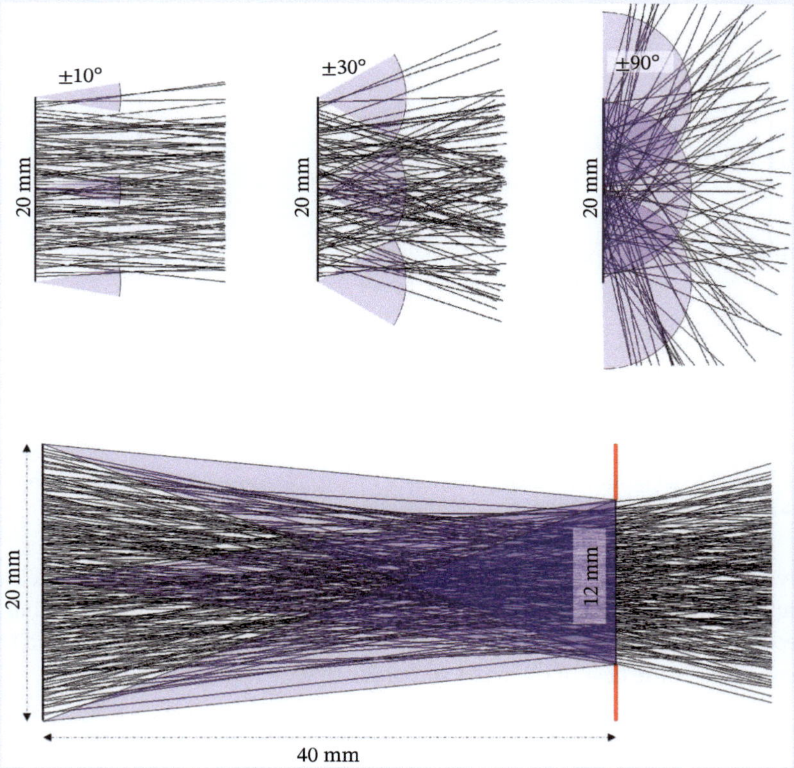

Figure 3.36 Sources with different pupils (aim regions). Top: telecentric, i.e. angular limitation, with ±10°, ±30° and ±90° (Lambertian). Bottom: Finite distance pupil, common in Köhler illumination configurations. All sources feature a diameter of $2\,r_S = 20\,\text{mm}$ and a total flux of 100 lm, but their directionality differs. The Köhler configuration limits the beam to a pupil of $d = 12$ mm at a distance of 40 mm.

For a better understanding, we set up some sources with the same size and flux, but different angular distributions (Figure 3.36). We then measure the on-axis illuminance E_T as a function of distance D from the source (well – we do not measure, we simulate!).

(Continued)

Knowledge Corner Topic 3.16: (Continued)

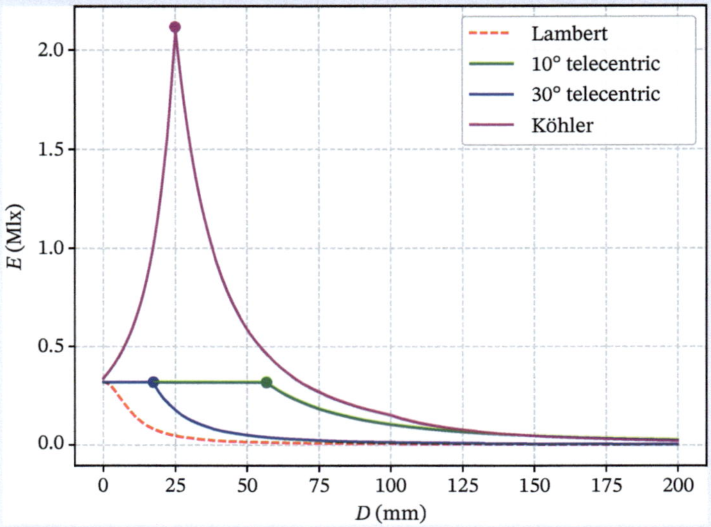

Figure 3.37 Axial illuminance as a function of the distance to source. The circles point to locations where the illuminance suddenly starts to drop with distance.

The results (Figure 3.37) show varying near-field behavior, followed by a decrease that appears to follow $1/D^2$, but how can we tell if it is really $1/D^2$ or just similar? Moreover, what would be the best location D_0 from where the distance should be measured to get the best fit? We first recognize that

$$E \propto \frac{1}{D^2} \quad \Rightarrow \quad \frac{1}{\sqrt{E}} \propto D \tag{3.144}$$

This allows us to use a linear regression to determine the slope a and y-axis intercept b for

$$\frac{1}{\sqrt{E}} = a\,D + b = a(D - D_0) \tag{3.145}$$

where $D_0 = -b/a$ is the x-axis intercept, which will turn out to be much more useful than the linear regression result for b.

With this linear regression model in mind, we plot $1/\sqrt{E}$ vs. D (see Figure 3.38).

For each curve, we can now visually determine the LPD, i.e. the point from where the curve is approximately linear. We see that, beginning at the LPD, the typical inverse square far-field dependence develops. The far field of a Lambertian source and of a ±30° beam starts very soon. For the ±10° source, the minimum distance is almost 60 mm, because at closer distance, the narrow telecentric beams from the source edge will not reach the target. The two kinks of the Köhler curve may be surprising at first. However, from larger distances, it is the smaller 12 mm pupil (at a distance of 40 mm from the actual source) that limits the beam, almost as if there were a Lambertian source. It is therefore appropriate that the Köhler far-field asymptote has a $D_0 = 40$ mm offset, whereas $D_0 = 0$ for the three other sources.

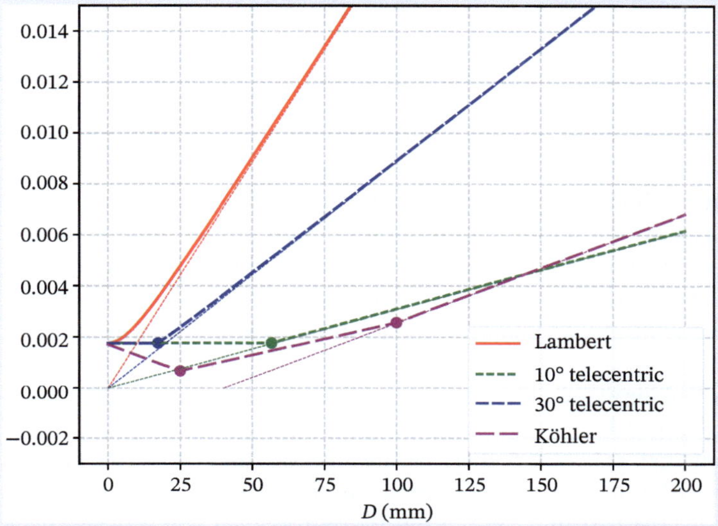

Figure 3.38 Plot of $1/\sqrt{E_0}$ as a function of distance, for the same four sources. The thin dashed lines show the far-field asymptotes. Their different slopes are due to the fact that our four sources all have the same flux and size, but different luminances and far-field intensities. The dots indicate sudden slope changes. A constant slope usually means a valid far field and the validity of the inverse square law. However, note the kink at $D = 100$ mm for the Köhler line (magenta), where the curve approaches the true asymptote: The Köhler line segment between 25 and 100 mm can deceive you into believing that convergence is already occurring.

We computed these results and curves from the formulas (and double-checked them by means of simulations – something we like to do because redundant information often is the key to recognizing errors early!). However, the intensity will usually be determined from measurements using a goniometer. In many cases, the measurement engineer will simply take measurements at a single fixed distance, trusting that the far-field assumption holds. However, it happens more often than we would like to admit that it is not clear where the far field really begins and where the "center of light (mass)", or the "virtual focus" [55] of the luminaire is located. Therefore, we recommend to perform illuminance measurements at several distances (the more, the better – redundant information again!), plot curves such as those in Figure 3.38 to visually determine the LPD, and apply linear regression (readily available in Excel®, Matlab®, Python and others, or even "by hand") to the data points with $D >$ LPD. We asked Werner Halbritter, the decade-long light measurement top expert at ams-OSRAM, and he told us that this is exactly what he does; in practice, he varies the lamp-to-sensor distance

(Continued)

Knowledge Corner Topic 3.16: (Continued)

around its nominal value by just as much as he can easily implement, and he finds that this variation is often sufficient to apply the $1/\sqrt{E}$ model in this way.

From the linear regression, we obtain the slope a and y-intercept b, from which we get

$$I_0 = \frac{1}{a^2} \qquad E_T = \frac{I_0}{(D - D_0)^2} \text{ with } D_0 = -b/a \tag{3.146}$$

This approach neatly solves all three of our problems: The far field begins where we are sure that $1/\sqrt{E_T}$ becomes linear. The intensity I_0 is given by the slope obtained from the linear regression. The x-axis intercept D_0 tells us from where we should measure the distance to the receiver, which is very useful to predict the illuminance when we deploy the actual source, for example in an office lighting situation or on a movie film set, or, as Werner Halbritter says, when he needs to use a calibration lamp at some nonstandard distance to provide a well-known illuminance.

We start by looking at the situations where the far field begins (excluding Lambertian), see Figure 3.39.

We conclude that, at least for otherwise homogenous luminance, the far field cannot begin earlier than at the distance where the receiver gets *rays from the whole source* for the first time. Watch out for pupils when you consider what "source" means here. For constant luminance within the beam, the LPD is a purely geometric condition.

For a source with an angular limitation $\pm\theta$, far field starts no earlier than

$$\text{LPD} = \frac{r_S}{\tan\theta} \tag{3.147}$$

For a Köhler-like source with a large pupil of $r_P > r_S$ at D_P, the far field can begin between source and pupil,

$$\text{LPD} = D_P \frac{r_S}{r_S + r_P} \tag{3.148}$$

When the pupil is smaller than the source, $r_P < r_S$, the far field begins behind the pupil, no earlier than

$$\text{LPD} = D_P \frac{r_S}{r_S - r_P} \tag{3.149}$$

since at large distances, the pupil is the effective source boundary, not the edge of the actual light source.

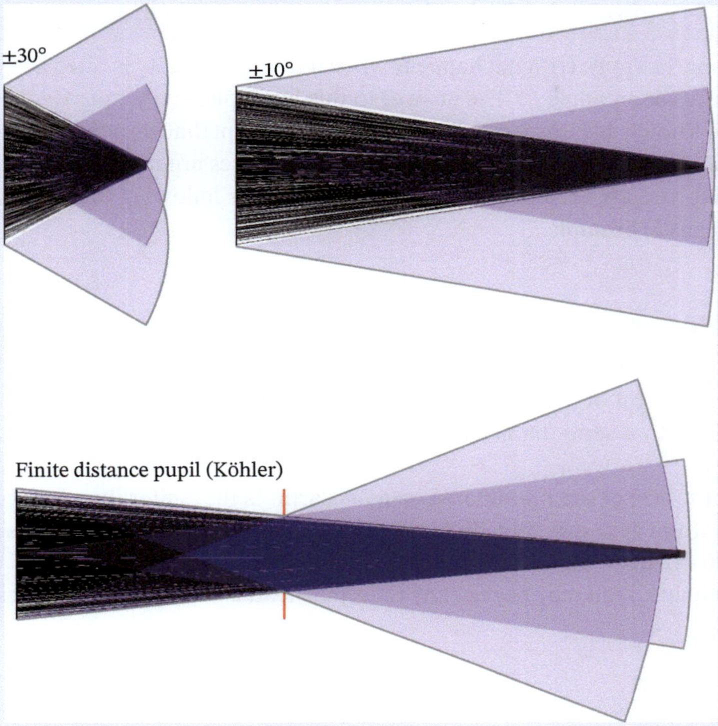

Figure 3.39 Rays from the source to a receiver at the limiting photometric distance (LPD) for sources of ±30° (top left), ±10° (top right), and a Köhler source.

Beyond these distances, the beam spreads according to the limiting angles.

When using formulations such as "cannot begin earlier than", "can begin at", or "no earlier than", we imply that in addition to seeing the whole source, the LPD must be large enough so that the small-angle approximation holds for the angular size (half angle α) of the source as seen from the LPD. The minimum LPD according to this latter requirement varies according to the precision you need: Here, the small-angle approximation error (see Knowledge Corner Topic 2.3) of your E estimate is $1 - \sin^2 \alpha / \tan^2 \alpha$, which is strictly valid for homogeneous luminance and circular source apertures and approximately valid otherwise.

Throughout the book, the inverse square law is applied in various instances:

- Gaussian beam irradiance and intensity, Eq. 2.87.
- Flux from an extended source to an area element such as shown in Figure 4.1.
- Knowledge Corner Topic 3.17.
- Exercise 4.43.
- A Lambertian source at the ceiling, Exercise 4.12.
- Headlights, Exercise 4.50.
- Exercise 3.13, sources of different luminance distributions.

(Continued)

Knowledge Corner Topic 3.16: (Continued)

Two final remarks apply: (i) What happens when our target area A_T is tilted away from the source at an angle β? The answer to this is simple: everything we discussed in the Knowledge Corner Topic remains valid, except that we need to multiply all E values by $\cos\beta$. (ii) If you feel that these sources are rather idealized and contrived, you are correct. See Exercise 4.43 and the following Simulation for a more realistic example.

Exercise 3.12: Orbiting reflector

A satellite at a height of 200 km above the Earth's surface unfolds an ideal flat-disk reflector with a semi-diameter of $r_S = 10\,\text{m}$ to illuminate a spot on Earth (Figure 3.40).

Calculate the size of the illuminated area on Earth. Is the center illuminance the same as if the area was directly hit by the Sun? Can you apply the inverse square law to calculate the light distribution on Earth?

For the estimates, you may neglect any reflector tilt and assume perpendicular incidence.

Figure 3.40 Reflector orbiting the Earth.

This idea was promoted by a startup, and there was some discussion on the internet regarding the applicability of the inverse square law to this case.

Solution 3.12

We start by calculating the flux on the $r_S = 10\,\text{m}$ reflector area ($A = 314\,\text{m}^2$) and with Eq. 3.224, we have

$$\Phi = EA = 111\,\text{klx} \cdot 314\,\text{m}^2 \approx 35\,\text{Mlm} \tag{3.150}$$

We may be tempted to apply the inverse square law, consider the disk as a Lambertian source, and use Eq. 3.110 to calculate the on-axis intensity

$$I_0 = \frac{\Phi}{\Omega_p} = \frac{35\,\text{Mlm}}{\pi\,\text{sr}} \approx 11\,\text{Mcd} \tag{3.151}$$

From this, we can then calculate the center illuminance on Earth via

$$E_0 = \frac{I_0\,\text{sr}}{D^2} = \frac{11\,\text{Mcd sr}}{(200\,\text{km})^2} = 0.275\,\text{mlx} \tag{3.152}$$

which is obviously far too small.

One problem is that there is actually no scattering from the reflector surface. The angular dimension of the reflected beam is determined by the Sun, yielding an illuminated disk (on Earth) of

$$R = D\tan\theta_S = 200\,\text{km} \cdot \tan 0.266° = 0.93\,\text{km} \tag{3.153}$$

with an area of $A_T = 2.7\,\text{km}^2$ and a corresponding illuminance of

$$E_T = \frac{\Phi}{A_T} = \frac{35\,\text{Mlm}}{\pi(930\,\text{m})^2} = 13\,\text{lx} \tag{3.154}$$

Still, this result appears a little small. Let's simulate (below).

What we did here is to calculate the illuminance on a purely geometrical basis. We could not apply the inverse square law because of missing requirements.

In summary, we need a larger reflector area to achieve an illuminance level comparable to that recommended for desk lighting. Or maybe a slightly concave reflector.

Dr. Wisenheimer:

Can diffraction at the reflector area have any effect on the spot size?

Diffraction at a circular aperture causes a far-field distribution with a first minimum at [56, Ch. 10.2.5]

$$\theta_D = 1.22\frac{\lambda}{2r_S} \tag{3.155}$$

which is orders of magnitude smaller than the angular extent of the Sun.

Simulation 3.5: Orbiting reflector

We simulate the situation and set up a disk source of $r_S = 10\,\text{m}$, with the Sun's angular size and set to a flux of $\Phi_S = \pi r_S^2 E_{\text{Sun}}$, and a receiver in a distance of 200 km (Figure 3.41).

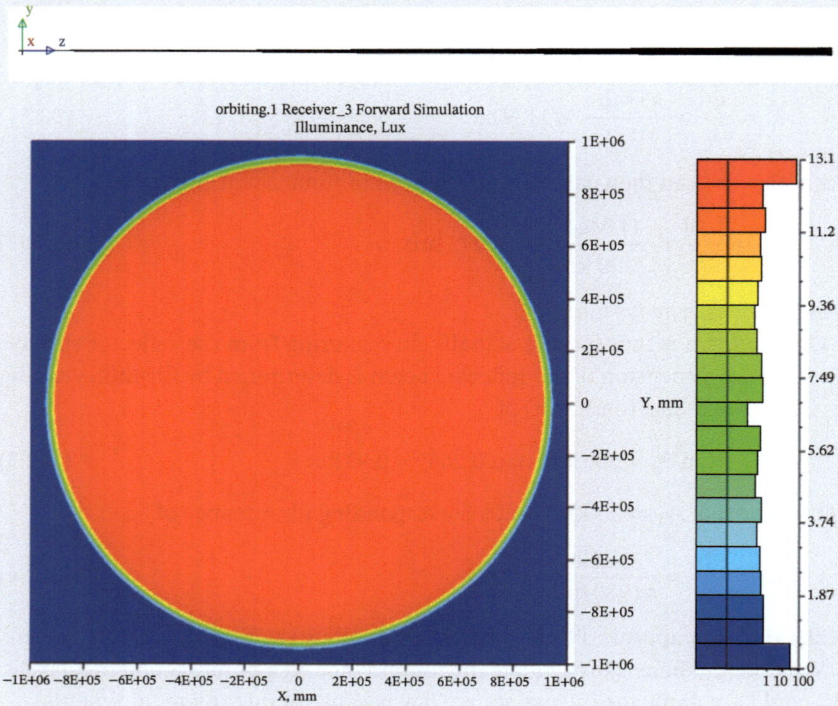

Figure 3.41 Sunlight from an orbiting reflector shining onto Earth. The simulation features a sunlike source of the size of the reflector disk, shining at a receiver on Earth. Bottom: Receiver illuminance.

At the receiver, we observe a constant illuminance of $E_T = 13\,\text{lx}$, exactly as calculated above.

Simulation file.

With a Lambertian scatterer of the same size as the reflector, we obtain a mere $E_T \approx 0.22\,\text{mlx}$, only slightly smaller than estimate above.

Dr. Wisenheimer:

We might be tempted to install a concave parabolic reflector instead of the flat one. However, as the target etendue (between the reflector area and a small area on Earth) is much smaller than the etendue spanned by reflector and the Sun, the former will always be filled, and we will obtain the same illuminance as in Figure 3.41.

Knowledge Corner Topic 3.17: Inverse square law (2)

In the following, we will derive the mathematics of receiver illuminance. We consider a meridional section of a rotationally symmetric arrangement. A source emits light of a luminance $L(r, \vartheta)$, the emission can be Lambertian, collimated, or focused (non-telecentric) as in Figure 3.42. The source area is $A_S = \pi r_S^2$, the light is measured with a small receiver at a distance D. ϑ is the direction of the ray that connects dA_S and A_T.

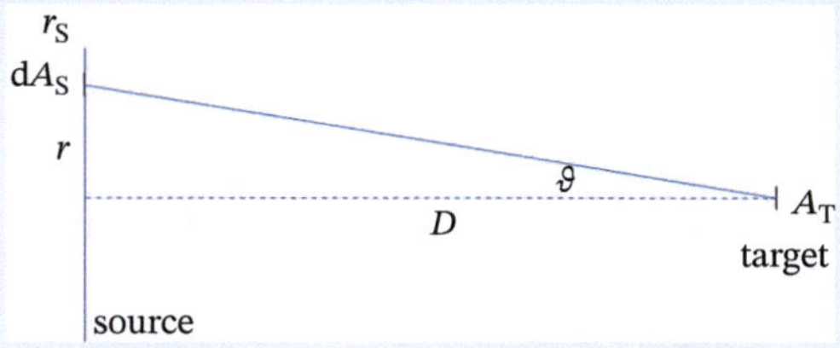

Figure 3.42 Source and target.

The flux on the receiver is then

$$\Phi = \iint L_S(r, \vartheta)\, dU_S \tag{3.156}$$

(the receiver picks luminance from within the target etendue portion of the source phase-space).

The integration in Eq. 3.155 of the two-dimensional function $L(r, \vartheta)$ is reduced to one dimension by the requirement that (cf. Figure 3.42)

$$r = D \tan \vartheta \tag{3.157}$$

An element on the source and the receiver form a light tube with the etendue

$$dU = \frac{dA_S A_T}{D^2} \cos^4 \vartheta(r)\ \text{sr} \tag{3.158}$$

(combining 3.42 and Eq. 3.190), as $D^2 \gg dA_S, A_T$. Because of rotational symmetry, we may consider dA_S as a ring element with

$$dA_S = 2\pi\, r\, dr \tag{3.159}$$

The differential of Eq. 3.156 is

$$dr = D \frac{d\vartheta}{\cos^2 \vartheta} \tag{3.160}$$

which leads us to (using Eq. 2.14)

$$dU = A_T 2\pi \sin \vartheta \cos \vartheta\, d\vartheta\ \text{sr} = A_T\, d\Omega_p \tag{3.161}$$

(Continued)

Knowledge Corner Topic 3.17: (Continued)

where the limit of the integration is Ω_p (the projected solid angle of the source as seen from the receiver). We can insert this into Eq. 3.155 to obtain

$$E_T = \frac{\Phi}{A_T} = \int L_S(r, \vartheta) \frac{2\pi r}{D^2 + r^2} \, dr \, sr = \int L_S(r, \vartheta) \, d\Omega_p(\vartheta) \tag{3.162}$$

Exercise 3.13: Inverse square law and source luminance distribution

Prove or debunk the inverse square law $E(D) \propto 1/D^2$ for

(a) a Lambertian source,

(b) a telecentric source of constant exitance E_0 with a $\cos^n$ intensity distribution (Knowledge Corner Topic 3.15), and

(c) a Köhler or light-tube source (Figure 3.43).

Quantity	Value
Source semi-diameter	$r_S = 4$ mm
Pupil semi-diameter	$r_f = 1$ mm
Pupil distance	$f = 5$ mm

Solution 3.13

(a) For a Lambertian source ($L_S = \text{const}$), we can integrate the angular term of Eq. 3.161 and get

$$E_T = L_S \Omega_p = L_S \, \pi \, sr \, \sin^2 \theta = L_S \pi \, sr \frac{r_S^2}{r_S^2 + D^2} = \frac{I_0 \, sr}{r_S^2 + D^2} \tag{3.163}$$

because of

$$I_0 = L_S A_S = L_S \, \pi r_S^2 \tag{3.164}$$

and clearly see the transition between near field ($D \ll r_S$) and far field ($D \gg r_S$). Hence, the inverse square dependence is only fulfilled in the far field, i.e. for a very small source.

(b) $\cos^n$ beam. The corresponding luminance is (Eqs. 3.134 and 3.132)

$$L_S(\vartheta) = E_0 \frac{n+1}{2\pi sr} \cos^{n-1} \vartheta \tag{3.165}$$

(with the constant luminous exitance E_0), leading to

$$\begin{aligned} E_T &= \int L_S(r,\vartheta)\,d\Omega_p(\vartheta) \\ &= E_0\frac{n+1}{2\pi}\int \cos^n\vartheta\,\sin\vartheta\,d\vartheta\,d\varphi = E_0\left(1-\cos^{n+1}\theta\right) \end{aligned} \tag{3.166}$$

with the integration limit (source size as seen from target)

$$\tan\theta = \frac{r_S}{D} \tag{3.167}$$

equivalent to

$$\cos\theta = \frac{1}{\sqrt{1+\tan^2\theta}} = \frac{D}{\sqrt{D^2+r_S^2}} \tag{3.168}$$

and we obtain

$$E_T(D) = E_0\left[1-\left(\frac{D}{\sqrt{D^2+r_S^2}}\right)^{n+1}\right] \tag{3.169}$$

For $D \gg r_S$, this can be approximated similar to Eq. 1.5, yielding

$$E_T(D) \approx E_0\frac{n+1}{2}\frac{r_S^2}{D^2} \tag{3.170}$$

Important: the far field of a $\cos^n$ beam follows the inverse square law.

(c) Light tube or Köhler source: see Figure 3.43.

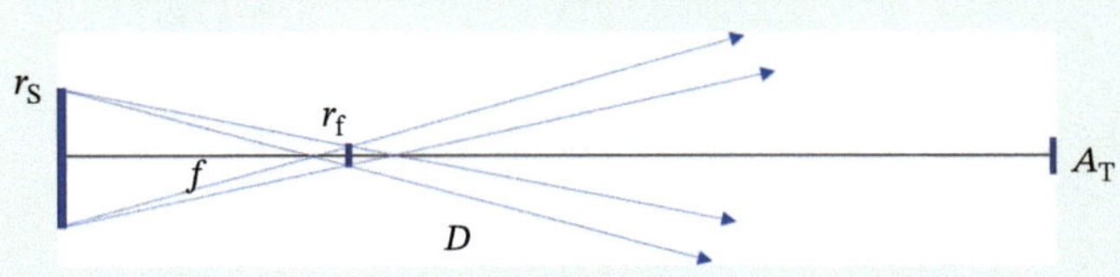

Figure 3.43 Source focused to a spot at $f = 4$ mm and target at $D = 12$ mm.

The target irradiance is given by Eq. 3.161

$$E_T(D) = L\Omega_P = L\pi\,\text{sr}\,\sin^2\theta(D) = L\frac{\tan^2\theta(D)}{1+\tan^2\theta(D)} \tag{3.171}$$

We use constant luminance but consider the arrangement by the angular limits in three different regions: Very close to the source disk, the contributing angular range is limited by the rear focus disk. Next, there is a region around the focus disk where it is limited by the source radius. Beyond that, its limit is again given by the focus disk.

(Continued)

Solution 3.13 (Continued)

The angles to apply in Eq. 3.170 are

$$\tan\theta = \begin{cases} \dfrac{r_f}{f-D} \; \forall \, D < f\dfrac{r_S}{r_S + r_f} \\ \dfrac{r_S}{D} \;\; \text{in between} \\ \dfrac{r_f}{D-f} \; \forall \, D > \dfrac{f r_S}{r_S - r_f} \end{cases} \tag{3.172}$$

Plots of the derived solutions are shown in the following simulation.

Simulation 3.6: Inverse square law and source luminance distribution

We validate the analytic expressions by comparing them to simulations (Figures 3.44 and 3.45).

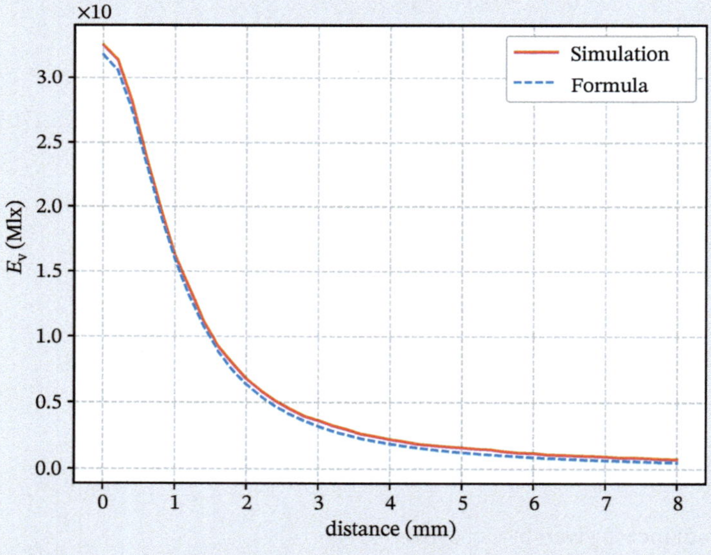

Figure 3.44 On-axis target illuminance as a function of receiver distance for a Lambertian source.

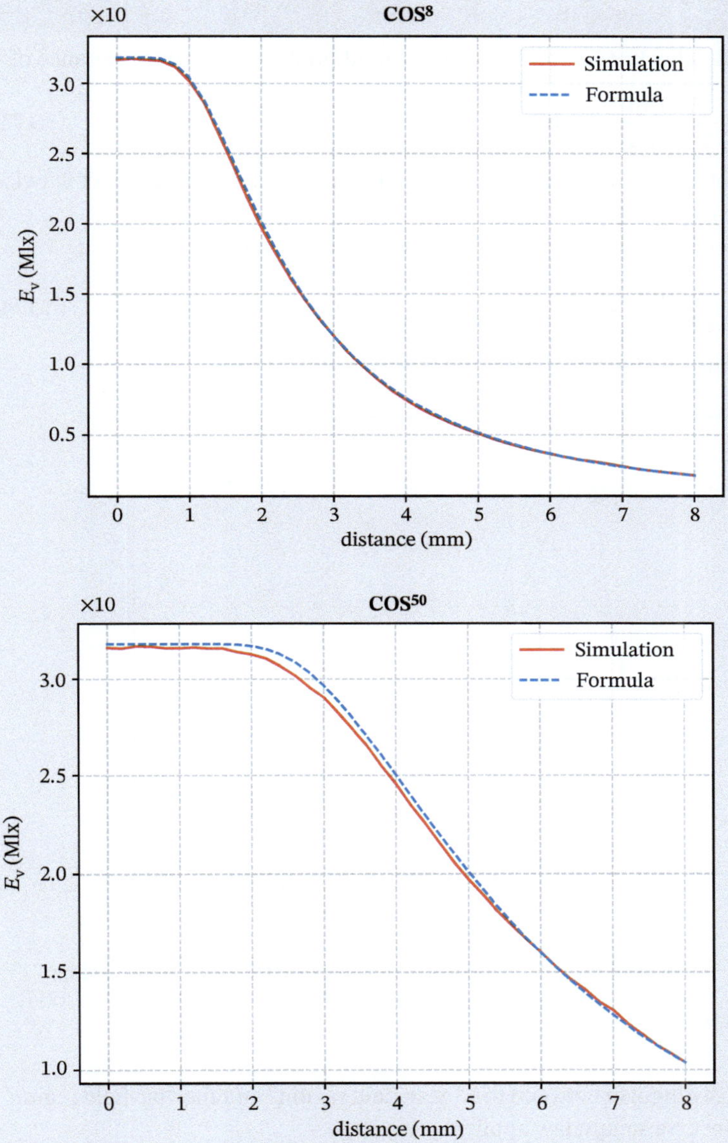

Figure 3.45 On-axis target illuminance as a function of receiver distance for $\cos^n$ sources.

For part (c), we have to take care to set up the source properly. We use a disk source with the focus disk as the aim area. The receiver is positioned close to the source so that the complete flux is reliably collected, then it is "defocused away" to scan the illuminance peak.

(Continued)

Simulation 3.6: (Continued)

We double-check: A source of 100 lm should produce an average exitance of

$$E_0 = \frac{100\,\text{lm}}{\pi\, 16\,\text{mm}^2} = 2\,\text{Mlx} \tag{3.173}$$

which is not exactly what we find in the simulation. The exitance peak is twice as large in the center.

In the focus area (the aim area) we expect an illuminance of

$$E_f = \frac{100\,\text{lm}}{\pi\, 1\,\text{mm}^2} = 32\,\text{Mlx} \tag{3.174}$$

which is confirmed by the simulation (Figure 3.46).

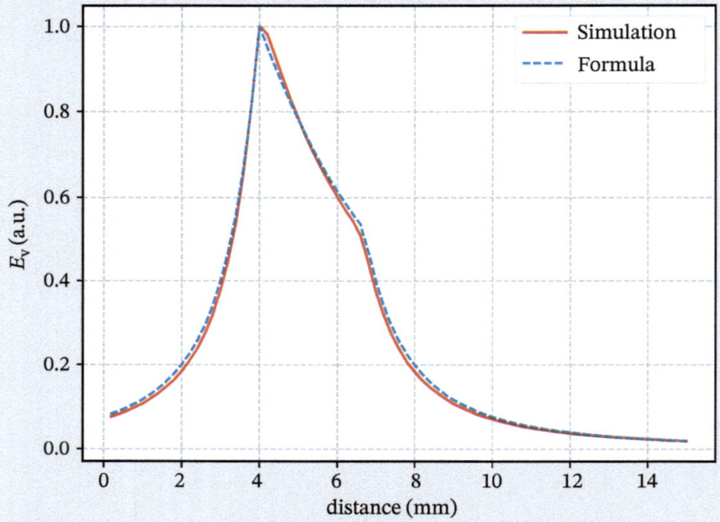

Figure 3.46 On-axis illuminance as a function of distance for a directional source ($f = 5$, $r_f = 1$, $r_S = 4$), showing a good correspondence between calculation (Eq. 3.171) and simulation.

We have to distinguish between the "near-field region" and the "far-field region", where the inverse square law applies.

Provided files dist_focus.xlsx and dist_focus.lts.

Exercise 3.14: Notebook screen

A 17″ notebook computer screen is specified with a rather typical luminance of $L = 250$ nit (Figure 3.47). We assume Lambertian emission of the screen (no screen gain).

Figure 3.47 Notebook screen. *Source:* Photos by the authors.

Calculate the luminous flux of the screen as a whole.

Solution 3.14

The area of the screen is (we assume a 4:3 aspect ratio with a factor of $\sqrt{3^2+4^2} = 5$ for the diagonal)

$$A = \left(\frac{17 \cdot 25.4\,\mathrm{mm}}{5}\right)^2 \cdot 3 \cdot 4 \approx 90\,000\,\mathrm{mm}^2 \tag{3.175}$$

With a projected solid angle of the hemisphere of π, we have an etendue of

$$U = \pi A = 0.28\,\mathrm{m}^2\mathrm{sr} \tag{3.176}$$

hence

$$\Phi = L \cdot U = 250\,\frac{\mathrm{cd}}{\mathrm{m}^2} \cdot 0.28\,\mathrm{m}^2\mathrm{sr} = 70\,\mathrm{lm} \tag{3.177}$$

Exercise 3.15: LCD Backlight

You are assigned the task to design an LCD backlight for a large cellphone with a planar light guide (Figure 3.48).

Figure 3.48 LCD backlight system. A BEF is a "brightness enhancement film" that reflects shallow incidence rays. These get scattered and reflected and, in best case, increase intensity around the panel normal direction. No BEF is used in the exercise.

Quantity	Value
Display area	140 mm · 70 mm
Target luminance	$L = 400$ nit
Optical transmission	70 %
LCD transmission	5 % (yes, it is that bad!)

(a) What luminous flux do you need from the LEDs? (b) The customer would like to use certain mid-power LEDs with 90 lm each. How many of these will you need? Into which potential (optical) trouble will you run?

Solution 3.15

(a) The etendue of the display is

$$U_T = \pi \, \text{sr} \cdot A \approx 30\,800 \, \text{mm}^2\text{sr} \tag{3.178}$$

and thus the luminous flux to be emitted is

$$\Phi_T = LU = 400 \, \text{cd/m}^2 \cdot 30\,800 \, \text{mm}^2\text{sr} = 12.3 \, \text{lm} \tag{3.179}$$

Therefore, we need an LED flux of

$$\Phi_S = \frac{12.3 \, \text{lm}}{70\% \cdot 5\%} \approx 350 \, \text{lm} \tag{3.180}$$

(b) This can be achieved by using *four* 90 lm LEDs. From the perspective of homogeneity, though, it would certainly be preferable to feed the display with (e.g.) 18 LEDs of 20 lm each.

Exercise 3.16: Illuminated retina

A human looks at the notebook screen from Exercise 3.14 ($L = 400$ nit) (Figure 3.49).

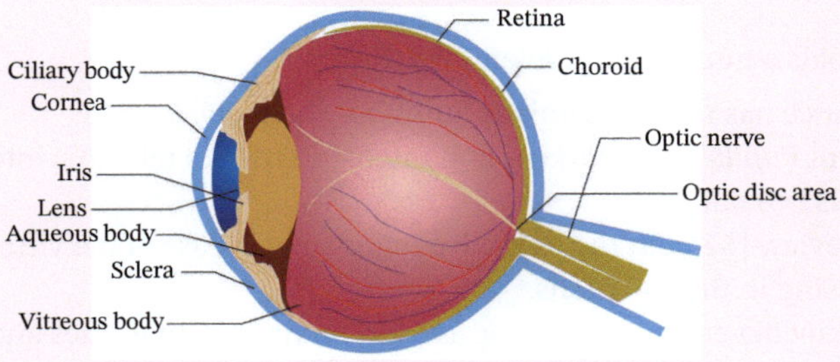

Figure 3.49 The human eye. A review of its "optical design" is given in [57].

Assume an eye pupil with diameter $2r_P = 4$ mm, a back focal length of 20 mm, and a refractive index of the vitreous humor of $n = 1.336$. Calculate the illuminance at the retina.

Solution 3.16

The etendue formed by the pupil and an area element dA at the retina is

$$U = \frac{n^2\pi^2 r_P^2 \, dA}{D^2} \tag{3.181}$$

As luminance is conserved on the way from the source to the retina, the flux is obtained by multiplying etendue with luminance (Eq. 3.108), and illuminance is obtained according to

$$E = \frac{LU}{dA} = L\left(\frac{n\,\pi\,r_P}{D}\right)^2 \mathrm{sr} = 400\,\frac{\mathrm{lm}}{\mathrm{m}^2}\left(\frac{1.336\cdot\pi\cdot 2\,\mathrm{mm}}{20\,\mathrm{mm}}\right)^2 = 70\,\mathrm{lx} \tag{3.182}$$

Sometimes the forms

$$E = L\left(\frac{n^2\pi A_P}{D^2}\right)\mathrm{sr} = L\left(\frac{n\pi}{2F/\#}\right)^2 \mathrm{sr} \tag{3.183}$$

with the pupil area $A_P = r_P^2\pi$ and the F-number $F/\# = D/2r_P$ are convenient as well.

Dr. Wisenheimer:

There is a quantity called *conventional retinal illuminance*, defined in [58],

$$T = L \cdot A_P \tag{3.184}$$

and is measured in *Trolands* (1 td = 1 μcd) [59, Sec. 845].

(Continued)

Solution 3.16 (Continued)

The value 1 td is obtained when an object of the luminance $L = 1$ nit is observed with an eye pupil area of $A_P = 1\,\text{mm}^2$ [60, Ch. 8]. The idea is related to the fact that the visual retinal stimulus is proportional to both object luminance and eye pupil area.

The whole topic holds a number of problems, however:

- Retinal illuminance has the same unit as luminous intensity.
- The German term *Pupillenlichtstärke* (pupil intensity) tries to take this into account (DIN 5031-6) but ...
- There is a nice review [61], but the refractive index (in this case of the vitreous body) is missing in the equations.
- The English Wikipedia entry is confusing as well, mixing up quantities and units.

Looking at our above equations, we conclude that

$$E = \frac{n^2\pi}{D^2}\text{sr} \cdot T \tag{3.185}$$

(cf. Eq. 4.7) and

$$T \approx E \cdot 70\,\frac{\text{mm}^2}{\text{sr}} \tag{3.186}$$

where the factor has a strange unit as well.

Exercise 3.17: Camera irradiance and the cos⁴ law

A camera takes a picture of a large Lambertian emitter at a distance D_0. Calculate the relative illuminance at the edge of the field for a field of view (FoV) of 2θ.

Solution 3.17

We consider an area $\text{d}A$ on the object (Figure 3.50). The etendue defined by this area and the entrance pupil of the lens is

$$U = \frac{\text{d}A' A'_\text{P}}{D^2} \tag{3.187}$$

Obliquity is taken into account by using

$$\text{d}A' = \text{d}A \cdot \cos\theta \tag{3.188}$$

$$A'_\text{P} = A_\text{P} \cdot \cos\theta \tag{3.189}$$

The distance is larger than the on-axis value D_0,

$$D = \frac{D_0}{\cos\theta} \tag{3.190}$$

When combining these expressions, we arrive at

$$U(\theta) = U_0 \cos^4\theta \tag{3.191}$$

with

$$U_0 = \frac{dA\,A_P}{D_0^2} \tag{3.192}$$

which defines the illuminance distribution on the target as

$$E(\vartheta) = \frac{LU(\theta)}{dA_T} = E(0)\cos^4\theta \tag{3.193}$$

with

$$E_0 = LM^2\frac{A_P}{D_0^2} \tag{3.194}$$

(M is the magnification).

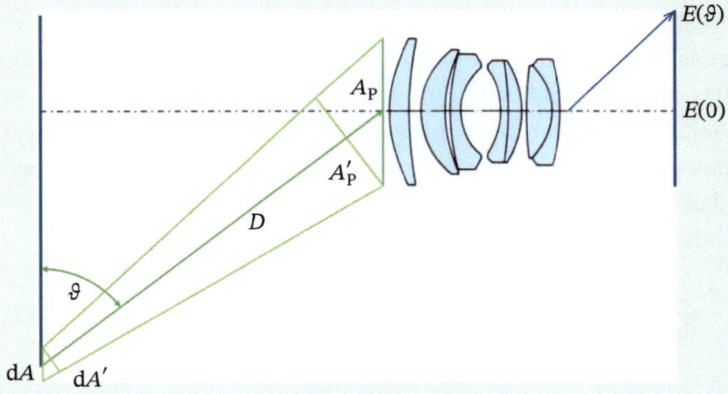

Figure 3.50 Light from a Lambertian emitter shining on a camera lens.

Dr. Wisenheimer:
The $\cos^4$ law [62] applies only to a Lambertian emitter and a small perfect lens. It does not apply, for example, to a torch light that shines into the entrance pupil of the camera lens or for Köhler illumination where the source is imaged onto the entrance pupil of a lens [63].

3.5 Black-body and Solar Radiation

Around 1900, physicists grappled with what Paul Ehrenfest dubbed the *ultraviolet catastrophe*: Trying to explain the thermal radiation spectrum on the basis of classical electrodynamics did not work. The classically predicted spectra would diverge to infinity in the ultraviolet. Ultimately, Max Planck [64] postulated that electromagnetic radiation comes in discrete energy packets, now known as *photons* or, more generally, as *quanta* (the plural of *quantum*). The fundamental importance of this suggestion (and the new theory he built around it, which culminated in his celebrated black-body radiation law) cannot be overestimated. Planck was not only a great physicist, he was also a good writer. His 1914 textbook "The Theory of Heat Radiation" [65] is quite worth reading even today, in our opinion.

Black-body radiation is important not just as a physical theory, but also for engineering illumination optics. Many important thermal sources (incandescent lamps, high-pressure Xenon lamps, the Sun) can be well approximated by black-body radiation. Moreover, black-body radiation pretty much defines what we call "white". Time to visit the Knowledge Corner café for a through discussion.

Knowledge Corner Topic 3.18: Black-body radiation

It is not an oxymoron – "black" bodies do radiate. You are certainly familiar with how things radiate when they are heated until they glow – burning charcoal, for example, which is really black when cold. When charcoal burns and glows, it is still black in the sense that it reflects very little of any incident light. Try it with a green laser pointer or a flashlight at your next barbecue! Moreover, there is an intimate, deep relation between how things look (their spectral reflectivity) and how exactly they glow. This relation is known as Kirchhoff's law, which we discuss in Knowledge Corner Topic 3.21.

The spectral radiance of a black body in vacuum is given by [38, 64, 66]

$$L_{\mathrm{e},\lambda}(\lambda, T) = \frac{\mathrm{d}L_{\mathrm{e}}(T)}{\mathrm{d}\lambda} = \frac{2hc^2}{\lambda^5\,\mathrm{sr}} \frac{1}{\exp\left(\frac{hc}{\lambda k_{\mathrm{B}} T}\right) - 1} \tag{3.195}$$

assuming an ideally emitting surface and with λ being the vacuum wavelength , T the absolute temperature, h being Planck's constant, k_{B} Boltzmann's constant, and c the speed of light in vacuum (Figure 3.51). The unit of $L_{e,\lambda}$ is $\mathrm{W/(m^2 sr\ nm)}$.

It is not trivial – quite the opposite, it is beyond the scope of this book – to discuss the proper use and meaning of the refractive index in these equations (see [67]). Equation 3.194 is valid when the black body is immersed in a medium, as long as we remember that "our" L is actually basic radiance, the flux per etendue, and λ is the vacuum wavelength. When in doubt, use frequency ν instead of wavelength, because frequency is independent of the medium (cf. Exercise 3.20).

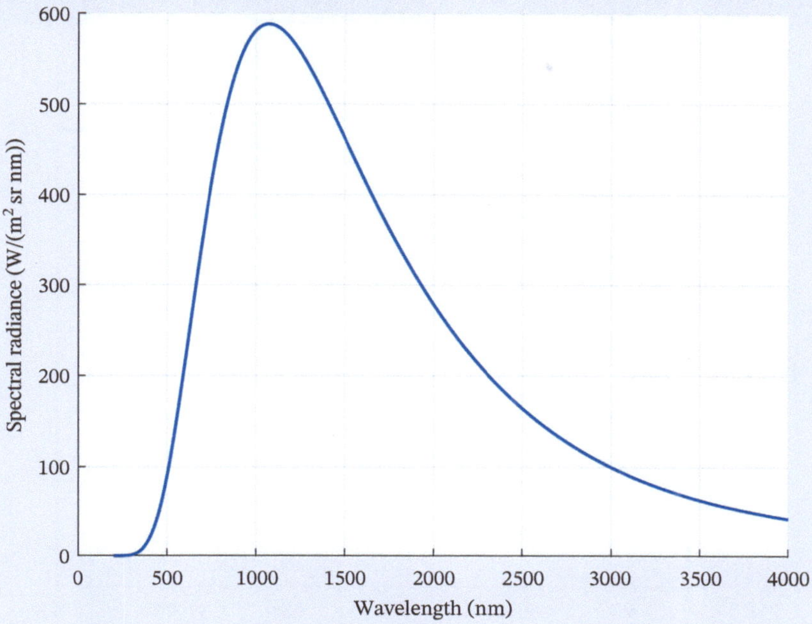

Figure 3.51 Black-body spectral radiance at $T = 2700\,\text{K}$.

With the help of the substitution $x = hc/\lambda k_B T$ and using

$$\int_0^\infty \frac{x^3\,\mathrm{d}x}{\mathrm{e}^x - 1} = \frac{\pi^4}{15} \tag{3.196}$$

we can integrate over all wavelengths to obtain the (total) (basic) radiance

$$L_e(T) = \int_0^\infty L_e(\lambda, T)\,\mathrm{d}\lambda = \frac{2\pi^4 k_B^4}{15 h^3 c^2}\,T^4 = \frac{\sigma T^4}{\pi\,\text{sr}} \tag{3.197}$$

and thus a radiant exitance (emitted power per unit surface area or radiance times the angular extent of a hemisphere $n^2\pi$ sr) of (cf. [68, Ch. 4])

$$E(T) = L_e n^2 \pi\,\text{sr} = n^2 \sigma T^4 \tag{3.198}$$

with

$$\sigma = \frac{2\pi^5 k_B^4}{15 h^3 c^2} = 5.67 \cdot 10^{-8} \frac{\text{W}}{\text{m}^2\text{K}^4} \tag{3.199}$$

This is the famous *Stefan–Boltzmann law* (honoring Ludwig Boltzmann and his 1866 thesis advisor Josef Stefan).

The luminance is obtained by calculating

$$L_v(T) = 683\,\frac{\text{lm}}{\text{W}} \int_0^\infty L_e(\lambda, T) V(\lambda)\,\mathrm{d}\lambda \tag{3.200}$$

which cannot be expressed analytically but has to be plotted (see Figure 3.52).

(Continued)

Knowledge Corner Topic 3.18: (Continued)

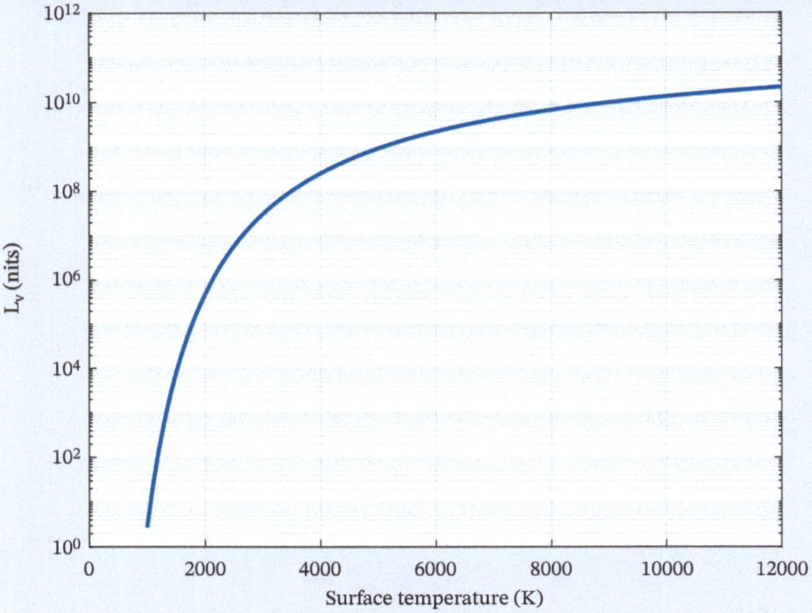

Figure 3.52 Black-body luminance.

Finally, we obtain the *luminous efficacy of thermal radiation* (see Knowledge Corner Topic 3.24) as

$$\eta = 683\,\frac{\mathrm{lm}}{\mathrm{W}}\,\frac{\displaystyle\int_0^\infty L_e(\lambda, T)v(\lambda)\,\mathrm{d}\lambda}{\displaystyle\int_0^\infty L_e(\lambda, T)\,\mathrm{d}\lambda} \tag{3.201}$$

which shows a maximum at around $T = 6500\,\mathrm{K}$. We provide the file CIE_planck.xlsx that contains a variety of related calculations. When you look for tabulated values for $V(\lambda)$, CIE kindly provides the official dataset for download.

Finally, we refer to additional calculations related to the Stefan–Boltzmann law in Exercise 4.28.

Exercise 3.18: Radiator

How much radiated power does a square radiator produce (surface area $1\,\mathrm{m}^2$, $60\,°\mathrm{C}$, ideally emitting black surface)? Calculate the radiant exitance. How much power does an identical radiator at a distance of 1 m receive?

Solution 3.18

With Eq. 3.197, we obtain a radiant exitance of

$$E = 5.7 \cdot 10^{-8} \frac{\mathrm{W}}{\mathrm{m^2K^4}} \cdot (273\,\mathrm{K} + 60\,\mathrm{K})^4 \approx 700\,\frac{\mathrm{W}}{\mathrm{m^2}} \tag{3.202}$$

and a radiance

$$L = \frac{E}{\pi\,\mathrm{sr}} = 223 \frac{\mathrm{W}}{\mathrm{m^2sr}} \tag{3.203}$$

The radiator and its twin together form a light tube. To calculate its etendue based on Eq. 3.47, we use $r = 0.5$ m and account for the square shape by introducing a factor of $4/\pi$ HANDLE WITH CARE for both sides:

$$U = \frac{\pi^2}{4}\left((\sqrt{2}-1)\,\mathrm{m}\right)^2 \mathrm{sr} \cdot \left(\frac{4}{\pi}\right)^2 = 4\left((\sqrt{2}-1)\,\mathrm{m}\right)^2 \mathrm{sr} \approx 0.7\,\mathrm{m^2sr} \tag{3.204}$$

Hence, the target will receive a power

$$\Phi = 223 \frac{\mathrm{W}}{\mathrm{m^2sr}} \cdot 0.7\,\mathrm{m^2sr} \approx 150\,\mathrm{W} \tag{3.205}$$

In the following, we will explore more things using a simple radiator.

Exercise 3.19: Radiance enhancer

Let's invent something! Rube approaches you with his prototype of an optical device that he claims can *increase radiance*.[a]

He wants you to use this device to concentrate the radiation from the radiator's surface from the previous exercise onto a 0.1 m × 0.1 m ideal receiver. To what temperature would this receiver heat up?

[a] Such things can happen. Not the radiance increase, of course, but people who invent devices to this (or similar) ends [139, 140].

Solution 3.19

If we could guide the full 700 W to the new target via a miraculous optical device, we would obtain an irradiance of

$$E = \frac{700\,\mathrm{W}}{0.01\,\mathrm{m^2}} = 70\,000\,\frac{\mathrm{W}}{\mathrm{m^2}} \tag{3.206}$$

(Continued)

Solution 3.19 (Continued)

According to Eq. 3.197, the receiver would therefore reach a temperature of

$$T = \sqrt[4]{\frac{E}{\sigma}} = \sqrt[4]{\frac{70\,000\,\mathrm{W/m^2}}{5.7 \cdot 10^{-8}\,\mathrm{W/m^2K^4}}} \approx 1050\,\mathrm{K} \tag{3.207}$$

Consequently, the radiance enhancer would allow for the generation of elevated temperatures. Obviously, we would immediately go ahead and connect it thermally to the source radiator, while harvesting some of the energy downstream!

Dr. Wisenheimer:

As you will already have guessed, the final implication is an indirect proof that the Second Law of thermodynamics will prevent us from increasing radiance by using an optical system [44].

The discussion of *luminance increase* in the literature [69] usually involves *recycling* [70] and does therefore not violate any conservation laws. Furthermore, there are *luminescent concentrators* that increase luminance by means of wavelength conversion [71, 72].

Here comes the ... well, our most important light source. We collect and present important data as a reference for various exercises in physics, solar collection and solar simulation.

Knowledge Corner Topic 3.19: The Sun

A warning ahead: data found in the literature differ to some degree [73], so you should not be surprised when you find slightly deviating figures on the internet.

Throughout the book, we refer to the figures we provide in the file sun.xlsx as well as the references therein, such as [74, 75]. Expect different figures coming up in an internet search (usually without accompanying original sources).

Geometry

Our star, called the Sun, is a nearly perfect sphere with a radius $r_S = 6.957 \cdot 10^8$ m [74].

The Earth orbit has a distance of one astronomical unit or

$$D_S = 1.494 \cdot 10^{11}\,\mathrm{m} \tag{3.208}$$

from the Sun. Consequently, we see the Sun under an angular cone of

$$\tan\theta_S = \frac{6.957 \cdot 10^8\,\mathrm{m}}{1.496 \cdot 10^{11}\,\mathrm{m}} \tag{3.209}$$

which leads to $\theta_S = 0.266°$ (half angle). Its surface area is

$$A_S = 4\pi\, r_S^2 = 6.08 \cdot 10^{18}\,\mathrm{m}^2 \tag{3.210}$$

and its projected solid angle as seen from Earth is

$$\Omega_{P,S} = \pi\,\mathrm{sr}\,\sin^2\theta = 6.8 \cdot 10^{-5}\,\mathrm{sr} \tag{3.211}$$

Radiometry
The average surface temperature of the Sun is 5772 K (exact value under discussion [73]). From the Stefan–Boltzmann law (Eq. 3.196), we obtain the radiance of the Sun's surface as

$$L_{e,S} = \frac{\sigma T^4}{\pi\,\mathrm{sr}} = 20\,\frac{\mathrm{MW}}{\mathrm{m^2 sr}} \tag{3.212}$$

and the total radiant flux

$$\Phi_{e,S} = L_S A_S \pi\,\mathrm{sr} = 3.83 \cdot 10^{26}\,\mathrm{W} \tag{3.213}$$

Its radiant intensity is

$$I_{e,S} = \frac{\Phi_{e,S}}{4\pi\,\mathrm{sr}} = 3.05 \cdot 10^{25}\,\frac{\mathrm{W}}{\mathrm{sr}} \tag{3.214}$$

The Sun's irradiance at the Earth's orbit is found via the inverse square law (Eq. 3.142) to be

$$E_{e,S} = \frac{I_{e,S}\,\mathrm{sr}}{D_S^2} = 1.36\,\mathrm{kW/m^2} \tag{3.215}$$

Our atmosphere's transmission reduces this value to

$$E_{e,E} = 1.0\,\mathrm{kW/m^2} \tag{3.216}$$

(the well known "solar constant"). Therefore, the radiometric atmospheric transmission is

$$\tau_e = \frac{E_{e,E}}{E_{e,S}} = 0.735 \tag{3.217}$$

Spectral
Black-body radiation at $T = 5772$ K features a luminous efficacy of radiation (Eq. 3.243) of $K = 92$ lm/W. In contrast, measured spectra provide a value of

$$K_S = 101\,\mathrm{lm/W} \tag{3.218}$$

(Continued)

Knowledge Corner Topic 3.19: (Continued)

for the extra terrestrial spectrum (see provided file). The Earth's atmosphere does slightly alter the spectrum to a different value of

$$K_E = 110\,\mathrm{lm/W} \tag{3.219}$$

We can calculate the Sun's total luminous flux based on Eqs. 3.212 and 3.217 to be

$$\Phi_{v,S} = \Phi_{e,S}K_S = 3.83 \cdot 10^{26}\,\mathrm{W} \cdot 101\,\frac{\mathrm{lm}}{\mathrm{W}} = 3.88 \cdot 10^{28}\,\mathrm{lm} \tag{3.220}$$

and its luminous intensity as

$$I_{v,S} = I_{e,S}K_S = 3.05 \cdot 10^{25}\,\frac{\mathrm{W}}{\mathrm{sr}} \cdot 101\,\frac{\mathrm{lm}}{\mathrm{W}} = 3.08 \cdot 10^{27}\,\mathrm{cd} \tag{3.221}$$

The photopic transmission of the atmosphere is

$$\tau_v = \frac{E_{v,E}}{E_{v,S}} = \frac{E_{e,E}K_E}{E_{e,S}K_S} = \tau_e\frac{K_E}{K_S} = 0.80 \tag{3.222}$$

Hence, the luminance of the Sun measured outside the Earth's atmosphere is

$$L_{v,S} = L_{e,S}K_S = 2.0\,\mathrm{Gnit} \tag{3.223}$$

but on Earth, we see a luminance of only

$$L_{v,E} = L_{e,S}K_S\,\tau_v = L_{e,S}\,\tau_e K_E = 1.63\,\mathrm{Gnit} \tag{3.224}$$

Finally, we obtain an illuminance of

$$E_{v,S} = 1.36\,\mathrm{kW/m^2} \cdot 101\,\mathrm{lm/W} = 137\,\mathrm{klx} \tag{3.225}$$

at the Earth's orbit and of

$$E_{v,E} = 1.0\,\mathrm{kW/m^2} \cdot 110\,\mathrm{lm/W} = 111\,\mathrm{klx} \tag{3.226}$$

on Earth, after passing through the atmosphere on a clear day.

Exercise 3.20: Solar spectrum

For a black-body emitter of $T = 5772$ K, calculate the wavelength of the maximum spectral radiance $L_e(\lambda)$ and the wavelength of the maximum spectral luminance $L_v(\lambda)$. Express the spectral radiance in terms of frequency and calculate the frequencies for the maxima.

Solution 3.20

We start with

$$L_e(\nu, T)\,d\nu = L_e(\lambda, T)\,d\lambda \tag{3.227}$$

where we insert

$$\lambda = \frac{c}{n\nu} \tag{3.228}$$

implying

$$|d\lambda| = \frac{c}{n^2\nu^2}\,|d\nu| = \frac{\lambda^2}{c}\,|d\nu| \tag{3.229}$$

and obtain from Eq. 3.194

$$L_e(\nu, T) = \frac{2hn\nu^3}{c^2}\frac{1}{\exp\left(\frac{h\nu}{k_B T}\right) - 1} \tag{3.230}$$

These functions are displayed in Figure 3.53.

The positions of the maxima are:

- For the spectral radiance at 502 nm and $3.4 \cdot 10^{14}\,\mathrm{s}^{-1}$, respectively.
- For the spectral luminance at 554 nm and $5.3 \cdot 10^{14}\,\mathrm{s}^{-1}$, respectively.

We could also introduce the variables [68]

$$X_P = \frac{hc}{n\lambda k_B T} \tag{3.231}$$

and investigate the function

$$\frac{X_P^5}{\exp X_P - 1} \tag{3.232}$$

which has a maximum at $X_P = 5$. Similarly, we introduce

$$Y_P = \frac{h\nu}{nk_B T} \tag{3.233}$$

and we investigate

$$\frac{Y_P^3}{\exp Y_P - 1} \tag{3.234}$$

with a maximum at $Y_P = 2.8$. Both values correspond well to the locations of the radiance maxima provided above. For luminance, we rely on the provided spreadsheet CIE_planck.xlsx.

We refer to Ref. [76] for the interpretation of the spectral distributions of light and on the use of wavelength, frequency or wavenumber as the independent variable.

Solution 3.20 (Continued)

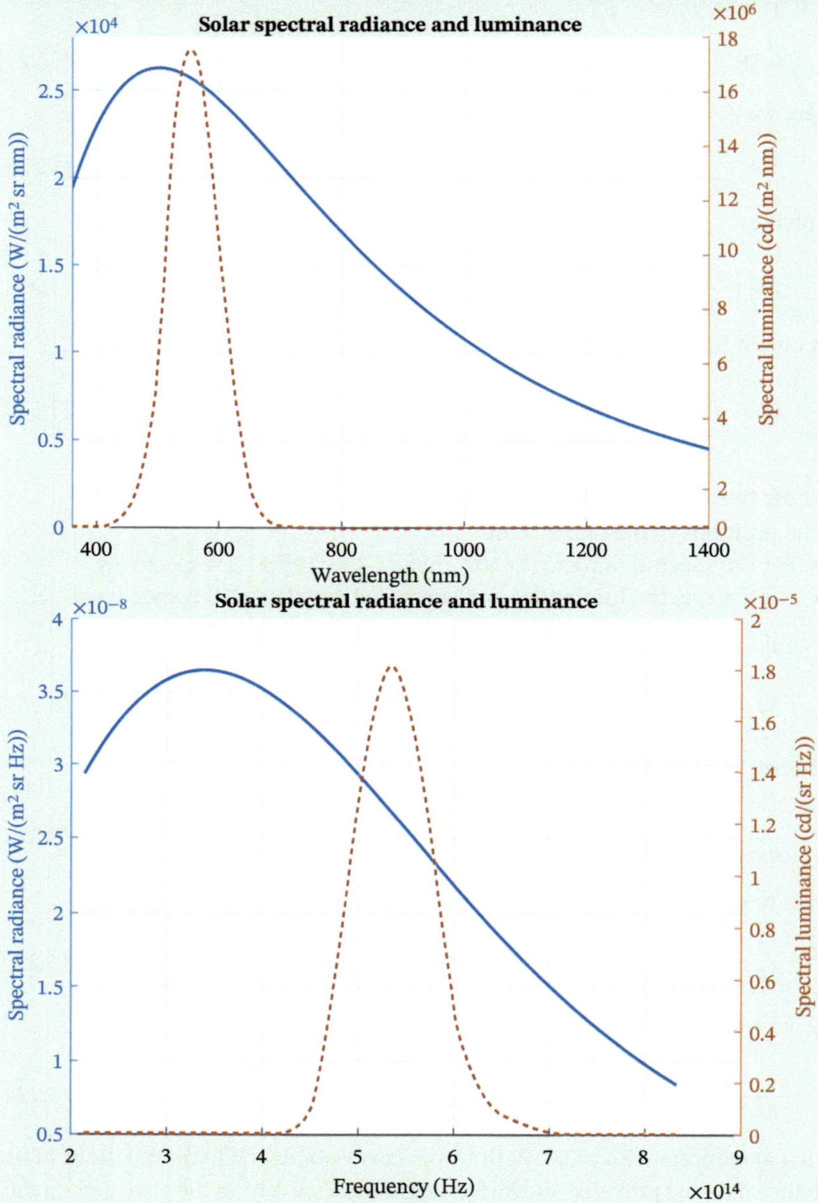

Figure 3.53 Spectral radiance and luminance of 5772 K black-body radiation in the wavelength and frequency domains.

Dr. Wisenheimer:

From Eq. 3.230 we derive

$$\lambda_{\max} T = \frac{hc}{5k_\mathrm{B}} = 2.88 \cdot 10^{-3}\,\mathrm{K\,m} \tag{3.235}$$

which is Wien's displacement law (for a discussion see [77]).

Knowledge Corner Topic 3.20: Chlorophyll absorption

In the previous exercise, we learned that solar radiation has a peak of its spectral radiance at 500 nm (Figure 3.54). Why does chlorophyll not absorb in the green where the highest spectral radiance is located? Why did plants choose red (and blue) in favor of green [78, 79]?

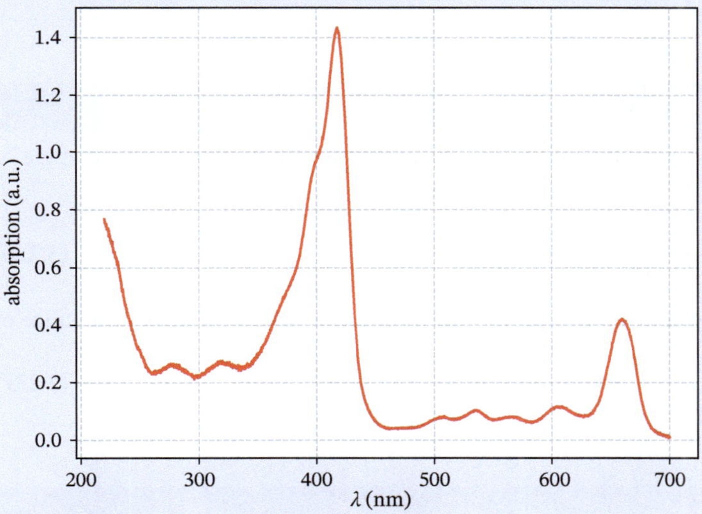

Figure 3.54 Absorption spectrum of Chlorophyll A in a solution [80].

First, we need to mention some reflection measurements where an almost negligible difference of green reflectance between green and brown leaves was found [81]. Apparently, the human vision with the $V(\lambda)$ peak in the green tends to emphasize the green color to some extent.

One textbook [78] explains absorption with the help of oscillators and resonance between the frequencies of incident light and the absorbing molecule.

(Continued)

Knowledge Corner Topic 3.20: (Continued)

What is important in this argument is the spectral radiance maximum in the frequency domain (possibly shifted to shorter wavelengths by water absorption in the atmosphere), causing an absorption peak in the red.

Current research suggests a different nature of the phenomenon, preferring stability at the expense of efficiency. The absorption in the red and blue spectral ranges might be less efficient, but is less affected by changing lighting conditions [82, 83].

Eye pigments work differently, and the human eye has its $V(\lambda)$ maximum just in the green because we are used to living in a world of green leaves.

Exercise 3.21: Cloudy sky

On a day with overcast sky, solar irradiance is reduced from Eq. 3.215 to $E = 100\,\mathrm{W/m^2}$. Calculate the radiance and luminance of the sky.

Solution 3.21

Irradiance is caused by light from a hemisphere with a projected solid angle of $\Omega_\mathrm{p} = \pi$ sr. Eq. 3.112 then yields a radiance of

$$L_\mathrm{e} = \frac{E}{\Omega_\mathrm{p}} = \frac{100\,\mathrm{W}}{\mathrm{m^2}\,\pi\,\mathrm{sr}} = 32\,\frac{\mathrm{W}}{\mathrm{m^2\,sr}} \tag{3.236}$$

Using data given in Knowledge Corner Topic 3.19, we obtain a sky luminance of

$$L_\mathrm{v} = L_\mathrm{e}K = 32\,\frac{\mathrm{W}}{\mathrm{m^2\,sr}} \cdot 118\,\frac{\mathrm{lm}}{\mathrm{W}} \approx 3.8\,\mathrm{knit} \tag{3.237}$$

Exercise 3.22: Lighting a fire with moonlight

Is it possible to light a fire with moonlight? The intuitive answer of most people is "no". This sounds reasonable, but you might want to explore for yourself. Here are some conditions: You are trying to ignite dry wood with an ignition temperature of 300 °C in air at ambient pressure. It is midnight, a very clear night in September, at sea level, with a full Moon above you, at a location near 45° latitude.

Solution 3.22

This is indeed possible, in several ways. One way involves wavelength conversion. For this, we use a long, cylindrical, extremely transparent and smooth glass rod doped with a clear, non-scattering fluorescent dye. The dye molecules will absorb short-wavelength (blue to green) light, emit longer-wavelength (yellow) light, which is not absorbed by the dye molecules and which is guided along the rod by TIR, potentially achieving a large irradiance at either end of the long rod. Simply putting a small piece of wood near one end of the rod would then do the job.

Other means of concentrating diffuse radiation can be easily contrived. There are some thermodynamic limits, which are discussed in detail by Ries [44]. A similar idea is to concentrate the moonlight for use in a huge optically pumped laser at a low-threshold setting (or to use many smaller such lasers), and then to focus the light from this laser onto a piece of wood. We are not sure if sufficiently low-threshold systems do exist currently, but there is no fundamental reason why they should not. Yet another idea is to use a large array of solar cells, which will create some electricity, and if the array is large enough, a simple resistive heater will be hot enough to ignite the wood.

The question becomes optically interesting when generation of heat is only allowed via absorption of photons coming directly from the Moon after possibly being redirected by passive optical elements such as mirrors and lenses. We then ask: Can we use a solar thermal power plant to light a fire at the central tower receiver by aiming the mirrors of the heliostat field toward the full Moon? With direct sunlight, this is of course easy – the central tower absorber will quickly become white hot. But with moonlight?

The Sun illuminates the Moon with an irradiance of about $E = 1360\,\mathrm{W\,m^{-2}}$ (the solar constant). On the ground, in our clear day at sea level at 45° latitude, this extraterrestrial irradiance is reduced by absorption in the atmosphere (cf. Knowledge Corner Topic 3.19) to the so-called "air mass 1.5" direct and circumsolar irradiance of about $1\,\mathrm{kW/m^2}$.

The overall reflectivity of the Moon, its so-called albedo, is rather low, around 10% [84]. We simplify the Moon to be a planar, Lambertian disc with a reflectivity of $\varrho = 10\%$. While the Moon is neither planar nor simply diffusely reflecting, this is in fact a conservative estimate due to a phenomenon called *opposition surge* [85] that makes surfaces with large-scale roughness (such as the Moon) reflect more strongly than a Lambertian surface; on a full Moon, no shadows are apparent, and an additional contribution comes from coherent backscattering. So, the Moon reflects light with a radiant exitance of $E\varrho \approx 136\,\mathrm{W\,m^{-2}}$. Assuming the Moon is spectrally constant gray, this reflected light will then undergo the same attenuation in the atmosphere (transmission is 73.5%, Eq. 3.216).

From nonimaging optics theory, we know that the limit for the concentration of radiation in air prevents the maximum irradiance at the tower receiver from

Solution 3.22 (Continued)

exceeding the radiant exitance of the source (in this case, the Moon), even when using a secondary concentrator, yielding $E_{max} = 100\,W\,m^{-2}$. This is clearly insufficient to ignite a piece of wood at the tower.

Is this our final word on this question? Maybe we should expand our field of view. To be able to do so, we review some fundamental knowledge about absorption and emission.

Knowledge Corner Topic 3.21: Kirchhoff's law of radiation

Let us heat a body to a certain temperature T and look at a certain location $\mathbf{r}$ on that body from a certain direction $\mathbf{k}$. From that location $\mathbf{r}$, the body will emit thermal radiation along $\mathbf{k}$ toward our eye. In general, the spectrum, or, more precisely, the spectral radiance $L_\lambda(\lambda, T, \mathbf{r}, \mathbf{k})$, will not be the same as the black-body spectrum $L_{\lambda,bb}(\lambda, T)$ at that temperature (which is independent of $\mathbf{r}$ and $\mathbf{k}$). Let us now aim a ray with a wavelength λ and a flux $\Phi_{\lambda,\text{incident}}$ at that same spot $\mathbf{r}$ from direction $\mathbf{k}$. The body will partially absorb this ray and convert it to heat, and will also partially transmit or reflect its power. We can now define the local *absorptivity* $\alpha(\lambda, \mathbf{r}, \mathbf{k})$ as the fraction

$$\alpha(\lambda, \mathbf{r}, \mathbf{k}) = \frac{\Phi_{\lambda,\text{absorbed}}}{\Phi_{\lambda,\text{incident}}} \tag{3.238}$$

of absorbed light, and the local *emissivity* ε as the ratio of emitted spectral radiance and black-body radiance according to

$$\varepsilon(\lambda, \mathbf{r}, \mathbf{k}) = \frac{L_{\lambda,\text{emitted}}(\lambda, T, \mathbf{r}, \mathbf{k})}{L_{\lambda,\text{bb}(\lambda,T)}} \tag{3.239}$$

Kirchhoff's law of Radiation [65, 86, 87] then states, perhaps surprisingly, that

$$\alpha(\lambda, \mathbf{r}, \mathbf{k}) = \varepsilon(\lambda, \mathbf{r}, \mathbf{k}) \tag{3.240}$$

Let this sink in for a moment. Two seemingly unrelated properties, namely (i) how exactly a hot body glows, and (ii) what color it has, are in fact two sides of the same coin. Something deep must be going on.

Now, absorptivity and thus also emissivity are limited to $0 \le \alpha \le 1$. An immediate consequence is that a perfectly black, fully absorbing body radiates the highest possible amount of thermal radiation, and that this radiation will precisely follow the black-body spectrum. In fact, this is where the "black-body" spectrum derives its name from.

But why are absorptivity and emissivity related in the first place? Kirchhoff argued as follows: Imagine that we can insulate the body at question, at a temperature T, perfectly – both against radiation and thermal conduction. We leave only a narrow radiation channel along our ray open, with a narrow-band spectral filter around a wavelength λ. Now we aim this radiation channel at a black-body radiator, the temperature of which we can adjust. When everything is set up, we measure the energy balance. Obviously, the Second Law requires that *if* there is any net flow of radiative energy, it must be from hot to cold, and because the two subsystems are at the same temperature, the net flow of radiation must be zero. Done!

Dr. Wisenheimer:

Black-body radiation is a really fundamental phenomenon. It is not only emitted by bodies we would usually call "hot". It's everywhere. As you sit in an office and read this, you are immersed in thermal, near-black-body radiation coming from the walls and the ceiling. When you read this on a park bench, there is long-wave black-body radiation coming from the sky. We cannot see it because its wavelength is in the millimeter range, but we can measure it: the famous cosmic microwave background, a remnant from the very early universe (mere microseconds after the Big Bang, theorists say). And the Hawking radiation emitted by Black Holes as they evaporate extremely slowly – the only thing that ever comes back from a Black Hole, according to Stephen Hawking – comes with precisely a black-body spectrum. It is really amazing just how fundamental it is.

But now let us get back to moonlight and wood.

Knowledge Corner Topic 3.22: Absorption spectrum engineering

Our wood is not black. It emits thermal radiation and loses heat via conduction to the air. We can put an AR-coated window onto the tower receiver's aperture, and evacuate the volume behind, to mitigate conduction losses. We can enclose the wood, and some ambient-pressure air, in a box that is hanging within the vacuum chamber. The box is perfectly thermally insulated, except for one outer planar absorber face that is thermally connected to the interior. Then, the interior of the box, with wood and air, will be in thermal equilibrium with the outside world to which it is connected radiatively via the absorber face. Then, the only loss mechanisms are the reflection of moonlight and thermal radiation at the absorber face. We have to make the face absorb more moon-reflected sunlight and emit less thermal radiation. To achieve this, we could add a secondary concentrator that will concentrate the moonlight onto the absorber face, so that if you would look back from the face, you would always see the Moon, either directly or via optics. The question is then: What is the maximum temperature at which the absorber face (and thus, because of perfect insulation, the complete interior of the box including the wood) is in thermal equilibrium with the radiation from the Moon?

To answer this question, we need to consult Kirchhoff's law of radiation (see Knowledge Corner Topic 3.21) and look at the spectrum in detail. In short,

(Continued)

Knowledge Corner Topic 3.22: (Continued)

Kirchhoff asserts that emissivity and absorptivity, both as a function of location, direction and wavelength, are equal. Since our absorber face is fully illuminated through the secondary concentrator, and the moonlight is assumed to be Lambertian, i.e. with a spectral radiance independent of location and direction, the dependence is only on wavelength. For each wavelength, a radiation temperature can be assigned to the moonlight's spectral radiance, namely, the temperature of a black-body radiator which emits the same spectral radiance at that wavelength. This temperature will vary with wavelength.

To get a feeling for radiation temperature, think of the following experiment. Assume you have a black-body radiator with a temperature you can freely adjust. Up to about 3000 K, you could use a microwave-heated cavity made of tungsten with a little hole in it; this hole is your black-body radiation emitter (the hole is black, since virtually nothing comes back from the cavity when you send light into the hole). Now you have a tunable narrow-band wavelength filter. For any wavelength, you look at the Moon and at the black-body radiator through the same filter, and adjust the cavity temperature until both appear equally bright (same spectral radiance). For the Moon, this temperature is larger, especially in the blue part of the spectrum; a glowing body must have a temperature in the kK (Kilokelvin) range to emit any visible amount of blue light.

Now, if the absorber temperature is higher than the moonlight's radiation temperature at a certain wavelength, the absorber should be perfectly reflecting and thus non-emitting at that wavelength. On the other hand, if the incoming radiation is "hotter" than the absorber, the absorber should be perfectly absorbing, because despite full emission, there will be a net heat gain at that wavelength and temperature. After settling to equilibrium, the absorber temperature, and thus the wood, will be at least at the threshold temperature. Since the Moon's radiation temperature is >1000 K for many wavelengths, the wood will finally start to burn. We are sorry, xkcd, but on this one, we think you are wrong. You cannot start a fire from moonlight just with a magnifying glass alone. When you add spectral filters and good insulation, you can.

Dr. Wisenheimer:

This is the well-known principle of *selective solar thermal absorbers* [88, 89], widely used in the solar thermal collector industry to reduce thermal radiative losses and thus increase solar thermal collector efficiency.

Accordingly, spectral considerations may play a decisive role in illumination optics. It is time to look deeper into the spectral properties of light.

3.6 Spectral Calculus, Color and Vision

3.6.1 Spectral Quantities

Knowledge Corner Topic 3.23: The $V(\lambda)$ curve

Radiant flux is the integral over spectral radiant flux:

$$\Phi_e = \int_{\lambda_{\text{min}}}^{\lambda_{\text{max}}} \Phi_{e,\lambda}(\lambda)\mathrm{d}\lambda \tag{3.241}$$

Luminous flux is derived from the spectral radiant flux via weighting with the eye sensitivity curve $V(\lambda)$ (Figure 3.55):

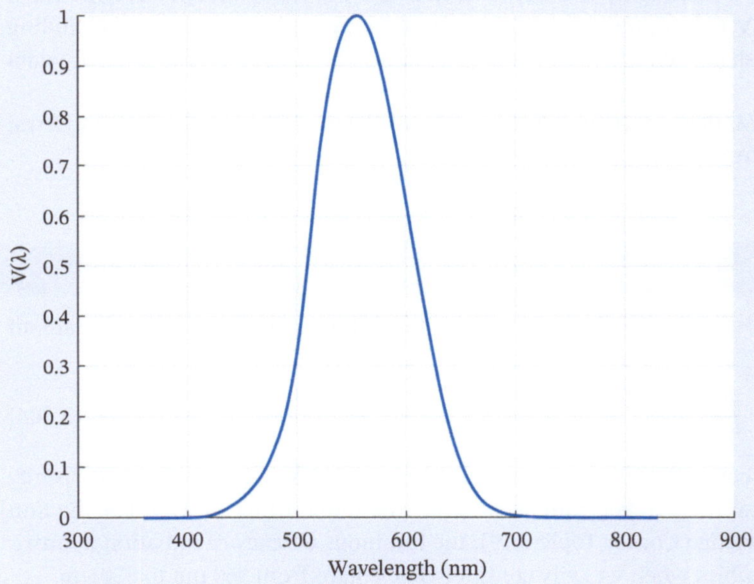

Figure 3.55 The photopic $V(\lambda)$ curve of eye sensitivity.

$$\Phi_\text{v} = K_\text{cd} \int_{360\,\text{nm}}^{830\,\text{nm}} V(\lambda)\Phi_{\text{e},\lambda}(\lambda)\,\mathrm{d}\lambda \tag{3.242}$$

multiplied with the maximum efficacy for monochromatic radiation at 555 nm

$$K_\text{cd} = 683\,\frac{\text{lm}}{\text{W}} \tag{3.243}$$

(one of the seven *defining constants* of the SI International System of Units). The full SI brochure is available for download from BIPM.

(Continued)

Knowledge Corner Topic 3.23: (Continued)

Dr. Wisenheimer:

It used to be a major pain to try and obtain the $V(\lambda)$ curve and many other standardized colorimetric spectra, such as the CIE 1931 color-matching functions or others. You either had to rely on doubtful secondary internet data or buy the (not exactly cheap) CIE standards with the spectra tabulated in the annex, and then copy the tables from the pdf annex into an editor to get machine-readable data.

Thankfully, this pain is gone. CIE heeded the call, and now provides many spectra for free download. And when CIE does something, they usually do it thoroughly and well. In addition to providing the spectra in a simple ASCII CSV format, there is a JSON metadata file for each spectrum, including md5 and sha256 checksums, sums of columns and a sample row, to ensure data integrity.

Julius' Matlab® spectrum library, freely available at Github, uses these spectral data sets from CIE.

Knowledge Corner Topic 3.24: Efficacies and efficiencies

The *spectral luminous efficacy of radiant flux* is the ratio of luminous and radiant flux

$$K = \frac{\Phi_\mathrm{v}}{\Phi_\mathrm{e}} \tag{3.244}$$

with K_cd being K for a single spectral line at 555 nm. It is helpful to specify the spectral range taken into account for the radiant flux as well. For the Sun (see Knowledge Corner Topic 3.19), the luminous efficacy of radiation approximately doubles when we only consider visible light from 360 nm to 830 nm.

The *luminous efficacy of a source* is its luminous flux per electrical power,

$$\eta_\mathrm{v} = \frac{\Phi_\mathrm{v}}{\Phi_\mathrm{el}} \tag{3.245}$$

Both luminous efficacies are measured in lm/W. For short, we often use the terms "spectral efficacy" and "luminous efficacy".

In contrast, the colloquial wall-plug efficiency (WPE) of a source, or officially its radiant efficiency is defined as

$$\eta = \frac{\Phi_\mathrm{e}}{\Phi_\mathrm{el}} \tag{3.246}$$

(radiant flux vs. electrical power, in percent).

Dr. Wisenheimer:

So what's this efficacy/efficiency business? Names of quantities are important! When you talk to me and use the word "efficiency", we both need to know exactly what you mean by that – otherwise we will not be able to communicate efficiently and precisely. And precise communication is of the utmost importance in science and engineering.

Efficiency is what you get in relation to what you put in, both in the same units, measured as a dimensionless fraction, or, equivalently, in percent. *Efficacy*, according to Merriam-Webster, is the "power to produce a desired result or effect", where power and effect usually do not have the same units; here, the units are W for power and lm for effect. Alas, some people use "wall-plug efficiency" (WPE) to denote η_v (in lm/W) rather than η (in %), as would be correct.

The conversion efficacy of a luminescent process is

$$\eta_c = \frac{\Phi_{v,conv}}{\Phi_{e,pump}} \tag{3.247}$$

(luminous flux of converted light vs. pumping radiant flux, unit lm/W).

Typical values are

- WPE of a blue LED chip, $\eta \approx 50\%$
- Luminous efficacy of a white LED, $\eta_v \approx 100 \ldots 200\,\text{lm/W}$
- Luminous efficacy of a red LED, $\eta_v \approx 60 \ldots 110\,\text{lm/W}$
- Conversion efficacy $\eta_c \approx 200\,\text{lm/}$of a yellow phosphor. However, at high pump irradiance and high local temperature, efficacy may drop to as low as $\eta_c \approx 40\,\text{lm/W}$ due to quenching.

Exercise 3.23: Constant spectrum

What is the spectral efficacy of radiation of a top-hat spectral distribution (*constant* spectral radiant flux) between 380…780 nm?

Solution 3.23

For a constant $\Phi_{e,\lambda}$, the ratio of Eqs. 3.241 and 3.240 is

$$\eta = 683\,\frac{\text{lm}}{\text{W}} \cdot \frac{\displaystyle\int_{380\,\text{nm}}^{780\,\text{nm}} V(\lambda)\,\mathrm{d}\lambda}{780\,\text{nm} - 380\,\text{nm}} \tag{3.248}$$

The integral is evaluated numerically, yields 107 nm, and we obtain

(Continued)

Solution 3.23 (Continued)

$$\eta = 683\,\frac{\text{lm}}{\text{W}} \cdot \frac{107\,\text{nm}}{400\,\text{nm}} \approx 183\,\frac{\text{lm}}{\text{W}} \tag{3.249}$$

which is quite a bit less than for typical white LEDs.

Exercise 3.24: Blue LED

Compare the radiant and luminous flux values of two blue and royal blue LED types for the wavelength and power minima and maxima. (LED size and package being equal otherwise.)

The LED datasheet states:

	Dominant or Peak	Wavelength	Min. Flux	Max. Flux
Blue	465 nm (min.)	485 nm (max.)	70 lm	112 lm
Royal blue	440 nm (min.)	455 nm (max.)	1.3 W	1.635 W

Which is the "brightest" bin in terms of radiance?

Solution 3.24

We note that the blue LED power is specified in lm and not in mW (in contrast to "royal" blue). The $V(\lambda)$ values for the respective wavelengths allow for an approximate translation of radiant and luminous flux values (approximate because only a single representative wavelength is used).

Wavelength	$v(\lambda)$	$\Phi_{e,min}$	$\Phi_{v,min}$	$\Phi_{e,max}$	$\Phi_{v,max}$
440 nm	0.023	1300 mW	*20.4* lm	1635 mW	*25.7* lm
455 nm	0.048	1300 mW	*42.6* lm	1635 mW	*53.6* lm
465 nm	0.074	*1384* mW	70 lm	*2216* mW	112 lm
485 nm	0.169	*606* mW	70 lm	*970* mW	112 lm

Numbers presented in *Italics* are *calculated*, the others are datasheet values.

Answer: The brightest bin (measured in lm) implies the highest radiance.

When Julius needs more precise values from general spectral computations, he likes to use his free, open-source Matlab spectrum library. Then, reading a typical blue spectrum from file, shifting it to $\lambda_{dom} = 465$ nm (which is not trivial), and computing the radiant flux when the luminous flux is 112 lm are just a few lines of simple code; all the rest is done by the library under the hood:

```
blue_spectrum = ReadLightToolsSpectrumFile('LERTDUW_S2WP_blue.sre');
blue_465nm_spectrum = ShiftToLdom(blue_spectrum, 465);
K = SpectralEfficacy(blue_465nm_spectrum);
blue_465_Phiv_lm = 112;
blue_465_Phi_W = blue_465_Phiv_lm / K

blue_465_Phi_W = 2.0560
```

We see that our 2216 mW estimate was a little optimistic, but the conclusion is still valid.

Open-source libraries with comparable functionality are also available for Python, e.g. `luxpy` (by Kevin Smet) or `colour-science` (by a group of contributors, https://www.colour-science.org/).

Dr. Wisenheimer:

For blue LEDs, LED vendors sometimes specify luminous flux in lm, and sometimes radiant flux in W or mW. For color, they sometimes specify the dominant wavelength, sometimes peak wavelength. They tend to choose lm and dominant wavelength for longer wavelengths, and W and sometimes peak wavelength for shorter wavelengths. Photometric lumens are appropriate when the LED is used in a stage light or architectural lighting as the only color, because in these cases, visual brightness impression is what counts. However, when used in a tunable lamp with other colors, lumen is most likely not the unit of choice. It is amazing how little blue luminous flux is needed for RGB white, where 100 lm are typically generated with about 70 lm green, 27 lm red and only 3 lm (!) blue. Omitting blue from the mix would make it hardly any less bright, but render the color a saturated yellow instead of white. It would be much preferable to see *color weight* as specification and grouping criterion (see Eq. 3.264), because for RGB color mixing, the color shift due to the blue contribution is way more important than its contribution to brightness.

Knowledge Corner Topic 3.25: Spectral weighting

There are many aspects of a spectrum that are calculated by weighting it with a specific action spectrum

$$\eta = \int w(\lambda)\Phi_{e,\lambda}\,d\lambda \tag{3.250}$$

The concept is rather important in color science (see below) but also widely used in the following areas:

- Human visual aspects and photoreceptor properties:
  - Photopic (human day vision): Eq. 3.241
  - Scotopic (human night vision)

(Continued)

Knowledge Corner Topic 3.25: (Continued)

- Mesopic [90] (adaptation between day and night, twilight vision). This is of special interest for nighttime driving and road lighting
- Melanopsin sensitivity [91, 92]. Melanopsin is a photopigment in the eye that can suppress melatonine synthesis when activated by blue light [93], important for so-called "human-centric lighting", which attempts to adapt illumination spectra to human circadian rhythms.

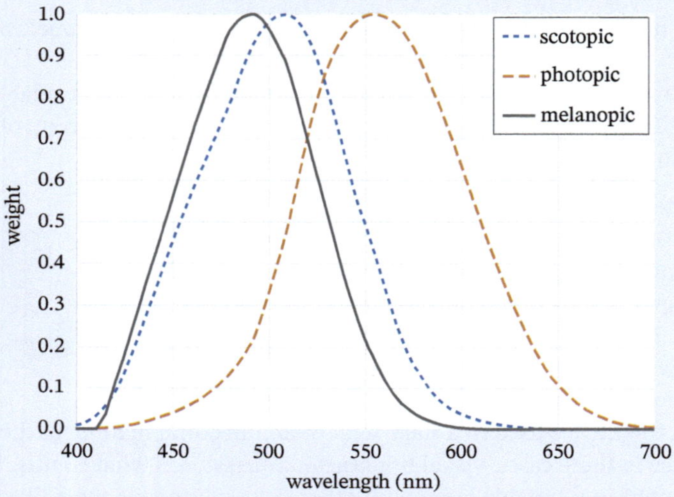

Figure 3.56 Photoreceptor sensitivities: scotopic, photopic (CIE 1924), and melanopic curves.

We took these curves from CVRL, melanopic sensitivity [92]. CVRL is another great source of information for spectra, especially for recent research, complementing the established standards at CIE.

- Human photobiological effects (harmful or beneficial) [94] (see Figure 3.57):
  - Photokeratitis and photoconjunctivitis (painful conditions of the eye caused by UV radiation, including arc eye and snow blindness) [95]
  - Pigmentation (skin darkening) [96]
  - Previtamin D synthesis [97]
  - Psoriasis (phototherapy for a noncontagious autoimmune skin disease) [98]
  - Photocarcinogenesis [99]
  - Erythema (skin reddening, sunburn) [96]
  - Blue-light hazard (blue content and eye safety of light sources) [100–102]
  - Bilirubin (see Knowledge Corner Topic 3.26)

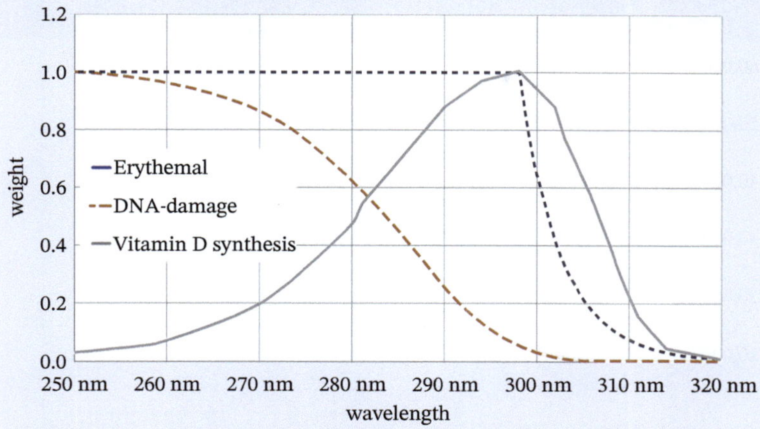

Figure 3.57 Sensitivity to UV radiation (after TEMIS.nl [97, 103, 104, 105]).

- Photobiologic effects of plants and horticulture [106] (Section 4.5).
  Grow-lighting is a "hot topic" with a lot of current research, especially regarding the features of LED lighting. We only mention a few aspects:
  - Chlorophyll synthesis [107] (cf. Figure 3.54)
  - Phototropism (growth) [108]
  - Flowering [109]
- Technical
  - Detector response
  - Film sensitivity
  - Photosensitizers [110] for paints, lacquers, or UV curable optical glue
  - Bleaching and weathering

Knowledge Corner Topic 3.26: Bilirubin treatment

Neonatal jaundice is a serious problem for babies who cannot yet metabolize bilirubin (a normal metabolic decomposition product from red blood cells) and develop a yellow to orange skin color. Fortunately, there is a cure: a phototherapy with blue light. The spectral effectiveness of light for bilirubin treatment is shown in Figure 3.58.

The absorption spectrum (Figure 3.58) translates to a spectral irradiance requirement of

$$E_\lambda = 35 \frac{\mu\text{W}}{\text{cm}^2\,\text{nm}} \tag{3.251}$$

averaged within the spectral range of 430–490 nm. This is governed by the EN 60601-2-50 standard [112, 113].

As of today, bilirubin lights are mostly equipped with blue LEDs.

(Continued)

Knowledge Corner Topic 3.26: (Continued)

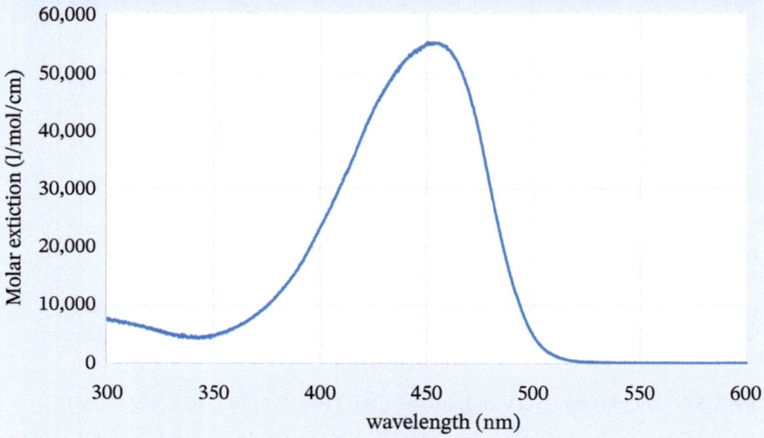

Figure 3.58 Bilirubin absorption spectrum [111].

Exercise 3.25: Bili light

Imagine a backlight of 40 cm · 60 cm illuminated by blue LEDs. (In the real world, we would take better care of our newborns and make the backlight a soft mattress made of light-guiding fibers.)

The plan is to use royal blue Luxeon Rubix LEDs with a nominal radiant flux of $\Phi = 1.5$ W. Assume a backlight efficiency of $\eta = 50\%$ and a WPE (Eq. 3.245) of $\eta = 50\%$ and calculate the necessary LED electrical power to meet the specification.

Solution 3.25

As a rough approximation, we estimate the total irradiance as

$$E = E_\lambda \cdot 60\,\text{nm} = 2.1\,\frac{\text{mW}}{\text{cm}^2} \tag{3.252}$$

The necessary radiant flux from the backlight is then

$$\Phi = EA = 2.1\,\frac{\text{mW}}{\text{cm}^2} \cdot 2400\,\text{cm}^2 \approx 5\,\text{W} \tag{3.253}$$

This means that we need to provide 20 W of LED electrical power. An array of twenty 1 W LEDs should be fine; more LEDs translate to saving electric energy and achieving a better homogeneity.

3.6.2 Color

In the following section, we provide a sketch of the science of colorimetry, intended to enable you to carry out some basic calculations, such as LED color mixing. Colorimetry, however, is a huge and fascinating field; we can barely scratch its surface here. In particular, we will not delve into the subject of the *color rendering quality* of light sources. Not because color rendering is not important (it is), but because the computations involved with it are beyond what can be done on paper or with spreadsheets, which is the focus of our exercises. We recommend Kevin Houser and Michael Royer's excellent webinar Advances in Characterizing Color Rendition of Light Sources.

In [114, Part II] you will find an excellent explanation of how the human eye processes color. Wyszecki and Stiles wrote the comprehensive textbook on color science [115]. It is somewhat outdated, but on the other hand, the science of colorimetry does not advance too quickly, and is always very much trying to be backward compatible. For full information on the CIE system, we refer to the book by Schanda [116]. One of us (Julius) is teaching a course on "Colorimetry for LED Users", twice a year online, and at selected SPIE conferences in person.

Knowledge Corner Topic 3.27: Basic colorimetry

The human eye has *three* types of color sensitive receptors. Mathematically, they reduce the information of the complete incident spectrum to three numbers by integration with three weighting functions (see Knowledge Corner Topic 3.25). Any color space for human vision will be three-dimensional (this is the first Grassmann law [117]; the other Grassmann laws deal with additive color mixing and linearity [116]).

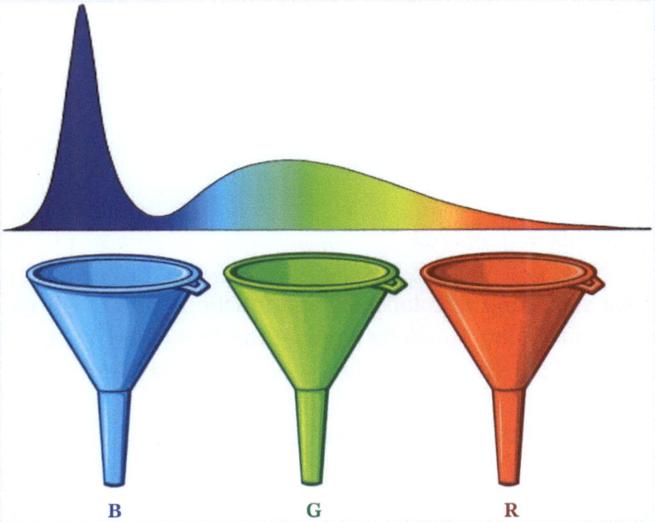

Figure 3.59 Photoreceptors reduce the information content of the spectrum.

(Continued)

Knowledge Corner Topic 3.27: (Continued)

The three primary receptor signals are combined by weighted addition (much green, less red, and very little blue) in the retina to create an overall brightness impression. This brightness impression varies with wavelength, and is strongest for 555 nm yellow-green. The CIE has published a standardized sensitivity curve, the so-called $V(\lambda)$ spectral luminous efficiency function, which is used to derive the photometric quantities (luminous flux, luminous intensity, illuminance, luminance) from the underlying spectra (see Knowledge Corner Topics 3.23 and 3.25).

The impression of color begins with similar neural computation in the retina: The green signal is subtracted from red, and the blue signal from the sum of green and red, to provide the red-green and blue-yellow color dimensions [114].

Color-matching functions

In 1931, a *standard colorimetric observer* was defined based on experimental investigations, where monochromatic primaries (red: 700 nm, green: 546.1 nm, blue: 435.8 nm) were used to match monochromatic test stimuli [118, 119]. The relative luminance was obtained as

$$L = R + 4.5907\,G + 0.0601\,B \tag{3.254}$$

with photometric weights. The so-called tristimulus values R, G, and B can be calculated by integrating a spectrum S_λ with *color-matching functions* (i.e. weights) $\bar{r}, \bar{g}, \bar{b}$ (derived from experiments with human subjects using these three primaries):

$$\begin{aligned} R &= \int \bar{r}(\lambda) S_\lambda(\lambda)\,\mathrm{d}\lambda \\ G &= \int \bar{g}(\lambda) S_\lambda(\lambda)\,\mathrm{d}\lambda \\ B &= \int \bar{b}(\lambda) S_\lambda(\lambda)\,\mathrm{d}\lambda \end{aligned} \tag{3.255}$$

Color-matching functions are dimensionless, and are scaled such that they cover equal areas

$$\int \bar{r}(\lambda)\,\mathrm{d}\lambda = \int \bar{g}(\lambda)\,\mathrm{d}\lambda = \int \bar{b}(\lambda)\,\mathrm{d}\lambda \tag{3.256}$$

To (i) avoid negative values in the color matching functions, and to (ii) use $V(\lambda)$ as one of the color-matching functions by requiring that $\bar{y}(\lambda) = V(\lambda)$, the following transformation of the tristimulus values was introduced [116]:

$$\begin{pmatrix} X \\ Y \\ Z \end{pmatrix} = \begin{pmatrix} 2.768892 & 1.751748 & 1.130160 \\ 1 & 4.590700 & 0.060100 \\ 0 & 0.056508 & 5.594292 \end{pmatrix} \begin{pmatrix} R \\ G \\ B \end{pmatrix} \tag{3.257}$$

The new tristimulus values (X, Y, Z) can again be calculated directly from a spectrum by weighting with corresponding *color-matching functions* $\bar{x}, \bar{y}, \bar{z}$ (see Figure 3.60):

$$\begin{aligned} &= \int \bar{x}(\lambda) S_\lambda(\lambda)\, \mathrm{d}\lambda \\ &= \int \bar{y}(\lambda) S_\lambda(\lambda)\, \mathrm{d}\lambda \\ &= \int \bar{z}(\lambda) S_\lambda(\lambda)\, \mathrm{d}\lambda \end{aligned} \tag{3.258}$$

which are displayed in Figure 3.59.

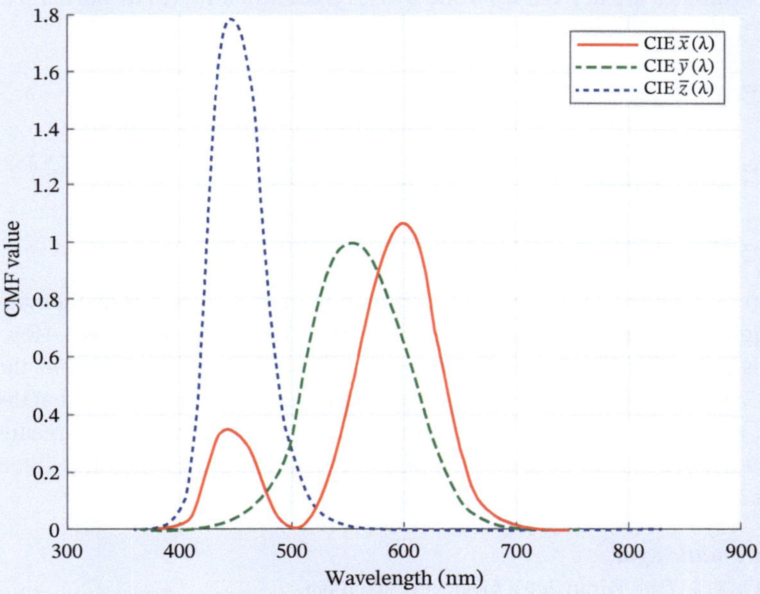

Figure 3.60 CIE 1931 $\bar{x}, \bar{y}, \bar{z}$ color-matching functions [116].

The color-matching functions are normalized for equal areas under the curves as well, meeting the requirement that a constant, flat spectrum (equal-energy stimulus) should yield equal tristimulus values [116]

$$X = Y = Z \tag{3.259}$$

which was realized by the choice of parameters in Eq. 3.256.

"The magnitudes of the XYZ components are proportional to physical energy, but their spectral composition corresponds to the color-matching characteristics of human vision" [120].

Because of [116, Ch. 3]

$$\bar{y}(\lambda) = V(\lambda) \tag{3.260}$$

(Continued)

Knowledge Corner Topic 3.27: (Continued)

we can state with Eqs. 3.241 and 3.257 that

$$\Phi_v = K_{cd} Y \tag{3.261}$$

when the spectrum under consideration has a meaning of spectral flux (W/nm). Corresponding relations exist for luminous intensity, illuminance and luminance.

Finally, we mention the physical unit of the tristimulus values: the W (Watt, as the radiant flux).

Color coordinates

Color coordinates are derived from the X, Y, Z tristimulus values by normalization:

$$\begin{aligned} x &= \frac{X}{X+Y+Z} \\ y &= \frac{Y}{X+Y+Z} \\ z &= \frac{Z}{X+Y+Z} \end{aligned} \tag{3.262}$$

Therefore, we have $x+y+z=1$. It is common to drop z, since not mathematically required due to normalization, keeping only x, y as color coordinates. However, it is geometrically instructive to view a (X, Y, Z) triple as a point in the abstract 3D color space. Geometrically, the condition $x+y+z=1$ means that the point (x, y, z) lies on the 1–1–1 plane. Computing color coordinates then means to *project* (X, Y, Z) toward the origin and to see where the projection line intersects the 1–1–1 plane.

Monochromatic light

For $S(\lambda) = \delta(\lambda)$ (the Dirac delta function), we have

$$\begin{pmatrix} X \\ Y \\ Z \end{pmatrix}(\lambda) = \begin{pmatrix} \bar{x}(\lambda) \\ \bar{y}(\lambda) \\ \bar{z}(\lambda) \end{pmatrix} \tag{3.263}$$

and the corresponding color coordinates are

$$\begin{pmatrix} x \\ y \\ z \end{pmatrix}(\lambda) = \frac{1}{\bar{x}(\lambda)+\bar{y}(\lambda)+\bar{z}(\lambda)} \begin{pmatrix} \bar{x}(\lambda) \\ \bar{y}(\lambda) \\ \bar{z}(\lambda) \end{pmatrix} \tag{3.264}$$

We plot the trajectory $(y(\lambda), x(\lambda))$ in the xy plane and obtain the “color shoe” in the xy color diagram (Figure 3.61).

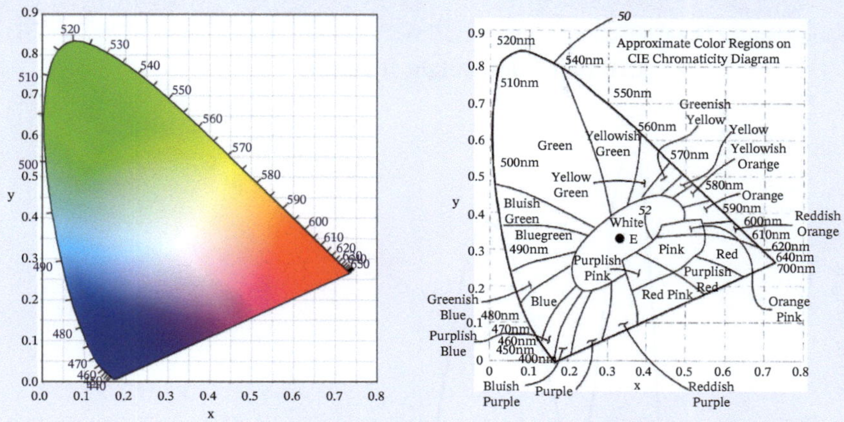

Figure 3.61 Color triangle: color coordinates of monochromatic light (left) and color regions (right) [121].

Color weight

The *color weight* is the sum of the tristimulus values:

$$W = X + Y + Z \tag{3.265}$$

and relates the primary to the respective color coordinate

$$\begin{aligned} X &= Wx \\ Y &= Wy \\ Z &= Wz \end{aligned} \tag{3.266}$$

which is quite convenient for many colorimetric computations.

We define the color weight function

$$\bar{w}(\lambda) = \bar{x}(\lambda) + \bar{y}(\lambda) + \bar{z}(\lambda) \tag{3.267}$$

which yields

$$W = \int \overline{w}(\lambda) S_\lambda(\lambda)\, \mathrm{d}\lambda \tag{3.268}$$

The function $\bar{w}$ characterizes the color mixing weight of a spectral distribution with emphasis on blue and red (Figure 3.62). The dip in the green explains why we have to put more green power into a color mix to achieve white light.

(Continued)

Knowledge Corner Topic 3.27: (Continued)

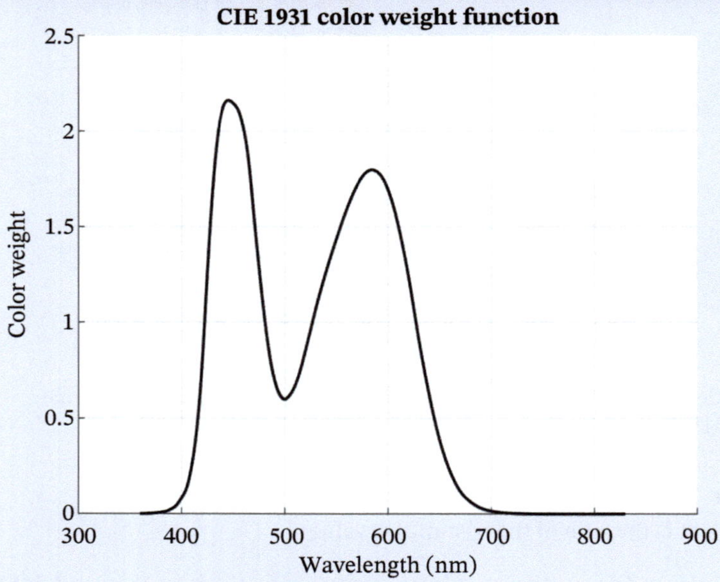

Figure 3.62 Color weight function $\overline{w}$.

Color mixing

According to Grassmann, additive color mixing is a linear combination. The addition of spectra

$$S_{12}(\lambda) = S_1(\lambda) + S_2(\lambda) \tag{3.269}$$

implies

$$\begin{pmatrix} X_{12} \\ Y_{12} \\ Z_{12} \end{pmatrix} = \begin{pmatrix} X_1 \\ Y_1 \\ Z_1 \end{pmatrix} + \begin{pmatrix} X_2 \\ Y_2 \\ Z_2 \end{pmatrix} \tag{3.270}$$

Geometrically, additive color mixing is a vector addition, concatenating arrows.

The color coordinate of the mix is

$$x_{12} = \frac{X_{12}}{W_{12}} = \frac{X_1 + X_2}{W_1 + W_2} = \frac{x_1 W_1 + x_2 W_2}{W_1 + W_2} \tag{3.271}$$

(and analogous for y and z). The actual weight for a mix of two normalized spectra is *color weight times radiant flux scaling*.

Gamut

Any system of primaries (for example, those cut out from white light by a projector's color wheel) forms a polygon in the xy color space containing all colors that can be obtained by mixing the primaries. This range is called the *color gamut* [122] (Figure 3.63). Often, there are three primaries, but sometimes four or more.

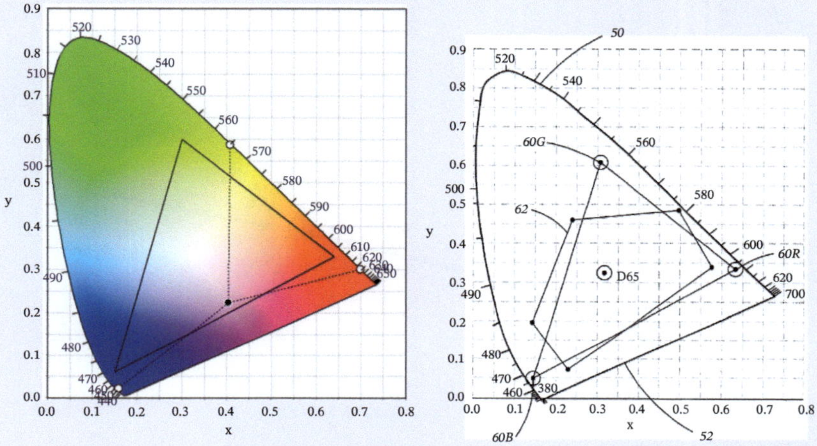

Figure 3.63 Left: gamut spanned by RGB LEDs compared to the Rec.709 standard [123]. Right: comparison to a gamut of five primaries [121].

For further information, see, for example, the Display Color Gamuts Shoot-Out.

Color distance

In order to distinguish two spectra or two samples, one would naively try to use the Euclidean distance in the xy color space. However: *Just noticeable chromaticity differences* (JNDs) can have different sizes in different regions of the color diagram [124]. As a first – and simple – way out of this dilemma, we will introduce one more coordinate system together with the transformation by which it is generated from the xy space.

The CIE 1976 UCS (uniform color space) within the CIELUV system [116] is defined by the transformation

$$u' = \frac{4x}{-2x + 12y + 3}$$
$$v' = \frac{9y}{-2x + 12y + 3} \tag{3.272}$$

The corresponding color diagram is shown in Figure 3.66 (right). Since this transformation is a projective transformation, straight lines in the xy space remain straight lines after transformation. In particular, the convex hull property of gamuts (mixed colors remain within the convex polygon spanned by the primaries) is preserved.

(Continued)

Knowledge Corner Topic 3.27: (Continued)

To express the perception of brightness, we need an auxiliary function (Figure 3.64)

$$f(x) = \begin{cases} \sqrt[3]{x} & \text{if } x > \left(\frac{24}{116}\right)^3 \\ \frac{841}{108}x + \frac{16}{116} & \text{below} \end{cases} \tag{3.273}$$

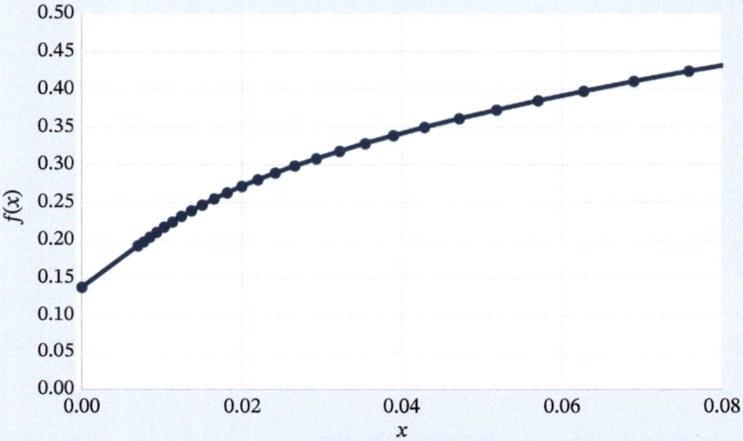

Figure 3.64 Auxiliary function with the linear region for small arguments.

and define a new quantity *lightness* as a measure of the visual perception of luminance,

$$L^* = 116 \cdot f\left(\frac{Y}{Y_n}\right) - 16 \tag{3.274}$$

where the index n is related to a reference white point. But how can this reference white be chosen? The underlying idea is that CIELUV is applied to *scenes*, such as an LCD monitor, a color print on a desk, or a dress of a model on stage. Diffuse reflectivity is implicitly assumed. Therefore, a natural choice of reference white presents itself: On a monitor, it corresponds to all RGB pixels either on full transmission or adjusted to calibration white, usually D65 (a standardized CIE daylight spectrum with a CCT (see below for CCT) of 6500 K). For a print or a fabric, the spectrum reflected by a perfectly white sample is selected.

In CIELUV, the color distance to reference white is expressed by the functions

$$u^* = 13L^*(u' - u'_n) \tag{3.275}$$

$$v^* = 13L^*(v' - v'_n) \tag{3.276}$$

For any two spectra, the distance between their color and lightness coordinates is then

$$\Delta u^* = u_1^* - u_2^* \tag{3.277}$$

$$\Delta v^* = v_1^* - v_2^* \tag{3.278}$$

$$\Delta L^* = v_1^* - v_2^* \tag{3.279}$$

The joint color and lightness difference ΔE_{uv}^* between the two spectra is the Euclidean distance

$$\Delta E_{uv}^* = \sqrt{(\Delta L^*)^2 + (\Delta u^*)^2 + (\Delta v^*)^2} \tag{3.280}$$

with the idea that when $\Delta E_{uv}^* \gtrsim 1$, the difference is just barely perceptible by well-trained observers with normal vision (1 JND).

The color distance of two sources of the same lightness is given by the distance in the *u'v'* diagram:

$$\Delta E_{uv}^* = 13L^*\sqrt{(u_1' - u_2')^2 + (v_1' - v_2')^2} \tag{3.281}$$

This quantity is important for tolerancing light sources (if you buy an LED lamp now and another of the same type some time later you would expect them not to show any noticeable color differences when operated side by side). As a consequence, high-end products are specified for their allowed color distance.

Dr. Wisenheimer:

There are more color distance metrics that work better than the unsuitable *xy* distance. The first was introduced by MacAdam [125], who published his famous ellipses in 1942; statements like "this LED is binned within a 3-step MacAdam ellipse" are quite common until today. At the same time as CIELUV, a similar but different color space called CIELAB was defined. All of these have their shortcomings, however; the ideal of "same computed distance means same perceived distance" has still not been reached. Today, the rather complicated CIEDE2000 metric [116] (based on CIELAB) is considered the gold standard. Julius' Matlab spectrum library and the OSRAM Sylvania Color Calculator both include a CIEDE2000 implementation.

Correlated color temperature (CCT)

What is white? And how can we best describe quantitatively what we call "cold white (bluish)" or "warm white (orangeish)" (Figure 3.65)? We cite a definition [126]:

> The correlated color temperature is the temperature of a black body having the chromaticity nearest the chromaticity associated with the given spectral distribution on a diagram where the (CIE 1931 standard observer

(Continued)

Knowledge Corner Topic 3.27: (Continued)

based) u', $2/3v'$ coordinates of the Planckian locus and the test stimulus are depicted [126].

There is a nice introduction: Guide to Light Color Temperature.

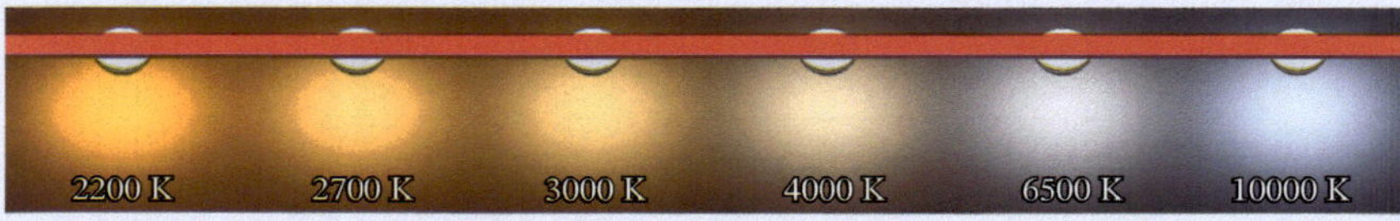

Figure 3.65 Lamps of different color temperatures (LightTools simulation). Have a look at the provided Simulation file.

It is straightforward to calculate the spectrum for a given temperature using Planck's formula, to weight it with the color-matching functions, and to obtain its coordinates in the *xy* color diagram (Figure 3.65). The reverse way – to calculate the best-matching black-body temperature from color coordinates – is harder.

We first recognize that all colors of black-body radiation must lie on a single 1D line in the *xy* diagram; after all, the spectrum is determined by a single number, the temperature T, which parameterizes this line that is known as the *Planck locus.*

It makes no sense to compute the best-matching temperature for arbitrary colors, such as saturated green, but only for colors that are more or less white. We have to somehow project our color point onto the Planck locus.

Since *xy* is very nonuniform, perceptually, we should perform this projection in a more uniform color space. At the time when Deane Judd proposed this approach, the best available CIE color space was the now-deprecated CIE 1960 UCS with its $u = u'$, $v = 2/3\, v'$ color coordinates. Owing to the desire for backward compatibility, the CIE 1960 UCS is used until today, solely to define what Judd called *correlated color temperature.*

The projection is achieved using so-called *Judd lines*, which are simply straight lines, perpendicular to the Planck locus in *uv* color space, and there is one such Judd line for each black-body temperature. To determine the color temperature of a lamp, you compute the *uv* coordinates of the lamp, find the unique Judd line on which that *uv* point lies, and look up the temperature to which this Judd line corresponds (see Figure 3.66).

The distance from the black-body curve in the *uv* space is termed d_{uv} [127]. Both CCT and d_{uv} form yet another coordinate system in the color space, in the vicinity of the black-body curve (anything beyond $d_{uv} > 0.05$ is considered nonwhite for CCT purposes).

If this sounds hard to achieve in practice, you may well use software such as the OSRAM Sylvania Color Calculator or Julius Matlab library; both have CCT and distance-to-Planck implemented.

If you do not have access to these numerical implementations: A polynomial-based, explicit transformation from color coordinates to CCT is described in [128] and demonstrated in CIE_planck.xlsx.

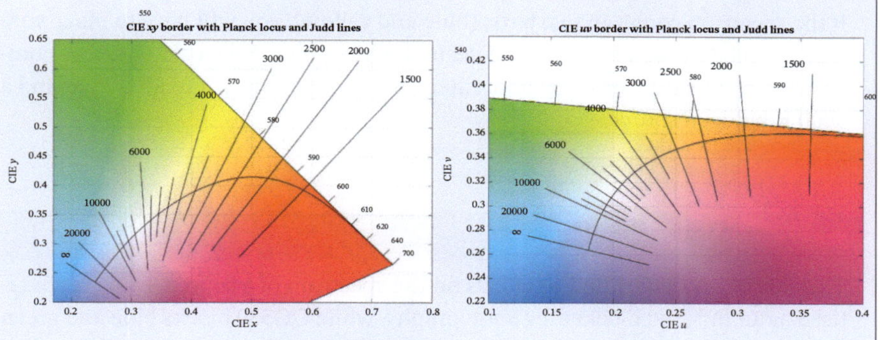

Figure 3.66 Both diagrams show the CIE "color shoe", with Planck locus and Judd lines, annotated with corresponding color temperatures. Left: CIE *xy* color space. Right: CIE *uv* color space. On the right, the Judd lines have length ± 0.05, the distance to the Planck locus where colors are still considered "white". On the left, the Judd lines are correctly transformed; their length corresponds to the *uv* diagram.

3.6.3 Color Filters

Color filters can alter the spectrum of a light source by selective reflection or absorption. They are used for technical purposes such as in machine vision or during ray file measurements (to measure the blue and yellow ranges of white light sources separately) or for the color correction of entertainment lights.

So-called dichroic filters consist of a stack of (many) thin layers on a substrate, often alternating layers of SiO_2 and TiO_2. Interference as a function of wavelength and angle of incidence then allows to split incident light in reflected and transmitted parts. Dichroic filter design means optimizing the layer materials and thicknesses to achieve the desired angle- and wavelength-dependent transmission/reflection characteristics.

Color filters made from dyed glass or plastic materials feature a wavelength-dependent absorption and remove specific spectral ranges from a beam of light. This function can be integrated in optical elements such as lenses.

Besides (passive) color filters, there are materials that can change the spectrum of light by luminescence, such as phosphors [129] and fluorescent solar collectors [130].

Exercise 3.26: Color correction

Download the Schott filter glass datasheets. Extract the BG47 and OG530 filter characteristics. Download the spectrum of a cold white (6500 K) LED of your choice, e.g. from Julius' Matlab spectrum library or from an LED vendor website. If the spectrum comes in two parts (blue and yellow), you will have to make sure that the relative weight yields the desired white point. Now calculate and compare the color shift caused by both filters, to both the cold-white LED light and a 3000 K halogen source.

Solution 3.26

The two filters have different effects on the spectrum of a source. BG47 is advertised as an infrared blocker for photography, while OG530 blocks blue and green light, thus looking orange (Figure 3.67).

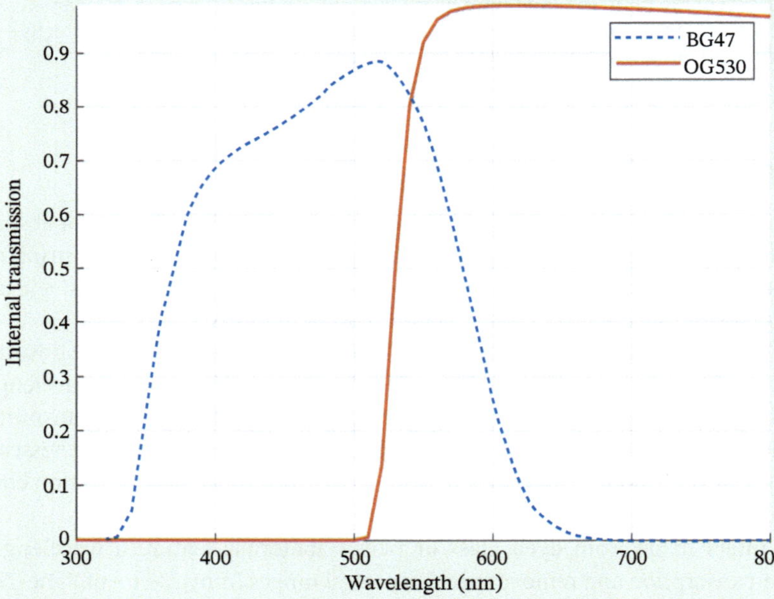

Figure 3.67 Transmission of two common color glass filters by Schott. The reference thickness for internal transmission is 1 mm.

We suggest to perform the calculations in the OSRAM Sylvania Color Calculator [131], using Julius' Matlab library, or by spreadsheet (provided for your convenience).

The proper way to analyze the effect of the filters is to prepare their transmission curves together with the source files, multiply them, and to calculate to color coordinates via Eqs. 3.257 and 3.261.

The resulting curves are shown in Figures 3.68 and 3.69.

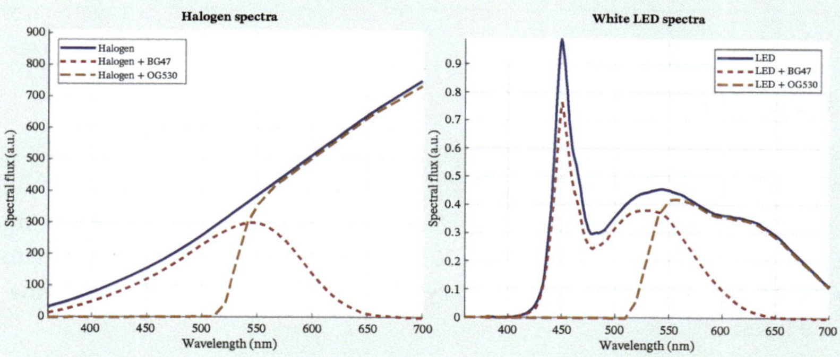

Figure 3.68 Effect of the color filters to the spectra: The BG filter causes a blue shift whereas the OG filter cuts all spectral components except of the red.

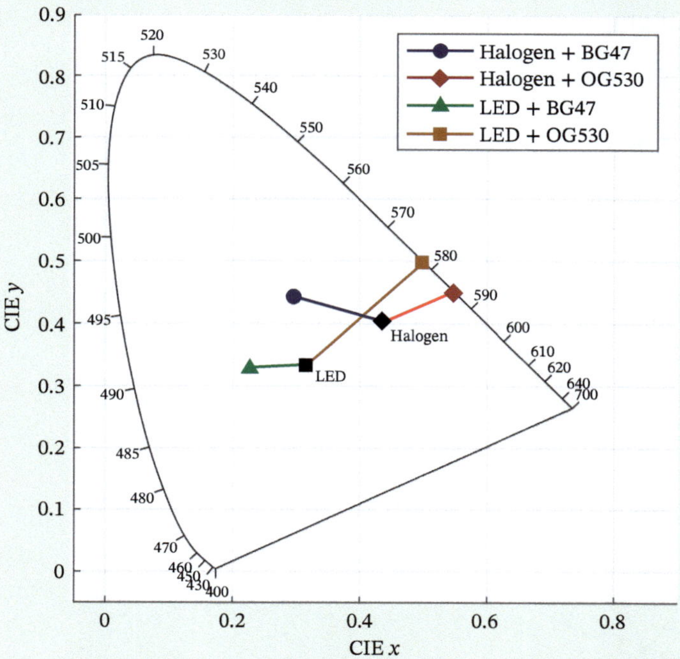

Figure 3.69 Effect of the color filters on the coordinates in the color diagram: The BG filter causes a shift to the left (larger for the halogen lamp, smaller for the LED), whereas the OG filter moves the color coordinate to the rim, creating a saturated orange.

(Continued)

Solution 3.26 (Continued)

Calculating the color coordinates of the filter curve itself is a futile endeavor. The result would be equivalent to using a constant spectrum, CIE standard illuminant E, which is just a hypothetical reference source of constant spectral power.

3.6.4 Light Mixing

Exercise 3.27: Three-line spectrum

Combine three spectral lines to obtain a (0.31, 0.33) white point with maximum luminous efficacy of radiation (Figure 3.70). You may need to use a spreadsheet to set up the spectral radiance function. Vary the wavelengths and individual powers of the lines.

Solution 3.27

First, download the file Three Lines Spectrum.

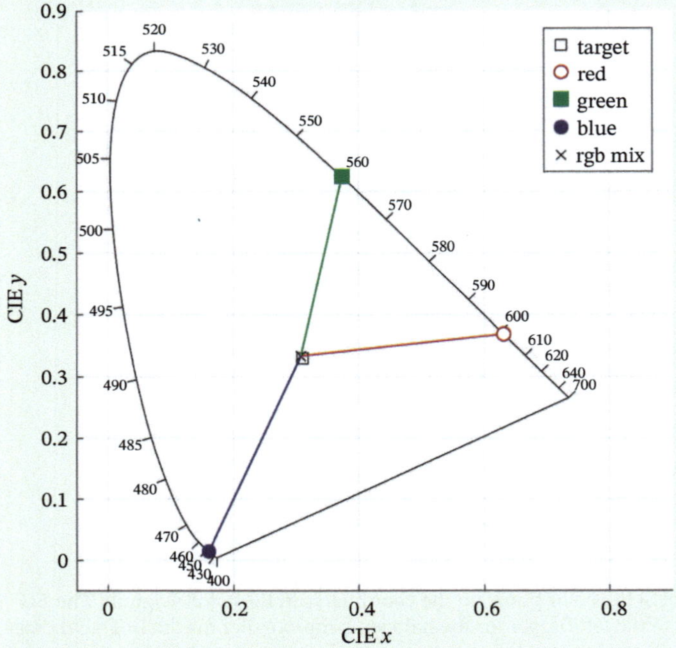

Figure 3.70 Mixing three monochromatic sources The combined radiation is at the target color coordinates and features $K = 412$ lm/W.

Vary the wavelengths and powers of the three single-line emitters. Optimize for highest luminous efficacy of radiation using the solver. The target color point acts as a constraint.

Your solution will look approximately like this:

- Red (601 nm, 1.1 W)
- Green (560 nm, 5.2 W)
- Blue (446 nm, 3.6 W).

You obviously need to provide more green light than red or blue. This is to compensate the green dip in the color weight function seen in Figure 3.62. Moreover, the green prefers longer wavelengths (toward yellow) to reduce the necessary radiant power to meet the target coordinates.

Dr. Wisenheimer:

A white point color mix is more efficient if done with only two lines [132]. However, in most real-world applications such as projectors, a mix of three primaries is preferable or simply required (cf. Exercises 3.31 and 4.45).

Exercise 3.28: Achromatic point

Calculate the color coordinates of a constant spectral distribution $S_\lambda(\lambda) = 1$ (CIE Standard Illuminant E) (this is a sequel to Exercise 3.23).

Solution 3.28

A constant spectrum yields by definition equal tristimulus values X, Y, and Z (equal horizontal sum of the matrix elements in Eq. 3.256), since the color-matching functions have been normalized to the same integral (area under the curve). Consequently, the color coordinates of a constant spectrum are (0.333, 0.333).

Exercise 3.29: Blue and yellow give white (1)

Mix blue light (450 nm, 1 W) with yellow light (570 nm, 1 W) and calculate the resulting color coordinates. Then increase the power of the yellow source (570 nm, 2 W). Which mix would you prefer for your living room?

Solution 3.29

We calculate the tristimulus values of the two monochromatic sources by looking up the color-matching functions from the spreadsheet.

Provided file: CIE_Planck.xlsx.

On this basis, we find the blue and yellow color coordinates and those of the mixtures.

	X	Y	Z	x	y
450 nm	0.34	0.38	1.77	0.157	0.018
570 nm	0.76	0.95	0	0.444	0.555
1:1	1.1	1.33	1.77	0.284	0.256
1:2	1.86	1.94	1.78	0.333	0.348

We find that the 1:2 result is close the black-body locus, while the 1:1 mixture will be distinctly bluish (Figure 3.71).

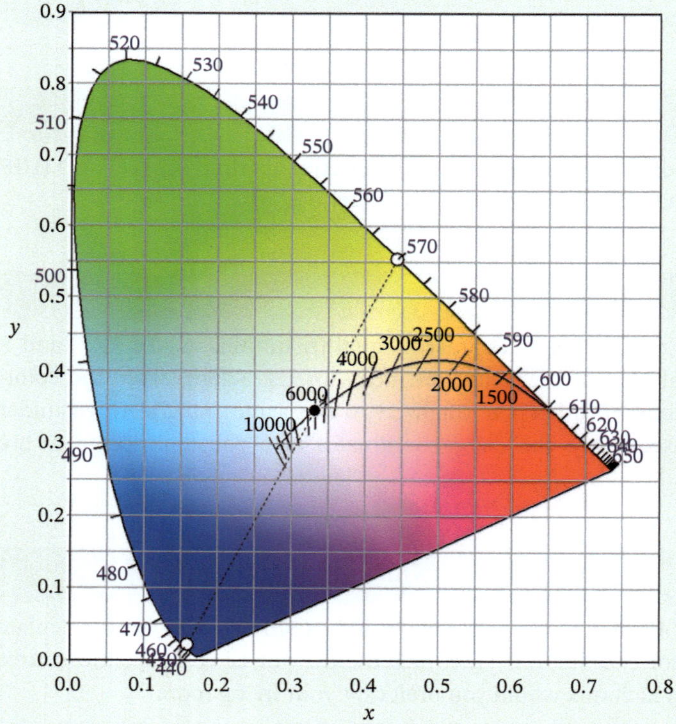

Figure 3.71 Mixing blue and yellow light. 1 W of blue plus 1.7 W of yellow light yield 2.7 W (≈ 1100 lm) of white light at 5700 K.

Exercise 3.30: Blue and yellow give white (2)

Now vary the wavelength of the blue source. Which wavelength in the blue region needs the least power to obtain white light when mixed with yellow light of 583 nm (obtaining a color on the black-body curve)?

Solution 3.30

For monochromatic lines of radiant flux P with wavelength λ, we have

$$W = \overline{w}P \tag{3.282}$$

where $\bar{w}$ is short for $\bar{w}(\lambda)$. We can write Eq. 3.270 as

$$x = \frac{x_{\mathrm{Y}}\overline{w}_{\mathrm{Y}}P_{\mathrm{Y}} + x_{\mathrm{B}}\overline{w}_{\mathrm{B}}P_{\mathrm{B}}}{\overline{w}_{\mathrm{Y}}P_{\mathrm{Y}} + \overline{w}_{\mathrm{B}}P_{\mathrm{B}}} \tag{3.283}$$

and similar for y. The necessary blue power is

$$P_{\mathrm{B}} = P_{\mathrm{Y}}\frac{\overline{w}_{\mathrm{Y}}}{\overline{w}_{\mathrm{B}}}\frac{x_{\mathrm{Y}} - x}{x - x_{\mathrm{B}}} = P_{\mathrm{Y}}\frac{\overline{w}_{\mathrm{Y}}}{\overline{w}_{\mathrm{B}}}\frac{y_{\mathrm{Y}} - y}{y - y_{\mathrm{B}}} \tag{3.284}$$

For a given blue wavelength, we would look up the color temperature of the mix by drawing a line in the color diagram. Then, we take the color coordinates of the blue wavelength and of the obtained color temperature from CIE_Planck.xlsx.

Furthermore, we look up the color weights $\bar{w}$ for yellow and this particular blue wavelength. Now, we can calculate P_{B} from Eq. 3.283.

As the color coordinates do not vary much (obvious for y), the minimum of P_{B} is related with the maximum of the color weight $\bar{w}$ at 446 nm (cf. Figure 3.62).

In a legacy three-chip projector, the light of a white source is split by dichroic filters into red, green, and blue primaries. After modulating each channel by an image provided by a liquid crystal display (LCD) or a digital micromirror device (DMD), the beams are then combined to form a color image.

Each photon produced by the source can be used in one channel only (or in none at all). The system is described by the two transition wavelengths B/G and G/R of the filter curves. The spectral split of white to the primaries is a tradeoff between various design goals:

- Highest white luminous flux (of course – any projector always lacks flux, and also ANSI lumen are always a major selling point)
- The flux has to be produced at a reasonable white point (you may make it as greenish as possible as long as customers do not complain HANDLE WITH CARE)
- Largest gamut (also known as spectral purity of the colors and promoted by a supplier initiative named 3LCD)

Exercise 3.31: RGB Split

Consider a three-chip projector operated with a tungsten halogen source of $T_C = 3000$ K. Assume that you can split its white light by arbitrary B/G and G/R spectral filter transition wavelengths. Calculate the R, G, and B color points and compare the gamut to the Rec.709 standard [123] (see Figure 3.63). Which choice of the transition wavelengths gives the highest luminous flux?

Solution 3.31

You may need to use a spreadsheet like RGB Split and set up the lamp spectrum as the spectral radiance of a black body (Eq. 3.194).

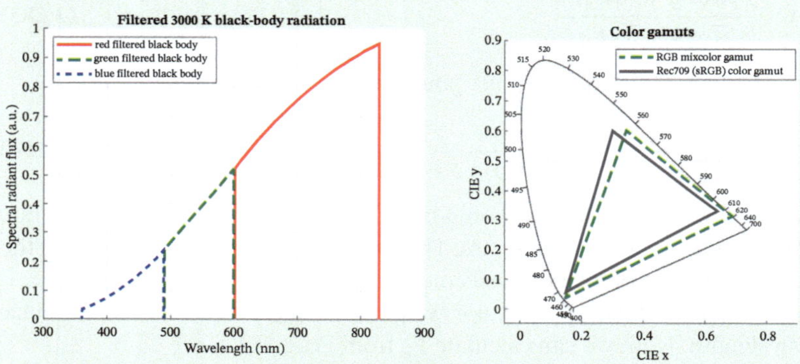

Figure 3.72 Three spectral intervals cut out from the spectrum of an incandescent lamp. The light from each part yields the color coordinates of a primary which form a gamut (green) that is then compared to the Rec.709 specification (gray).

You will see what happens to the gamut and total flux (total Y) for different spectral limits (Figure 3.72). You will learn that there is always a tradeoff between color saturation and high flux (spectral luminous efficacy at a given white point).

Dr. Wisenheimer:

In a lamp-based, time-sequential (one-chip) projector, the filters are arranged on a *color wheel*. The situation is a bit different, since the spectral widths of the segments can overlap in this case. The segment size is an important parameter, but segments must not be too small because of the limited switching speed of the pixels. Sometimes, there is also an additional white segment which increases the flux at the expense of color saturation.

Exercise 3.32: Remote phosphor

Consider an array of blue LEDs, completely enclosed by a hemispherical yellow phosphor shell as shown in Figure 3.73 (cf. [133, Figure 3]) The phosphor shell is thin, but thick enough to provide *full conversion* of the incident pump radiation.

In contrast to most white LEDs, where a phosphor layer is attached to a blue chip, this is a different way to realize phosphor conversion.

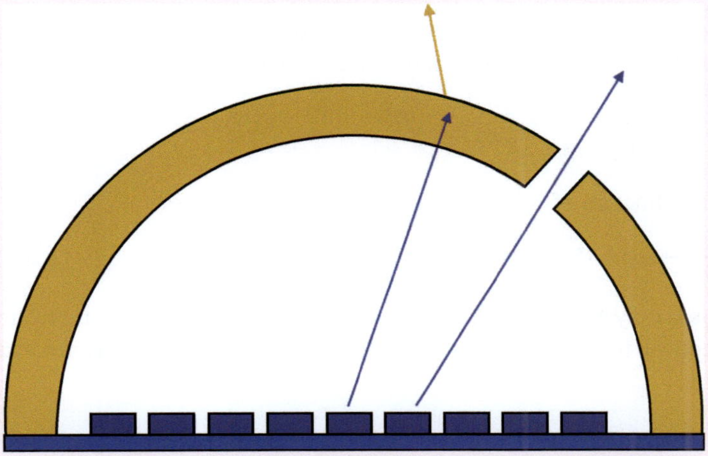

Figure 3.73 Remote phosphor shell on top of an array of blue LED chips. A hole is shown that allows some blue rays to pass.

If you drill some holes into the phosphor shell, the converted yellow light will mix with some transmitted blue light and yield white light (we assume perfect mixing in the far field).

What total hole area do you need to create in order to get white light on the black-body locus?

Quantity	Value
Phosphor conversion efficiency	$\eta_c = 30\%$
LED chip wavelength	$\lambda = 450$ nm
Converted light	$\lambda = 570$ nm

You may use a free tool such as the LED Color Calculator by John Selverian already mentioned above.

Solution 3.32

If we denote the fraction of holes in the phosphor shell by $0 < h < 1$, we have

$$\Phi_B = \Phi_0 h \tag{3.285}$$

and

$$\Phi_Y = \Phi_0(1 - h)\eta_c \tag{3.286}$$

We know from Exercise 3.29 that a factor of around 1.7 between the yellow and the blue radiant flux yields a color exactly on the black-body locus. This means that with

$$1.7h = (1 - h)\eta_c \tag{3.287}$$

we obtain a hole fraction of

$$h = \frac{\eta_c}{1.7 + \eta_c} = 0.15 \tag{3.288}$$

Of course, this is only a rough estimate. The holes will have side walls, there are multiple reflections in the cavity, and more. However, this estimate may serve as a starting point.

Exercise 3.33: LED color mix (1)

You have a white LED and a blue LED with the following data:

Quantity	Value
White LED color coordinates	$(x_W, y_W) = (0.31,\ 0.32)$
White LED luminous flux	$\Phi_W = 100\,\text{lm}$
Blue LED dominant wavelength	$\lambda_{\text{dom}} = 450\,\text{nm}$
Blue LED luminous flux	$\Phi_B = 100\,\text{lm}$

(a) Compute the X, Y, Z tristimulus values of the LEDs.
(b) Assume that you use both LEDs at full power with suitable color mixing optics attached.[b] Compute the x,y color coordinates of the mixed color.
(c) Where are these color coordinates located in the CIE xy "color shoe"? Can you believe this or is something wrong?

[b]Meaning a "side-by-side" combination into a target etendue at least as large as the combined etendues of both LEDs.

Solution 3.33

By now, we know that color weight is the key to simplifying color mixing calculations. We tend to use color weight measured in W, as outlined above, taking $K_{cd} = 683\,\mathrm{lm/W}$ into account. In fact, the ratios are the same when omitting K_{cd} to save a few steps, which is equivalent to measuring the color weight in lm. However, this is misleading and not very tidy, because the unit lm should be strictly reserved for luminous flux.

(a) We need all X, Y, Z tristimulus values for both sources. Remembering that $W = X/x = Y/y = Z/z = X + Y + Z$ (Eq. 3.264), we start by calculating the color weight of the white LED

$$W_\mathrm{W} = \frac{Y_\mathrm{W}}{y_W} = \frac{\Phi_\mathrm{W}}{K_\mathrm{M} y_\mathrm{W}} = \frac{100\,\mathrm{lm}}{683\,\frac{\mathrm{lm}}{\mathrm{W}} \cdot 0.32} = 0.457\,\mathrm{W} \tag{3.289}$$

The X tristimulus value is then

$$X_\mathrm{W} = x_\mathrm{W} W_\mathrm{W} = 0.31 \cdot 0.457\,\mathrm{W} = 0.142\,\mathrm{W} \tag{3.290}$$

The third color coordinate of the white LED is

$$z_\mathrm{W} = 1 - x_\mathrm{W} - y_\mathrm{W} = 0.37 \tag{3.291}$$

Hence,

$$Z_\mathrm{W} = z_\mathrm{W} W_\mathrm{W} = 0.37 \cdot 0.457\,\mathrm{W} = 0.169\,\mathrm{W} \tag{3.292}$$

Isn't it miraculous how easy and straightforward the color weight makes such a calculation?

Next, we address the blue LED. Take the values $(x_\mathrm{B}, y_\mathrm{B}) = (0.16, 0.02)$ at 450 nm from the provided file CIE_Planck.xlsx, or simply pick them from the xy color horseshoe diagram. The blue color weight is then

$$W_\mathrm{B} = \frac{Y_\mathrm{B}}{y_\mathrm{B}} = \frac{\Phi_\mathrm{B}}{K_\mathrm{M} y_\mathrm{B}} = \frac{100\,\mathrm{lm}}{683\,\frac{\mathrm{lm}}{\mathrm{W}} \cdot 0.02} = 7.32\,\mathrm{W} \tag{3.293}$$

This is much larger than the value for the white LED. The X tristimulus value of the blue LED is

$$X_\mathrm{B} = x_\mathrm{B} W_\mathrm{B} = 0.16 \cdot 7.32\,\mathrm{W} = 1.17\,\mathrm{W} \tag{3.294}$$

Therefore, we have

$$z_\mathrm{B} = 1 - x_\mathrm{B} - y_\mathrm{B} = 0.82 \tag{3.295}$$

and obtain

$$Z_\mathrm{B} = z_\mathrm{B} W_\mathrm{B} = 0.82 \cdot 7.32\,\mathrm{W} = 6.0\,\mathrm{W} \tag{3.296}$$

Solution 3.33 (Continued)

(b) Color-mixing both LEDs yields

$$X = X_W + X_B = 1.31\,\mathrm{W} \tag{3.297}$$

$$Y = Y_W + Y_B = \frac{200\,\mathrm{lm}}{683\,\frac{\mathrm{lm}}{\mathrm{W}}} = 0.293\,\mathrm{W} \tag{3.298}$$

$$Z = Z_W + Z_B = 6.169\,\mathrm{W} \tag{3.299}$$

The color weight of the mix, needed to compute its color coordinates, is thus

$$W = W_W + W_B = 7.78\,\mathrm{W} \tag{3.300}$$

Finally, we arrive at

$$x = \frac{X}{W} = \frac{1.31\,\mathrm{W}}{7.78\,\mathrm{W}} = 0.168 \tag{3.301}$$

$$y = \frac{Y}{W} = \frac{0.293\,\mathrm{W}}{7.78\,\mathrm{W}} = 0.037 \tag{3.302}$$

which is a location very close to the blue LED.

Blue LEDs contribute very little flux, but a lot of color. This is consistent with how the retina works [114]: For the brightness channel, computed directly in the retina, mostly the red- and green-sensitive cones contribute. The signal from the blue-sensitive cones contributes only to the yellow-blue channel. In a typical RGB mixed spectrum, and also in a typical white LED, the blue light contributes about 20% of the radiant flux, but only 3–4% of the luminous flux. You can see this also from the huge color weight of the blue LED.

Exercise 3.34: Tunable bicolor white

The simplest way to implement a white light source with a tunable color temperature is by mixing the light from cold-white and warm-white phosphors (Figure 3.74).

Since the Planck locus is curved, the color coordinates of the mix of both spectral distributions will end up slightly below the black-body curve, i.e. the mixed color will slightly tend to magenta. Calculate the *uv* distance to the Planck locus for the case of equal color weight.

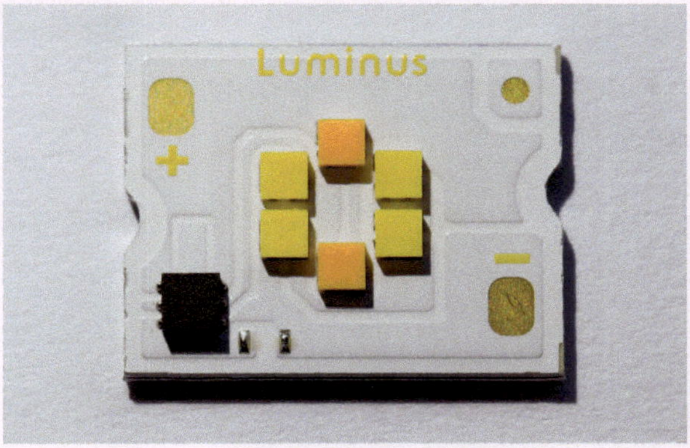

Figure 3.74 Tunable white LED module with cold white at 6500 K and warm white at 2700 K.

Solution 3.34

This task will require the use of a suitable software. Interactively, we recommend John Selverian's OSRAM Color Calculator [131]. You would create a 2700 K and a 6500 K black-body spectrum, save both as user spectra, compute and assign luminous fluxes via color weights (simply use the CIE y value – can you see why?) and look at the result ($d_{uv} = -0.0063$).

With Excel, you might use our provided file CIE_Planck.xlsx, enter the two color temperatures and obtain the color coordinate of their middle point (at $(x_M, y_M) = (0.364, 0.353)$) slightly below the black-body curve. Next, calculate the CCT of that point (approximately 4300 K) and its color coordinates ($(x, y) = (0.368, 0.369)$). A transformation of both to the uv system (also in the file) gives the same uv distance.

Last but not least, we present a solution using Julius' Matlab library. Here, we enter the two color temperatures and obtain from the spectra the color coordinates $(x_C, y_C) = (0.314, 0.324)$ (at 6500 K) $(x_W, y_W) = (0.460, 0.411)$ (at 2700 K). See Figure 3.75 for details and for the result.

Dr. Wisenheimer:

A few more lines of code would have revealed that the CIEDE2000 distance between the slightly magenta-ish mixed color and the same CCT, same flux black-body spectrum is about 9.5: a noticeable difference (theoretically, a CIEDE2000 value of 1 is "just noticeable", but in practice, the threshold is more like 2 or 3). High-end tunable white sources will need more than two internal source colors.

(Continued)

Solution 3.34 (Continued)

Exercise 3.34: distance to Planck of bicolor white LED mix

```
pl = PlanckLocus(); % a struct with Planck locus data
xy6500 = pl.xy_func(6500); % 6500K white point
xy2700 = pl.xy_func(2700); % 2700 K white point
% equal color weight mix is simply arithmetic mean
xymix = (xy6500 + xy2700)/2; % mixed color white point
[CCT,duv] = CCT_from_xy(xymix(1),xymix(2)); % mixed CCT and duv
fprintf("6500K x/y = (%0.3f,%0.3f)\n"...
      + "2700K x/y = (%0.3f,%0.3f)\n" ...
      + "Mixed x/y = (%0.3f,%0.3f)\n"...
      + "CCT = %0.0f K, duv = %0.4f",...
      xy6500(1),xy6500(2), xy2700(1),xy2700(2),...
      xymix(1),xymix(2), CCT,duv);
```

```
6500K x/y = (0.313,0.324)
2700K x/y = (0.460,0.411)
Mixed x/y = (0.387,0.367)
CCT = 3753 K, duv = -0.0063
```

```
% some plotting code ...
```

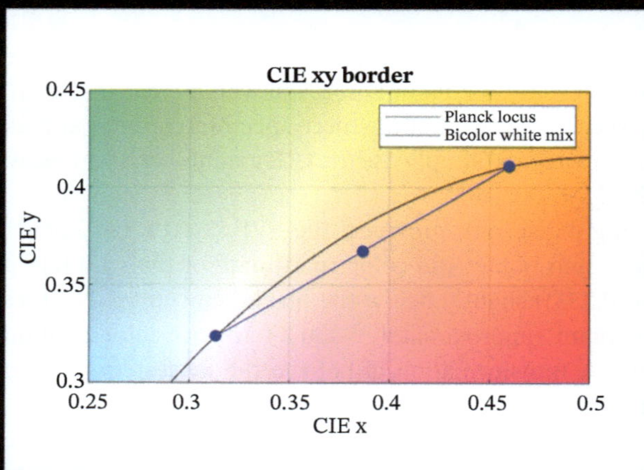

Figure 3.75 Solution to this exercise with a few lines of Matlab code, using Julius' library.

Exercise 3.35: LED color mix (2)

Highly efficient warm white light is the holy grail for indoor lighting.[c] We should prefer warm white (or even yellow!) LEDs in street lighting to avoid light pollution – which is stronger for bluish light [134] – and to reduce stress imposed onto insects. Another example is theater lighting: theatrical fixtures are in special need of both high-luminance and high-CRI warm-white sources.

Unfortunately, LEDs based on pumping warm-white phosphors are not very efficient because of conversion losses that are considerably larger than for cold-white LEDs. Therefore, there were many attempts to produce warm-white radiation by mixing of a variety of LED types.

Mint or lime LEDs are phosphor-converted blue LEDs (Figure 3.76) that deliver a greenish white in the region slightly above the black-body locus (Figure 3.77) [135]. They produce light more efficiently than green LEDs, and their emission can be beneficially mixed with light from red LEDs, creating a color in the warm-white region of the black-body curve with high efficacy and potentially also a high color-rendering index. This concept is known as "brilliant-mix" [136]. (One problem of this approach is that single-color red LEDs show a stronger efficiency drop with temperature, necessitating more complex electronics to compensate.) Assume a red LED at 625 nm and a mint LED at (0.333, 0.473). Which luminous flux ratio between the two sources is needed to hit the black-body curve?

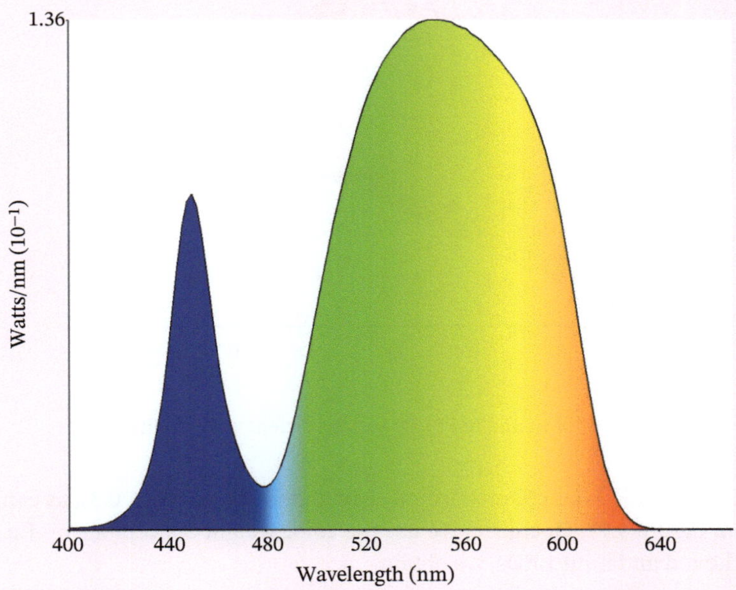

Figure 3.76 Typical spectrum of a phosphor-converted mint LED.

---

[c]True for western countries. In Asia, people rather prefer cold white light at home.

Solution 3.35

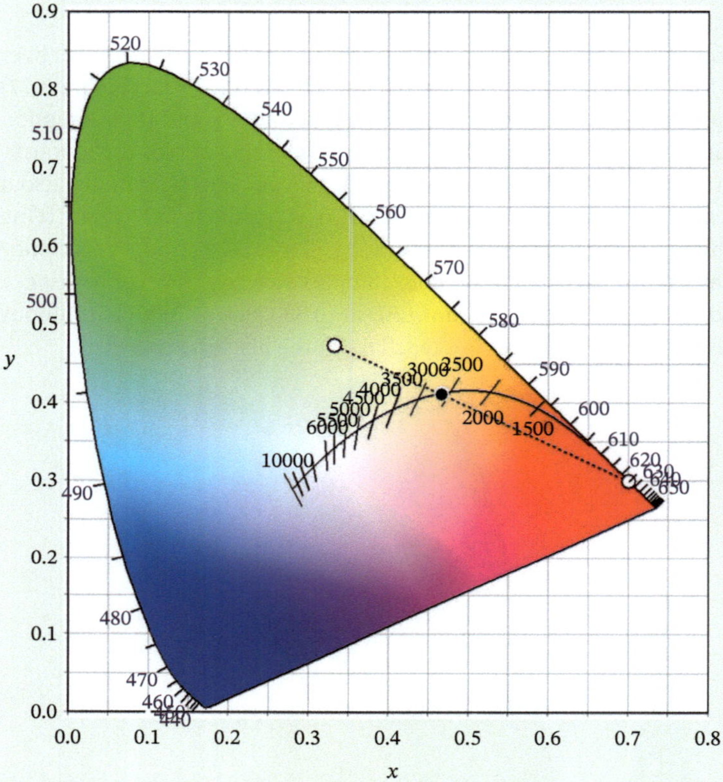

Figure 3.77 Mixing light of mint and red LEDs yields efficient warm white (left).

At 625 nm, the red LED has color coordinates of almost exactly (0.7, 0.3), as can be seen from Figure 3.77. Of course, we use the color-weight concept again, Eq. 3.264. For the red and mint LEDs, we obtain

$$W_r = \frac{Y_r}{y_r}, \qquad W_m = \frac{Y_m}{y_m} \tag{3.303}$$

The color coordinates of the light mix are (using Eq. 3.270)

$$\begin{aligned} x &= \frac{x_r W_r + x_m W_m}{W_r + W_m} \\ y &= \frac{y_r W_r + y_m W_m}{W_r + W_m} \end{aligned} \tag{3.304}$$

yielding the following system of equations for the color weights matching the color coordinates

$$W_r(x - x_r) + W_m(x - x_m) = 0$$
$$W_r(y - y_r) + W_m(y - y_m) = 0 \tag{3.305}$$

These equations define an implicit linear function in x, z, connecting the color coordinates of both sources,

$$\frac{x - x_r}{y - y_r} = \frac{x - x_m}{y - y_m} \tag{3.306}$$

This is equivalent to

$$x(y_r - y_m) + y(x_m - x_r) = x_m y_r - x_r y_m \tag{3.307}$$

which, with the given values, evaluates to

$$0.173\,x + 0.367\,y = 0.231 \tag{3.308}$$

This is a line connecting the color points of both LEDs. The line crosses the black-body curve at approximately 2600 K, with $(x, y) = (0.468, 0.409)$. Eq. 3.304 then becomes

$$-0.232W_r + 0.135W_m = 0$$
$$0.109W_r - 0.064W_m = 0 \tag{3.309}$$

both yielding a color-weight ratio of

$$W_r = 0.582W_m \tag{3.310}$$

and thus with Eq. 3.302

$$Y_r = 0.582\frac{y_r}{y_m}Y_m = 0.582\frac{0.3}{0.473} \cdot Y_m = 0.37\,Y_m \tag{3.311}$$

with a detailed calculation provided in Brillantmix.xlsx.

Keep in mind that this is the photometric or luminous flux ratio.

Dr. Wisenheimer:

Using additional blue LEDs, one can increase the color temperature of the light mix while keeping the efficacy and the CRI at high levels. Phosphor-converted green [137] LEDs can be used instead of mint as well, which is of special interest for high-luminance applications (Figure 3.78) [138]. Nowadays, mint LEDs are sometimes marketed as a white "Horti" (horticulture supplement) types.

(Continued)

Solution 3.35 (Continued)

This technology, depending on vendor, is known as Brilliant Mix or True White.

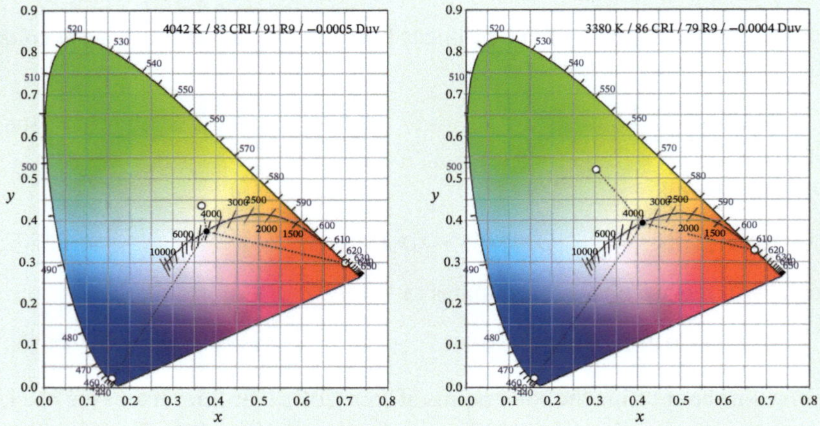

Figure 3.78 Color mix based on mint, red, and blue (left), and amber, blue and converted green (right).

References

1. Czapski, S. (1904). *Grundzüge der Theorie der optischen Instrumente nach Abbe.* Leipzig: Johann Ambrosius Barth.
2. Chaves, J. (2016). *Introduction to Nonimaging Optics*, 2e. CRC Press.
3. Nolte, D.D. (2010). The tangled tale of phase space. *Physics Today* 63 (4): 33–38. https://doi.org/10.1063/1.3397041.
4. Wolf, K.B. (2004). *Geometric Optics on Phase Space.* Berlin: Springer.
5. Reeb, O. (1942). Mitteilung des Fachausschusses "Optische Lichttechnik" der Deutschen Lichttechnischen Gesellschaft. *Zeitschrift für Instrumentenkunde* 62 (3).
6. Horan, J.J. (1975). The A-omega question. *Applied Optics* 14 (9): 2033. https://doi.org/10.1364/AO.14.002033.
7. Steel, W.H. (1974). Luminosity, throughput, or etendue? *Applied Optics* 13 (4): 704. https://doi.org/10.1364/AO.13.000704.
8. Goodman, D.S. (1993). The F-word in optics. *Optics and Photonics News* 4 (4): 38. https://doi.org/10.1364/OPN.4.4.000038.
9. Pratchett, T., Stewart, I., and Cohen, J.S. (2013). *The Science of Discworld.* London: Ebury.
10. Herkommer, A.M. (2014). The Lagrange invariant: a bridge between imaging and illumination design. In: *EOS Annual Meeting*, Berlin.

11 Rehn, H. and Muschaweck, J. (2023). Phase space diagrams in illumination optics. In: *International Optical Design Conference 2023*, Vol. 12798, 22–31, SPIE Press.

12 Liouville, J. (1838). Note sur la Théorie de la variation des constantes arbitraires. *Journal de Mathématiques Pures et Appliquées* 3: 342.

13 Jacobi, C.G.J. (1866). *Vorlesungen über Dynamik. Nebst fünf hinterlassenen Abhandlungen desselben*. Berlin: Georg Reimer. https://www.zvab.com/erstausgabe/Vorlesungen-Dynamik-Nebst-f\%C3\%BCnf-hinterlassenen-Abhandlungen/20271205461/bd.

14 Boltzmann, L. (1970). Weitere Studien über das Wärmegleichgewicht unter Gasmolekülen. In: *Kinetische Theorie II: Irreversible Prozesse Einführung und Originaltexte*, (ed. S.G. Brush), WTB Wissenschaftliche Taschenbücher, 115–225. Wiesbaden: Teubner. https://doi.org/10.1007/978-3-322-84986-1_3.

15 Brooker, G. (2002). Modern classical optics. *Oxford Master Series in Atomic, Optical and Laser Physics*, 1e. Oxford: Oxford University Press.

16 Marcuse, D. (1977). Loss analysis of single-mode fiber splices. *Bell System Technical Journal* 56 (5): 703–718. https://doi.org/10.1002/j.1538-7305.1977.tb00534.x.

17 Howell, J.R. (2023). *Configuration Factors*. Amazon Publishing. https://www.amazon.de/Configuration-Factors-John-Howell/dp/B0BZFJS9PD/ref=monarch_sidesheet_image.

18 Zimmer, H.G. (1970). *Geometrical Optics*. Springer.

19 Rehn, H. and Muschaweck, J.A. (2022). What light tubes are good for: connecting theories. *Proceeding of SPIE*, Vol. 12220. https://doi.org/10.1117/12.2631710.

20 Gross, H. (2005). *Handbook of Optical Systems*. Weinheim: Wiley.

21 Hottel, H.C. (1931). Radiant heat transmission between surfaces separated by non-absorbing media. *Transactions of the American Society of Mechanical Engineers* 53: 265–273.

22 Winston, R., Miñano, J.C., and Benitez, P. (2005). *Nonimaging Optics*. Academic Press.

23 Banda, S. (2002). *Lichttechnische Berechnungen: Grundlagen – Verfahren – Eigenschaften*. Renningen: Expert-Verlag.

24 Muschaweck, J. and Rehn, H. (2022). Design elements of illumination optics. *Components and Systems*. In preparation.

25 Muschaweck, J. and Rehn, H. (2019). Illumination design patterns for homogenization and color mixing. *Advanced Optical Technologies* 8 (1). https://doi.org/10.1515/aot-2018-0051.

26 Cassarly, W.J. (2014). IODC 2014 illumination design problem: the cinderella lamp. In: *Proceedings of SPIE, 9293*. https://doi.org/10.1117/12.2075913.

27 Rehn, H. and Muschaweck J. (2020). Étendue estimation for non-trivial geometry. In: *DGaO Proceedings*, Vol. 121.

28 Lambert, H. (1760). *Photometria Sive de Mensura et Gradibus Luminis, Colorum et Umbrae*. Augsburg: Klett.

29 Rehn, H. (2024). How to fill a light tube. In: *Illumination Optics VII*, Vol. 13022, 56–59, Bellingham, QA: SPIE Press.

30 Brick, P. and Rehn, H. (2018). Illumination device and method for producing an illumination device. https://patents.google.com/patent/WO2018197636A3/de?oq=WO+2018\%2f197636.

31 Hopkins, H.H. (1960). Optical system having cylindrical rod-like lenses, US3257902. https://patents.google.com/patent/US3257902A.

32 Rehn, H. (2016). Optical design for LED luminaires featuring Köhler illumination. In: *10th International Conference on Optics-Photonics Design & Fabrication*, Ravensburg-Weingarten.

33 Saleh, B.E.A. and Teich, M.C. (2007). Fundamentals of photonics. *Wiley Series in Pure and Applied Optics*. Hoboken, NJ: Wiley.

34 Heisenberg, W. (1927). Über den anschaulichen Inhalt der quantentheoretischen Kinematik und Mechanik. *Zeitschrift für Physik*, 43 (3): 172–198. https://doi.org/10.1007/BF01397280.

35 Keller, U. (2019). Interferenz und Kohärenz. https://ethz.ch/content/dam/ethz/special-interest/phys/quantum-electronics/ultrafast-laser-physics-dam/education/lectures/physik_II/nacht_der_welle/Lecture_QE_Keller_Kap4.pdf.

36 Kennard, E.H. (1927). Zur Quantenmechanik einfacher Bewegungstypen. *Zeitschrift für Physik* 44 (4): 326–352. https://doi.org/10.1007/BF01391200.

37 Ozawa, M. (2020). Heisenberg's original derivation of the uncertainty principle and its universally valid reformulations. http://arxiv.org/abs/1507.02010.

38 Boyd, R.W. (1983). *Radiometry and the Detection of Optical Radiation*. New York: Wiley.

39 Arecchi, A.V., Koshel, R.J., and Messadi, T. (2007). *Field Guide to Illumination*, Bellingham WA: SPIE Press.

40 McCluney, R. (2014). *Introduction to Radiometry and Photometry*, 2e. Boston: Artech House.

41 Nicodemus, F.E. (1963). Radiance. *American Journal of Physics* 31 (5): 368–377. https://doi.org/10.1119/1.1969512

42 Zalewski, E.F. (1995). Radiometry and photometry. In: *Handbook of Optics*, Vol. 2. New York: McGraw-Hill.

43 Clausius R. (1864). Ueber die Concentration von Wärme- und Lichtstrahlen und die Gränzen ihrer Wirkung. *Annalen der Physik* 197: 1–44.

44 Ries, H. (1982). Thermodynamic limitations of the concentration of electromagnetic radiation. *Journal of the Optical Society of America* 72 (3): 380. https://doi.org/10.1364/JOSA.72.000380.

45 Hagemann, V., Seidl, A., and Weidmann, G. (2020). Static ceramic phosphor assemblies for high power high luminance SSL-light sources for digital projection and specialty lighting. In: *Light-Emitting Devices, Materials, and Applications XXIV*, 53. San Francisco, CA: SPIE Press. https://doi.org/10.1117/12.2544000.

46 Russell, S. (2012). *The Architecture of Light*, 2e. Conceptnine.

47 Schouten, J.A. and Struik, D.J. (1921). Über das theorem von Malus-Dupin und einige verwandte theoreme in einer ein-dimensionalen Mannigfaltigkeit mit beliebiger quadratischer Maßbestimmung. *Rendiconti del Circolo Matematico di Palermo* 45 (1): 313–331. https://doi.org/10.1007/BF03018145.

48 Marle, C.M. (2014). A direct proof of Malus' theorem using the symplectic structure of the set of oriented straight lines. http://arxiv.org/abs/1409.0381.

49 Weatherburn, C.E. (1927). On congruences of curves. *Tohoku Mathematical Journal, First Series* 28: 114–125.

50 Born, M. and Wolf, E. (1999). *Principles of Optics*, 7e. Cambridge University Press.

51 Muschaweck, J. and Wiesmann, C. (2013). Extracting light out of LEDs. *Advanced Optical Technologies* 2 (4), 291–300. https://doi.org/10.1515/aot-2013-0027.

52 Reeb, O. (1952). Zur Frage der photometrischen Grenzentfernung. *Optik* 9.

53 Helbig, E. (1972). *Grundlagen der Lichtmesstechnik.* Leipzig: Akademische Verlags-Ges Geest & Portig.

54 Dotreppe, G.M., Audenaert, J., Scheir, G.H. et al. (2023). Off-axis limiting photometric distance of Lambertians and narrow beams. *Lighting Research & Technology* 56 (6): 550–569. https://doi.org/10.1177/14771535231208933.

55 Muschaweck, J. (2023). Virtual focus of a ray data source. *Optical Engineering* 62 (07). https://doi.org/10.1117/1.OE.62.7.070401.

56 Hecht, E. (2010). *Optics.* San Francisco: Addison-Wesley.

57 Artal, P. (2016). The eye as an optical instrument. In: *Optics in Our Time*, 285–297. Cham: Springer International Publishing. https://doi.org/10.1007/978-3-319-31903-2_12 .

58 Troland, L.T. (1917). On the measurement of visual stimulation intensities. *Journal of Experimental Psychology* 2 (1): 1–33. https://doi.org/10.1037/h0071652.

59 International Electrotechnical Vocabulary. (2023). https://www.electropedia.org/iev/iev.nsf/welcome?openform.

60 Miller, J.L., Schwertz, K., Friedman, E.J. et al. (2020). *Photonics Rules of Thumb*, 3e. SPIE Press. https://doi.org/10.1117/3.2553485. https://www.spiedigitallibrary.org/ebooks/PM/Photonics-Rules-of-Thumb-Third-Edition/eISBN-9781510631762/10.1117/3.2553485.

61 Thibos, L.N., Lopez-Gil, N., and Bradley, A. (2018). What is a troland? *Journal of the Optical Society of America* A 35 (5): 813. https://doi.org/10.1364/JOSAA.35.000813.

62 Max, R. (1945). The $\cos^4$ law of illumination. *Journal of the Optical Society of America* 35 (4): 283. https://doi.org/10.1364/JOSA.35.000283.

63 Rehn, H. (2024). Fundamental laws revisited: the $\cos^4$ law of radiometry. In: *DGaO-Jahrestagung*, Aachen. https://www.dgao-proceedings.de/download/125/125_b17.pdf.

64 Max P. (1901). Über das Gesetz der Energieverteilung im Normalspectrum. *Annalen der Physik* 309 (3): 553–563. https://doi.org/10.1002/andp.19013090310.

65 Planck, M. and Masius, M. (1914). *The Theory of Heat Radiation.* Philadelphia: P. Blakiston's Son & Co.

66 Wolfe, W.L. and Wolfe. W. (2024). *Optical Radiometry and Applications.* Bellingham, WA: SPIE Press.

67 Hartmann, J. (2006). Correct consideration of the index of refraction using blackbody radiation. *Optics Express* 14 (18): 8121. https://doi.org/10.1364/OE.14.008121.

68 Kittel, C. and Kroemer, H. (1980). *Thermal Physics*, 2e. 28. print edition, New York, NY: Freeman.

69 Bierhuizen, S., Krames, M., Harbers, G. et al. (2007). Performance and trends of high power light emitting diodes, 66690B. https://doi.org/10.1117/12.735102.

70 Fu, L. (2006). *Increasing the Brightness of Light Sources.* PhD thesis. Marburg. https://d-nb.info/982573987/34.

71 Batchelder, J.S., Zewai, A.H., and Cole, T. (1979). Luminescent solar concentrators 1: theory of operation and techniques for performance evaluation. *Applied Optics* 18 (18): 3090. https://doi.org/10.1364/AO.18.003090.

72 Brukilacchio, T.J. (2008). Light emitting diode illumination system, US8098375B2. https://patents.google.com/patent/US8098375B2.

73 Michael, P.R., Johnston, D.E., and Moreno, W. (2020). A conversion guide: solar irradiance and lux illuminance. *Journal of Measurements in Engineering* 8 (4): 153–166. https://doi.org/10.21595/jme.2020.21667.

74 Prša, A., Harmanec, P., Torres, G. et al. (2016). Nominal values for selected solar and planetary quantities: IAU 2015 resolution B3. *The Astronomical Journal* 152 (2): 41. https://doi.org/10.3847/0004-6256/152/2/41.

75 Gueymard, C.A. (2004). The sun's total and spectral irradiance for solar energy applications and solar radiation models. *Solar Energy* 76 (4): 423–453. https://doi.org/10.1016/j.solener.2003.08.039.

76 Martinsons, C., Behar-Cohen, F., Bergen, T. et al. (2024). Reconsidering the spectral distribution of light: do people perceive watts or photons? *Lighting Research & Technology* 56 (8): 886–899. https://doi.org/10.1177/14771535241246060.

77 Marr, J.M. and Wilkin, F.P. (2012). A better presentation of Planck's radiation law. *American Journal of Physics* 80 (5): 399–405. https://doi.org/10.1119/1.3696974.

78 Meschede, D. and Gerthsen, C. (2015). *Gerthsen Physik*, 25e. Berlin [u.a.]: Springer Spektrum.

79 Kume, A. (2017). Importance of the green color, absorption gradient, and spectral absorption of chloroplasts for the radiative energy balance of leaves. *Journal of Plant Research* 130 (3): 501–514. https://doi.org/10.1007/s10265-017-0910-z.

80 Strain, H.H., Thomas, M.R., and Katz, J.J. (1963). Spectral absorption properties of ordinary and fully deuteriated chlorophylls a and b. *Biochimica et Biophysica Acta* 75: 306–311. https://doi.org/10.1016/0006-3002(63)90617-6.

81 Virtanen, O., Constantinidou, E., and Tyystjärvi, E. (2022). Chlorophyll does not reflect green light – how to correct a misconception. *Journal of Biological Education* 56 (5): 552–559. https://doi.org/10.1080/00219266.2020.1858930.

82 Arp, T.B., Kistner-Morris, J., Aji, V. et al. (2020). Quieting a noisy antenna reproduces photosynthetic light-harvesting spectra. *Science* 368 (6498): 1490–1495. https://doi.org/10.1126/science.aba6630.

83 Ortega, R.P. (2020). Why are plants green? To reduce the noise in photosynthesis. https://www.quantamagazine.org/why-are-plants-green-to-reduce-the-noise-in-photosynthesis-20200730.

84 Wu, Y., Wang, Z., Cai, W., et al. (2018). The absolute reflectance and new calibration site of the moon. *The Astronomical Journal* 155 (5): 213. https://doi.org/10.3847/1538-3881/aabaf5.

85 Lachish, U. (2018). The Sun and the Moon a Riddle in the Sky. https://arxiv.org/abs/1808.01024.

86 Kirchhoff, G. (1860). Ueber das Verhältniss zwischen dem Emissionsvermögen und dem Absorptionsvermögen der Körper für Wärme und Licht. *Annalen der Physik und Chemie* 185 (2): 275–301. https://doi.org/10.1002/andp.18601850205.

87 Willers, C.J. (2013). *Electro-Optical System Analysis and Design: A Radiometry Perspective*. Number 236, Bellingham, WA: SPIE Press.

88 Bermel, P., Lee, J., Joannopoulos, J.D. et al. (2012). Selective solar absorbers. *Annual Review of Heat Transfer* 15. https://doi.org/10.1615/AnnualRevHeatTransfer.2012004119.

89 Zhang, J., Wang, C., Shi, J. et al. (2022). Solar selective absorber for emerging sustainable applications. *Advanced Energy and Sustainability Research* 3 (3): 2100195. https://doi.org/10.1002/aesr.202100195.

90 Stockman, A. and Sharpe, T. (2006). Into the twilight zone: the complexities of mesopic vision and luminous efficiency. *Ophthalmic and Physiological Optics* 26 (3): 225–239. https://doi.org/10.1111/j.1475-1313.2006.00325.x.

91 Thapan, K., Arendt, J., and Skene, D.J. (2001). An action spectrum for melatonin suppression: evidence for a novel non-rod, non-cone photoreceptor system in humans. *The Journal of Physiology* 535 (1): 261–267. https://doi.org/10.1111/j.1469-7793.2001.t01-1-00261.x.

92 Lucas, R.J., Peirson, S.N., Berson, D.M. et al. (2014). Measuring and using light in the melanopsin age. *Trends in Neurosciences* 37 (1): 1–9. https://doi.org/10.1016/j.tins.2013.10.004.

93 Alkozi, H.A. (2019). Melatonin and melanopsin in the eye: friends or foes? *Anales de la Real Academia Nacional de Farmacia* 85 (1): 49–59.

94 Internationale Beleuchtungskommission. (ed). (1993). *CIE Collection in Photobiology and Photochemistry: Abstracts of CIE Publications Prepared within Division 6*. Number 106 in Technical Report / International Commission on Illumination, Vienna: CIE Central Bureau. https://cie.co.at/publications/cie-collection-photobiology-and-photochemistry.

95 Cullen, A.P. (2002). Photokeratitis and other phototoxic effects on the cornea and conjunctiva. *International Journal of Toxicology* 21 (6): 455–464. https://doi.org/10.1080/10915810290169882.

96 Schmalwieser, A.W., Wallisch, S., and Diffey, B. (2012). A library of action spectra for erythema and pigmentation. *Photochemical & Photobiological Sciences* 11 (2): 251–268. https://doi.org/10.1039/c1pp05271c.

97 Webb, A.R., Terenetskaya, I.P., Holick, M.F. et al. (2023). Previtamin D action spectrum: challenging CIE towards a standard. *Lighting Research & Technology* 55 (4–5): 487–497. https://doi.org/10.1177/14771535221122937.

98 Parrish, J.A. and Jaenicke, K.F. (1981). Action spectrum for phototherapy of psoriasis. *The Journal of Investigative Dermatology* 76 (5): 359–362. https://doi.org/10.1111/1523-1747.ep12520022.

99 Forbes, P.D., Cole, C.A., and deGruijl, F. (2021). Origins and evolution of photocarcinogenesis action spectra, including germicidal UVC. *Photochemistry and Photobiology* 97 (3): 477–484. https://doi.org/10.1111/php.13371.

100 Bullough, J.D. (2000). The blue-light hazard: a review. *Journal of the Illuminating Engineering Society* 29 (2): 6–14. https://doi.org/10.1080/00994480.2000.10748312.

101 Hessling, M., Koelbl, P.S., and Lingenfelder, C. (2018). LED illumination - a hazard to the eye? *Optik & Photonik* 13 (4): 40–44. https://doi.org/10.1002/opph.201800029.

102 Leontopoulos, M., Leontopoulos, S., and Knoop, M. (2023). Consideration of blue light hazard for virtual reality head mounted displays. *Lighting Research & Technology* 55 (7–8): 759–771. https://doi.org/10.1177/14771535221145801.

103 TEMIS – Daily UV index and UV dose data from satellite. https://www.temis.nl/uvradiation/UVarchive/uvncfiles.php.

104 Webb, A.R., Slaper, H., Koepke, P. et al. (2011). Know your standard: clarifying the CIE erythema action spectrum. *Photochemistry and Photobiology* 87 (2): 483–486. https://doi.org/10.1111/j.1751-1097.2010.00871.x.

105 Bernhard, G., Mayer, B., Seckmeyer, G. et al. (1997). Measurements of spectral solar UV irradiance in tropical-Australia. *Journal of Geophysical Research: Atmospheres* 102 (D7): 8719–8730. https://doi.org/10.1029/97JD00072.

106 Ashdown, I. (2020). Specifying LED colors for horticultural lighting. *FIRES,* 11. https://www.ies.org/fires/specifying-led-colors-for-horticultural-lighting.

107 Cinque, G., Croce, R., and Bassi, R. (2000). Absorption spectra of chlorophyll a and b in Lhcb protein environment. *Photosynthesis Research* 64 (2/3): 233–242. https://doi.org/10.1023/A:1006467617697.

108 Brown, C.S., Schuerger, A.C., and Sager, J.C. (1995). Growth and photomorphogenesis of pepper plants under red light-emitting diodes with supplemental blue or far-red lighting. *Journal of the American Society for Horticultural Science* 120 (5): 808–813. https://doi.org/10.21273/JASHS.120.5.808.

109 Ahamadi, L., Matloobi, M., and Motallebi-Azar, A. (2023). Light source spectrum influences long-term flowering cycles and visual appearance in African violet (Saintpaulia ionantha Wendl.). *Lighting Research & Technology,* 1477153523 1183884. https://doi.org/10.1177/14771535231183884.

110 Dubbelman, T.M.A.R. and Spikes, J.D. (1994). Photosensitizers. In: *Photobiology in Medicine*, 61–70. Boston, MA: Springer US. https://doi.org/10.1007/978-1-4899-1313-5_4.

111 Agati, G. and Fusi, F. (1990). New trends in photobiology recent advances in bilirubin photophysics. *Journal of Photochemistry and Photobiology B: Biology* 7 (1): 1–14. https://doi.org/10.1016/1011-1344(90)85138-M.

112 Cremer, R.J., Perryman, P.W., and Richards, D.H. (1958). Influence of light on the hyperbilirubinæmia of infants. *The Lancet* 271 (7030): 1094–1097. https://doi.org/10.1016/S0140-6736(58)91849-X.

113 Tan, K.L. (1991). Phototherapy for neonatal jaundice. *Clinics in Perinatology* 18 (3): 423–439. https://doi.org/10.1016/S0095-5108(18)30506-2.

114 Elliott, A. and Fairchild, M. (2015). *Handbook of Color Psychology*. Cambridge: Cambridge University Press.

115 Wyszecki, G. and Stiles, W.S. (2000). *Color Science: Concepts and Methods, Quantitative Data and Formulae,* 2e. New York: Wiley.

116 Schanda, J. (ed.) (2007). *Colorimetry: Understanding the CIE System*, Hoboken, NJ: Wiley.

117 Grassmann, H. (1853). Zur Theorie der Farbenmischung. *Annalen der Physik und Chemie* 165 (5): 69–84. https://doi.org/10.1002/andp.18531650505.

118 Wright, W.D. (1929). A re-determination of the trichromatic coefficients of the spectral colours. *Transactions of the Optical Society* 30 (4): 141–164. https://doi.org/10.1088/1475-4878/30/4/301.

119 Guild, J. (1931). The colorimetric properties of the spectrum, 1. *Philosophical Transactions of the Royal Society A* 230: 681–693.

120 Poynton, C. Frequently asked questions about color. https://poynton.ca/PDFs/ColorFAQ.pdf

121 Dowling, K.J. (2010). Color management methods and apparatus for lighting devices. https://patents.google.com/patent/US7710369B2/en.

122 Masaoka, K. and Nishida, Y. (2015). Metric of color-space coverage for wide-gamut displays. *Optics Express* 23 (6): 7802. https://doi.org/10.1364/OE.23.007802.

123 ITU-R. (2015). BT.709: Parameter values for the HDTV standards for production and international programme exchange. https://www.itu.int/rec/R-REC-BT.709-6-201506-I/en.

124 MacAdam, D.L. (1942). Visual sensitivities to color differences in daylight*. *Journal of the Optical Society of America* 32 (5): 247–274. https://doi.org/10.1364/JOSA.32.000247.

125 MacAdam, D.L. (1942). Visual sensitivities to color differences in daylight*. *Journal of the Optical Society of America* 32 (5): 247. https://doi.org/10.1364/JOSA.32.000247.

126 International Lighting Vocabulary. (2011). https://cie.co.at/publications/international-lighting-vocabulary.

127 Smet, K., Royer, M., Baxter, D. et al. (2023). Recommended method for determining the correlated color temperature and distance from the planckian locus of a light source. *LEUKOS* 20 (2): 1–15. https://doi.org/10.1080/15502724.2023.2248397.

128 McCamy, C.S. (1992). Correlated color temperature as an explicit function of chromaticity coordinates. *Color Research & Application* 17 (2): 142–144. https://doi.org/10.1002/col.5080170211.

129 Schlotter, P., Schmidt, R., and Schneider, J. (1997). Luminescence conversion of blue light emitting diodes. *Applied Physics A* 64 (4): 417–418. https://doi.org/10.1007/s003390050498.

130 Goetzberger, A. and Greubel, W. (1977). Solar energy conversion with fluorescent collectors. *Applied Physics* 14.

131 Selverian, J. (2017). ColorCalculator software. https://www.researchgate.net/publication/320930275_ColorCalculator_software.

132 MacAdam, D.L. (1950). Maximum attainable luminous efficiency of various chromaticities. *Journal of the Optical Society of America* 40 (2): 120. https://doi.org/10.1364/JOSA.40.000120.

133 Wiesmann, C., Linkov, A., and Muschaweck, J. (2012). Estimating the performance of remote phosphor SSL devices by simulations. *Proceeding of SPIE* 85502. https://doi.org/10.1117/12.2001607.

134 Kyba, C., Ruby, A., Kuechly, H. et al. (2020) Direct measurement of the contribution of street lighting to satellite observations of nighttime light emissions from urban areas. *Lighting Research & Technology* 147715352095846. https://doi.org/10.1177/1477153520958463.

135 LUMILEDS. (2021). Lime and Mint LEDs: Unique green options for boosting efficiency, quality, and value of color-mixing luminaires. https://lumileds.com/company/blog/lime-and-mint-leds-unique-green-options-for-boosting-efficiency-quality-and-value-of-color-mixing-luminaires.

136 LEDs Magazine. (2011). Osram Opto unveils Brilliant-Mix LED mixing concept, https://www.ledsmagazine.com/architectural-lighting/retail-hospitality/article/16697672/osram-opto-unveils-brilliantmix-led-mixing-concept.

137 Philips. (2013). Lighting world first: Philips breaks 200 lumens per watt barrier. https://www.philips.com/consumerfiles/newscenter/main/design/resources/pdf/Inside-Innovation-Backgrounder-Lumens-per-Watt.pdf.

138 Laabs, H., Mehl, O., and Rehn, H. (2016). High-end white light sources fiber optics technology for medical applications. *LED Professional*. https://www.led-professional.com/resources-1/articles/high-end-white-light-sources-fiber-optics-technology-for-medical-applications.

139 Kokoschko, R. (1998). Vorrichtung zur Lichtsammlung, -Konzentrierung und -Leitung von direkter und diffuser Strahlung. EP0875724A2. https://patents.google.com/patent/EP0875724A2/.

140 Koulechoff, J. (2017). Device and method for increasing the energy density of radiation.WO2018113814A2. https://patents.google.com/patent/WO2018113814A2.

4

Practice

> In theory, there is no difference between theory and practice. In practice, there is.[1]

We agree with this statement. One reason for this is that in theory, you can get away with things because you can make suitable assumptions or neat approximations. In practice, reality will eventually catch up with you, and sometimes haunt you. The most dangerous situations occur when a theoretical result or a simulation is just a little too good to be true. Let's look at some examples.

We start with some rather basic illumination calculations, later extending our view to specialty lighting and optical instruments.

4.1 Light Sources

Exercise 4.1: Light bulb

A 40 W incandescent lamp with a bulb diameter of $d = 50$ mm is operating with an efficacy of $\eta = 10$ lm/W (a realistic value, unfortunately). Assume an ideally scattering spherical bulb realizing isotropic intensity and homogeneous luminance. Small, nearly planar surface patches of the bulb will then qualify as Lambertian emitters (see Knowledge Corner Topic 3.12). Calculate the luminous intensity I_v of the bulb and the surface exitance E_v and luminance L_v of the bulb surface.

1 Benjamin Brewster, in The Yale Literary Magazine, Vol 47, p. 202, 1882

Engineering Illumination Optics: Case Studies and Exercises, First Edition. Henning Rehn and Julius Muschaweck.
Published 2026 by Wiley-VCH GmbH.

Solution 4.1

The total flux of the bulb is

$$\Phi_v = \eta \cdot P_{el} = 400\,\text{lm} \tag{4.1}$$

The whole lamp is modeled as radiating isotropically, with an intensity of

$$I_v = \frac{\Phi}{\Omega} = \frac{400\,\text{lm}}{4\pi\ \text{sr}} = 32\ \text{cd} \tag{4.2}$$

The bulb surface area is

$$A = 4\pi r^2 = 4\pi \cdot (25\,\text{mm})^2 \approx 7850\,\text{mm}^2 \tag{4.3}$$

Luminous exitance is the flux per surface area:

$$E_v = \frac{\Phi_v}{A} = \frac{400\,\text{lm}}{7850\,\text{mm}^2} = 0.05\,\frac{\text{lm}}{\text{mm}^2} \approx 50\,\text{klx} \tag{4.4}$$

and surface luminance is

$$L_v = \frac{\Phi_v}{U_S} = \frac{E_v}{\Omega_p} = \frac{50\,\text{klx}}{\pi\ \text{sr}} \approx 16\,\frac{\text{kcd}}{\text{m}^2} = 16\,\text{knit} \tag{4.5}$$

Dr. Wisenheimer:

Theory and practice: An HDR photo of such a bulb would show just how inhomogeneous the luminance really is. The *average* value is correct, though!

Exercise 4.2: Light bulb vs. moon

At what distance from an observer do we have to place the bulb of the previous exercise to appear as bright as the full Moon?

Solution 4.2

As Dr. Wisenheimer will remark in Exercise 5.2, using the fuzzy word “bright” in a radiometric question is probably not a good idea. Let’s see!

We compare the visual impression based on retinal illuminance for both sources. Consider a distant circular source with luminance L_S, such that light is coming toward the eye from a small cone with half angle θ and projected solid angle Ω_p. The etendue of this light tube is (small-angle approximation)

$$U_T = A_P\Omega_p = A_P\pi\ \text{sr}\sin^2\theta \tag{4.6}$$

with the eye's pupil area A_P. The flux in this light tube is, of course, $\Phi = L_S U_T$. The retinal image radius is $r_T = f_{eye} \tan\theta$ (first order imaging, cf. Exercise 3.16), and the retinal image area is $A_T = r_T^2 \pi$, using the eye's focal length f_{eye}.

We obtain the retinal irradiance E_T from

$$E_T = \frac{\Phi}{A_T} = \frac{U_T L_S}{\pi r_T^2} = \frac{A_P \pi \text{ sr} \sin^2\theta}{\pi f_{eye}^2 \tan^2\theta} L_S \approx \frac{A_P}{f_{eye}^2} L_S \text{ sr} \tag{4.7}$$

Since A_P and f_{eye} are properties of the eye, not the source, retinal illuminance scales only with source luminance (as long as the iris diameter does not change, and the source does fill the small $\pm\theta$ cone).

In particular, retinal illuminance *does not depend on distance*!

The etendue concept leads us to the conclusion that *the question was indeed misleading.*

Here I am to help you!

Accordingly, we must compare luminances. The luminous exitance of the Moon is calculated from solar irradiance as

$$E_M = K_v E_S \varrho = 101 \frac{\text{lm}}{\text{W}} \cdot 1.36 \frac{\text{kW}}{\text{m}^2} \cdot 10\,\% = 13.7\,\text{klx} \tag{4.8}$$

with the solar irradiance, the luminous efficacy (Eq. 3.217) of solar radiation in free space (Knowledge Corner Topic 3.19), and the Moon's albedo introduced in Exercise 4.19. Assuming Lambertian scattering, the luminance is then

$$L_M = \frac{E_M}{\pi \text{ sr}} = 4360\,\text{nit} \tag{4.9}$$

which is less than that of the lamp. The lamp will appear brighter, even if we wrongly consider it homogeneous!

Exercise 4.3: Light bulb in a parabolic reflector

The bulb from Exercise 4.1 is placed at the focal point of a rather large parabolic reflector (Figure 4.1). The parameters are as follows:

Quantity	Value
Bulb semi-diameter	$R = 25$ mm
Bulb flux	$\Phi_S = 400$ lm
Parabola focal length	$f = 250$ mm
Parabola front diameter	$d = 8f = 2$ m

(Continued)

Exercise 4.3: (Continued)

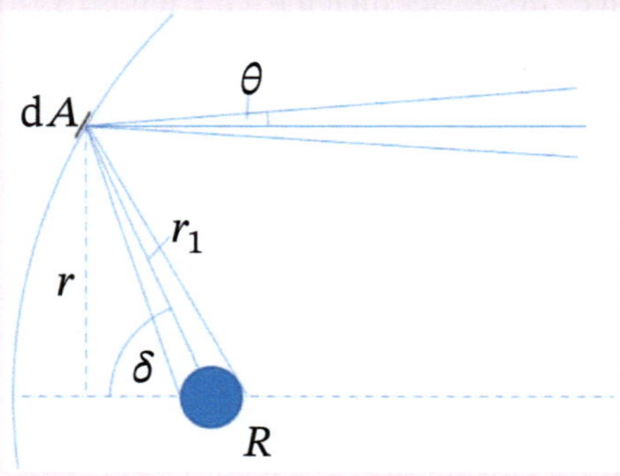

Figure 4.1 Bulb inside a parabolic reflector (cf. Figure 2.32).

Calculate the luminous intensity of the reflected light in the direction of the reflector axis.

Solution 4.3

The calculation is easy since only the intensity on the axis is asked for. It does not matter that the cones from different locations on the reflector are of different sizes.

The bulb luminance is

$$L = \frac{\Phi_S}{U} = \frac{\Phi_S}{4\pi R^2 \cdot \pi\,\mathrm{sr}} = \frac{400\,\mathrm{lm}}{\pi^2\,\mathrm{sr} \cdot 2500\,\mathrm{mm}^2} = \frac{51}{\pi}\,\mathrm{knit} \tag{4.10}$$

The reflector's front area is

$$A = \pi(4f)^2 = \pi\,\mathrm{m}^2 \tag{4.11}$$

Hence, the intensity is

$$I = LA = 51\,\mathrm{kcd} \tag{4.12}$$

Theory and practice: The peak, center luminance of the actual lamp might well be ten times larger, and the same factor would apply to the on-axis intensity!

Of course, we compare this to a simulation.

Simulation 4.1: Light bulb in a parabolic reflector

We set up the source and the parabolic reflector ($\rho = 2f$, vertex at $-f$, diameter $8f = 2$ m).

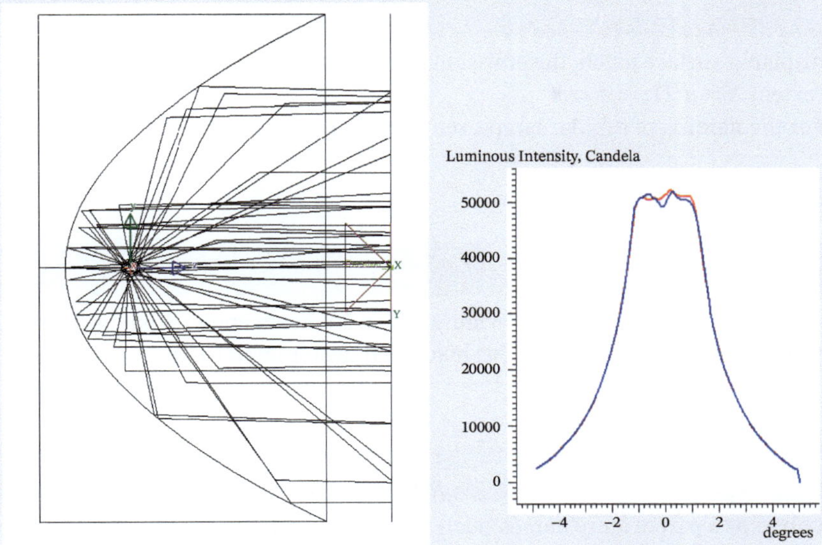

Figure 4.2 LightTools simulation and result.

We obtain 52 kcd (Figure 4.2), pretty close to the calculated estimate above. We provide the simulation file for your reference.

Next, we consider a different source.

Exercise 4.4: Fluorescent tube

A T5 lamp ($l = 1450$ mm, $d = 16$ mm) produces $\Phi = 5000$ lm out of 50 W electrical power. Calculate its surface luminance, assuming homogeneous emission from the phosphor layer.

Solution 4.4

The surface area is

$$A = \pi\, d\, l = 73\,000\,\text{mm}^2 \tag{4.13}$$

and its luminance is obtained via

$$L_\text{S} = \frac{\Phi_\text{S}}{U_\text{S}} = \frac{\Phi_\text{S}}{A \cdot \pi\,\text{sr}} = \frac{5000\,\text{lm}}{73\,000\,\text{mm}^2 \cdot \pi\,\text{sr}} = 22\,\text{knit} \tag{4.14}$$

Can we really compute the etendue of a cylinder in air by simply multiplying its area by π sr? Yes, we can. Angular extent is *local*, and locally, from each small,

(Continued)

Solution 4.4 (Continued)

near-planar surface patch, the emission goes into the full hemisphere with angular extent $X = n^2\Omega_p = \pi$ sr.

For the naming of tubular lamps, see Knowledge Corner Topic 4.1.

Let us consider another example.

Exercise 4.5: MR16 lamp

Calculate the etendue of an MR16 lamp with a 30° beam angle. What is the maximum possible flux of an MR16 30° halogen reflector lamp running at 2700 K?

Knowledge Corner Topic 4.1: Lamp diameter naming

Numbers as a part of lamp names, such as in T5 or MR16, are derived from a base unit of one-eighth of an inch.

"MR16" means "Multifaceted Reflector" 16 with a diameter of

$$d = 16 \cdot (1/8'') = 16 \cdot 3.175\,\text{mm} = 50.8\,\text{mm} = (2'') \tag{4.15}$$

This corresponds to a reflector front circumference of 16 cm. Again the number 16, but this is no coincidence since $\pi \approx 25.4/8$.

A "Par64" lamp features a parabolic reflector of a diameter of about 20 cm.

A "T5" tubular lamp has a diameter of $5 \cdot (1/8'') = 5 \cdot 3.175\,\text{mm} \approx 16\,\text{mm}$.

A "T1¾" radial LED has a diameter of $1.75 \cdot (1/8'') = 1.75 \cdot 3.175\,\text{mm} = 5.55\,\text{mm}$.

Solution 4.5

The etendue of an MR16 lamp with a 30° beam angle is

$$U = A \cdot X = (25.4\,\text{mm})^2\pi \cdot (\sin 15°)^2\pi\,\text{sr} = 426\,\text{mm}^2\text{sr} \tag{4.16}$$

Assuming that the filament of the halogen bulb is a black-body emitter at 2700 K, we get a luminance of 12 Mnit from Eq. 3.197. Since the filament is a coil (with gaps seen from the side), and the tungsten surface is gray and not black, the *effective luminance* (Eq. 3.92) will be rather *half* of that value (not a proof but our experience, see also Knowledge Corner Topic 3.21).

Consequently, the upper flux limit will be in the order of

$$\Phi = L \cdot U = 6\,\text{Mnit} \cdot 426\,\text{mm}^2\text{sr} \approx 2500\,\text{lm} \tag{4.17}$$

which would require at least a 150 W nominal power halogen burner, operated at 2700 K with slightly reduced power.

But in reality, a burner as strong as this cannot be operated in such a small volume. The highest commercially available flux in this lamp size is rather about $50\,\text{W} \cdot 20\,\text{lm/W} = 1000\,\text{lm}$.

In practice, there are limits other than merely etendue conservation.

Dr. Wisenheimer:

In the preceding exercises you encountered typical luminance values for general lighting. Filaments operate in the order of $L \approx 10$ Mnit at 2700 K. Some specialty halogen lamps feature up to 60 Mnit at 3300 K, at the cost of a shorter lifetime.

In contrast, *translucent* lamps or luminaire surfaces will deliver around $L \approx 20$ knit, as we have seen in Exercise 4.1.

LED retrofits for general lighting feature very similar low luminance values, much as their legacy predecessors; see next exercise.

Exercise 4.6: Incandescent lamp efficacy

An incandescent lamp with a filament temperature of $T = 2700$ K emits 89% of its radiated power beyond 780 nm.

To increase efficacy, one could apply a reflective coating (a short-pass filter) on the bulb (e.g. using IRC technology, and there are even some more recent activities [1]).

With such a coating, the infrared spectral content would be recycled at the filament, which would reduce the necessary electric power to heat it to the desired temperature.

Assume that a share of $r = 50\%$ of the IR portion is reflected and subsequently absorbed by the filament and calculate the efficiency increase as less electrical power is needed to reach the same visible radiant flux Φ_{vis}. Neglect any electric losses.

Solution 4.6

Initially, electrical power is converted to visual and infrared radiant flux.

$$P_{\text{el},1} = \Phi_{\text{vis}} + \Phi_{\text{IR},1} \tag{4.18}$$

The efficiency (useful visible radiant flux per electrical power) is

$$\eta_1 = \frac{\Phi_{\text{vis}}}{P_{\text{el},1}} \approx 10\% \tag{4.19}$$

With the bulb coating, we have

$$P_{\text{el},2} = \Phi_{\text{vis}} + \Phi_{\text{IR},2} = P_{\text{el},1} - r\Phi_{\text{IR},1} \tag{4.20}$$

(Continued)

Solution 4.6 (Continued)

The improved efficiency is

$$\eta_2 = \frac{\Phi_{vis}}{P_{el,2}} = \frac{\Phi_{vis}}{\Phi_{vis} + (1-r)\Phi_{IR,1}} = \frac{\Phi_{vis}}{\Phi_{vis} + (1-r)(P_{el,1} - \Phi_{vis})} \tag{4.21}$$

hence,

$$\eta_2 = \eta_1 \frac{1}{\eta_1 + (1-r)\,(1-\eta_1)} \tag{4.22}$$

With $r = 50\%$, we have

$$\eta_2 = \eta_1 \cdot \frac{2}{\eta_1 + 1} \tag{4.23}$$

and with $\eta_1 = 10\%$, we get $\eta_2 = 18\%$.

Theory and practice: This approach only works when the IR light is precisely reflected onto the filament, which is rather difficult to achieve; anything but a perfect sphere with the filament having precise point symmetry relative to the center is hard to imagine. The larger the sphere, the more difficult the tolerances become – the smaller the sphere, the hotter the inner surface, which is not good for coatings.

Nowadays, incandescent lamps are history. Consequently, we follow the triumphant progress of LEDs.

Exercise 4.7: CoB LED luminance

You have a "Chip on Board light engine" (as in Figure 3.19):

The diameter of the light-emitting surface (LES) is $2r_S = 10$ mm, the total flux is $\Phi_S = 1200$ lm. Assume that the LED is a Lambertian emitter, i.e. has constant luminance over area and angle.

a. Compute the (average) illuminance on a screen directly above the light engine.

b. Calculate the average luminance of the light engine.

c. Calculate the luminous intensity of the naked light engine perpendicular to its plane.

d. Estimate the illuminance at 10 mm distance, perpendicular above the light engine center.

e. How much less is this illuminance compared to what you would expect from the $1/r^2$ relation (Eq. 3.142) at the same distance of 10 mm (as if the light engine were a point source)?

Solution 4.7

a. The illuminance on a screen directly above the LED (luminous exitance) is simply

$$E_S = \frac{\Phi_S}{A_S} = \frac{1200\,\text{lm}}{\pi\, 25\,\text{mm}^2} \approx 15\,\frac{\text{lm}}{\text{mm}^2} = 15\,\text{Mlx} \tag{4.24}$$

which is much *brighter* (see Exercise 5.2) than a surface illuminated by direct sunlight.

b. The luminance is flux per etendue for Lambertian emitters, so

$$L_S = \frac{\Phi_S}{U_S} = \frac{\Phi_S}{\pi \text{sr} \cdot A_S} = \frac{1200\,\text{lm}}{\pi^2\, 25\,\text{mm}^2\text{sr}} \approx 5\,\frac{\text{cd}}{\text{mm}^2\text{sr}} = 5\,\text{Mnit} \tag{4.25}$$

which is substantially *less* than the luminance of single high-power LED chips.

c. The luminous intensity *on axis* is luminance times area,

$$I_0 = L_S \cdot A_S = \frac{\Phi_S}{U_S} \cdot A_S = \frac{1200\,\text{lm}}{\pi\,\text{sr}} \approx 400\,\text{cd} \tag{4.26}$$

d. There are different approaches to calculate the true illuminance at the given distance.

d1. Using the concept of a *light tube* filled with constant luminance: We consider a tiny circular area element (radius R) at the target plane and calculate the etendue between this area and the LED area using Hottel's formula (Eq. 3.36). With (e.g.) $R = 0.1$ mm, $r = 5$ mm, and $D = 10$ mm, we obtain $U = 0.0197\,\text{mm}^2\text{sr}$. Then,

$$E_0 = \frac{\Phi_S}{A_S} = \frac{L_S U_S}{\pi R^2} \approx 3.14\,\text{Mlx} \tag{4.27}$$

d2. Using the fact that for constant luminance, illuminance is luminance times angular extent: From the target center, we see the LED under a cone angle of

$$\tan\theta = \frac{r_S}{D} = \frac{5\,\text{mm}}{10\,\text{mm}} \tag{4.28}$$

giving $\theta = 26.6°$. This allows us to find the projected solid angle from Eq. 2.12, $\Omega_p = 0.628$ sr, and to obtain the target illuminance

$$E_0 = L_S \Omega_p = 3.14\,\text{Mlx} \tag{4.29}$$

Unsurprisingly, both values are equal. If we apply the inverse square law (Eq. 3.142) in spite of the rather short distance, we overestimate the illuminance by more than 25%:

$$E_0 = \frac{I_0}{D^2}\text{sr} = \frac{400\,\text{cd}}{100\,\text{mm}^2}\text{sr} = 4\,\text{Mlx} \tag{4.30}$$

(Continued)

Solution 4.7 (Continued)

Theory and practice: Real LEDs are not Lambertian. Play with our $\cos^n$ calculator! For Lambertian, the exponent is $n = 1$. Set the exponent to, say, 1.1 and 0.9 (both not unusual for such LEDs). Look at the collimation strength I_0/Φ as well as the intensity distribution curve. You will see that at the same flux, intensity and therefore on-axis illuminance will vary by about ±7%. There are too many digits in 3.14 Mlx!

Exercise 4.8: LED extraction efficiency

The widely used LED material GaN has a refractive index of $n = 2.5$. Assume a chip size of 1.25 mm · 1.6 mm. The thickness of thin-film LEDs is negligible as there are only a few μm between the bottom Ag mirror and the exit face (see Figure 4.3). The Ag mirror is rather rough and strongly scattering. The top surface is assumed polished and smooth. This is not the case in actual devices, but a useful assumption for now.

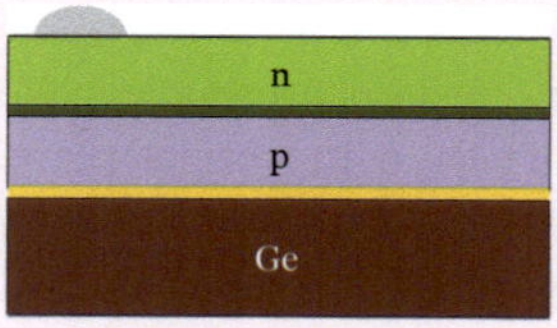

Figure 4.3 Thin-film LED chip with active region (junction, dark gray), mirror (yellow), substrate (dark brown) and top contact (light gray). Adapted from [2]. The drawing is not to scale: The n–p part above the mirror is just a few μm thick.

- **a.** A screen S1 is placed inside the LED chip, above the junction, but below the LED–air interface. Compute the etendue of the full phase space volume of S1.
- **b.** Compute the critical angle θ_c for total internal reflection from LED chip to air.
- **c.** Compute the normal incidence Fresnel reflectivity of the GaN–air interface.
- **d.** Now consider the "directly usable" ray bundle that is emitted from the junction/mirror subsystem, consisting only of rays that are within θ_c. These are the rays which can leave the LED directly. Compute the etendue of this "direct" ray bundle.
- **e.** A screen S2 is placed in air, directly above the LED chip. Compute the etendue of the full phase-space volume of S2. Comment on the relation between this full phase space in air and the "direct" ray bundle.

Solution 4.8

a. The etendue inside the LED, on screen S1, is:

$$U_{\text{in}} = AX = An_{\text{in}}^2 \Omega_{\text{p}} = 2\,\text{mm}^2 \cdot 6.25 \cdot \pi\,\text{sr} = 39\,\text{mm}^2\text{sr} \tag{4.31}$$

b. The critical angle for total internal reflection inside the chip results from

$$\sin\theta_{\text{C}} = \frac{n_{\text{out}}}{n_{\text{in}}} = 0.4 \tag{4.32}$$

which yields $\theta_{\text{C}} \approx 24°$. That's a small TIR escape cone! Rays beyond that angle of incidence will be reflected and subsequently recycled or absorbed. (Modern InGaN LEDs have very large recycling rates.)

c. Fresnel reflectivity at perpendicular incidence on the interface is [3]

$$\rho = \frac{(n_{\text{in}} - 1)^2}{(n_{\text{in}} + 1)^2} \approx 18\% \tag{4.33}$$

That is a huge reflection loss! For typical low-index glasses, we have only $\rho \approx 4\%$.

d. The etendue inside the LED of a ray bundle with $\pm\theta_{\text{C}}$ is:

$$U_{\text{in},\theta_{\text{C}}} = An_{\text{in}}^2 \Omega_{\text{p}} = 2\,\text{mm}^2 \cdot 6.25\pi \sin^2\theta_{\text{C}}\,\text{sr} = 6.28\,\text{mm}^2\text{sr} \tag{4.34}$$

e. The etendue at the screen S2 on top of the chip is simply

$$U_{\text{out}} = An_{\text{out}}^2 \Omega_{\text{p}} = 2\,\text{mm}^2 \cdot \pi\,\text{sr} = 6.28\,\text{mm}^2\text{sr} \tag{4.35}$$

We see that $U_{\text{in}} = U_{\text{out}}$ – in other words, we see etendue conservation in action! The rays inside the chip, on screen S1, within $\pm\theta_{\text{C}}$, are precisely (by definition) the rays that completely fill the phase space outside, on screen S2.

Simulation 4.2: Fresnel losses

We simulate a source inside an LED medium (or glass) to get familiar with the Fresnel formula (Figure 4.4).

Provided file.

(Continued)

Simulation 4.2: (Continued)

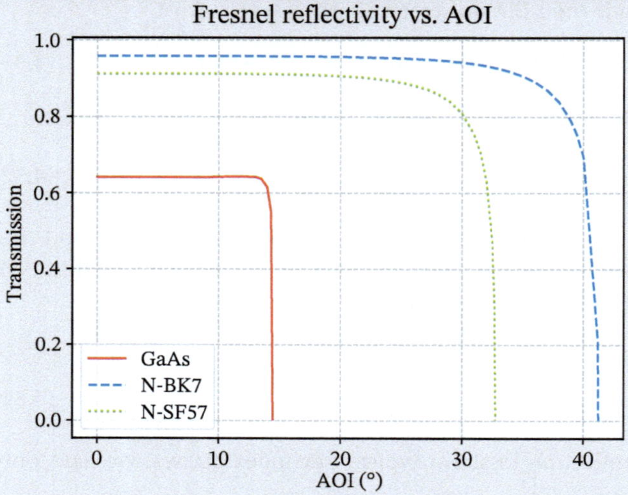

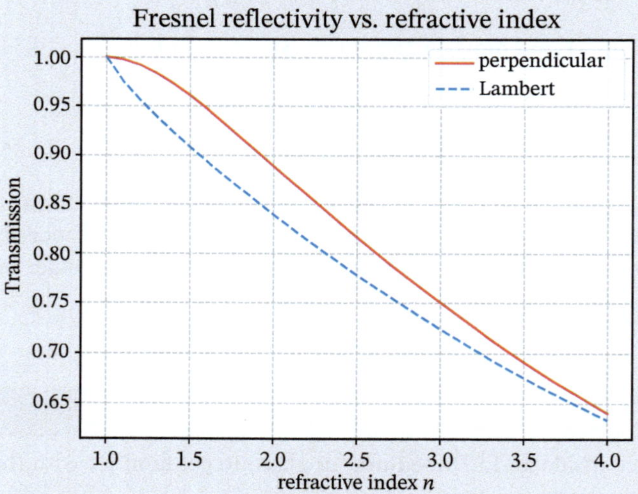

Figure 4.4 Top: Transmission out of a material as a function of the inside AOI (calculated using Eq. 4.33 and the Fresnel equations. Bottom: Transmission for rays perpendicular to the surface and average transmission for an Lambertian source embedded in a material of given refractive index.

Exercise 4.9: LED dome (1)

Now, we will take a look at a differently packaged LED type: the variant with a dome (Figure 4.5). Here, an almost hemispherical lens made of silicone or glass is attached (i.e. immersed) to the chip(s).

We consider an LED with a variety of white chips: LED with dome.

Figure 4.5 Left: Geometric model of a domed four-chip LED with generic sources inside the dome and detector directed to the LED. As usual, the CAD rendering does not consider refraction. Center: True color view (detector focused into the LED). Right: Photorealistic rendering taking into account the refraction at the dome surface. The enlarged virtual image of the chips is obvious.

Assume a size of $1.25 \cdot 1.6\,\text{mm}^2$ for each of the four chips and an immersed clear silicone dome with $n = 1.41$ and a radius of $r = 2.5$ mm.

a. Calculate the etendue of the light leaving the chip (e.g. on a screen just above the chip).

b. Calculate the etendue of the LED ray bundle leaving the dome, e.g. on a hemispherical screen just outside in air. Compare the angular cones to those of bare chips of same area!

c. Finally, assume the same LED dome geometry, but with the silicone dome covered with a phosphor material, emitting diffuse converted light into all directions [4, 5]. Calculate the etendue of the diffuse light leaving the dome. Which screen is most convenient for this purpose?

Solution 4.9

a. The etendue at a screen on top of the chip but already inside the silicone is

$$U_{\text{sil}} = AX = 4\,A_{\text{chip}} n_{\text{sil}}^2\, \Omega_{\text{p}} = 8\,\text{mm}^2 \cdot 1.41^2 \pi \text{sr} = 50\,\text{mm}^2\text{sr} \tag{4.36}$$

compared to that of a bare (in air) chip area

$$U_{\text{bare}} = 8\,\text{mm}^2 \cdot \pi\,\text{sr} = 25\,\text{mm}^2\text{sr} \tag{4.37}$$

This states an important rule of thumb for LED domes: They increase source etendue by a factor of two – with all the consequences.

b. The U_{sil} etendue value is conserved to a screen outside the dome, since the dome is far enough away to avoid TIR. Since the area of the latter is larger, there must be an angular distribution smaller than a that of a Lambertian.

$$\Omega_{\text{p}} = \frac{U_{\text{sil}}}{A_{\text{dome}}} = \frac{U_{\text{sil}}}{2\pi\, r^2} = \frac{50\,\text{mm}^2\text{sr}}{39\,\text{mm}^2} = 1.25\,\text{sr} \tag{4.38}$$

(Continued)

Solution 4.9 (Continued)

corresponding to a cone angle of $\approx \pm 40°$. Is that real? See the following simulation.

c. The phosphor cover causes Lambertian emission from each area element of a screen outside the dome.

$$U_{\mathrm{ph}} = A_{\mathrm{dome}}\Omega_{\mathrm{p}} = 2\pi r^2 \cdot \pi\,\mathrm{sr} = 123\,\mathrm{mm}^2\,\mathrm{sr} \tag{4.39}$$

which is much larger than that of the dome LED. Care has to be taken when using remote phosphors in etendue limited illumination designs as such phosphor shells blow up the etendue a lot. Consequently, the use of remote phosphors may require sophisticated optical systems as shown in [6, 7] instead of simple shells.

Simulation 4.3: LED with dome

To confirm the above result (b), we set up the referenced LED with its ray data and look at it through a $d = 0.2$ mm pinhole just outside the dome (Figure 4.6). We are lucky – in our example, all the rays originate just outside the dome.

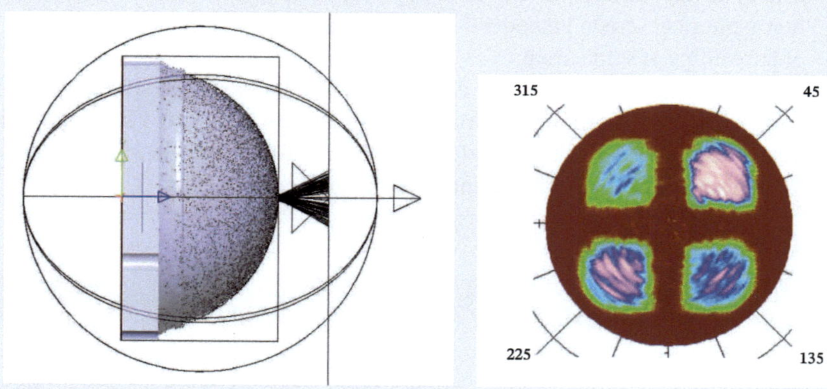

Figure 4.6 Pinhole on a dome to inspect the local cone angle. The angular distribution (slightly smaller than the rim of 45°) confirms the result of the above solution.

The result confirms our above estimate (b). A simulation file is provided.

Here I am to help you!

When I look right into the pinhole (in a simulation or physically implemented in the laboratory), I can gather useful information about the local angular distribution, color, and exitance [8].

Exercise 4.10: To dome or not to dome?

For overall efficacy (in luminous flux per electric power), should we use a $1 \cdot 1\,\text{mm}^2$ square LED with a dome lens or rather a $1.4 \cdot 1.4\,\text{mm}^2$ square LED without a dome lens? Both LEDs would be operated with 1 W. (Exercise suggested by Matthias Walser).

Solution 4.10

There is not a single clear answer, unfortunately, because several effects are competing.

The current vs. voltage curve for any LED is fundamentally a current *density* vs. voltage curve: Two parallel LEDs have twice the current, but the same voltage. The smaller LED with half the area will have higher voltage and therefore less current (about 5–10%) at the same electric power. But it is current that determines light output, ideally with one photon per electron. High luminance is always accompanied by smaller efficacy. A little relief might come from so-called *droop*: Quantum efficiency decreases with increasing current, and 10% less current means less than 10% drop in photon flux.

On the other hand, the dome will “pull” more photons out of the chip, because in the dome, there is twice as much angular extent, and photons find their way out easier. This may more than offset the current droop in single-color LEDs. In white LEDs, the reasoning is not so clear: When the photons can get out easier, the blue photons will not stay in the phosphor platelet as long, and will have less chance to be converted to yellow by the phosphor particles. If we could just add a dome to a white LED, we would experience a blue shift. LED vendors must therefore use denser or thicker phosphor platelets for domed white LEDs to offset this blue shift. This will counteract the outcoupling gain due to the dome.

The etendue of both LEDs is nearly identical: $U = An^2\pi$ with $n \approx 1.41$ for typical silicone domes. There is no difference in the etendue fundamentals. However, the virtual image (by the dome lens) of the chip is enlarged and aberrated, and the dome lens is usually not tightly toleranced. Both effects make etendue limited designs more difficult. And when your optical architecture requires optics close to the LED, a dome might be in your way. A TIR lens would probably work well with a domed LED, though.

So the bottom line is: it is hard to tell if a domed LED will have higher efficacy or not. Practical optical design considerations are probably more important in most cases.

4.2 Lighting

Exercise 4.11: Constant Illuminance

Which intensity distribution of a point source at a ceiling height D_0 gives a constant illuminance on the floor?

Solution 4.11

Let θ be the angle of a ray to the vertical, and let $D(\theta)$ be the distance between the point source and the point on the floor to which this ray points. Then, $D_0 = D\cos\theta$ (draw the triangle to see why). The inverse square law (Eq. 3.142) then yields

$$E(\theta) = \frac{I(\theta)}{D^2}\mathrm{sr}\cdot\cos\theta = \frac{I(\theta)}{D_0^2}\mathrm{sr}\cdot\cos^3\theta \tag{4.40}$$

Hence, we need

$$I(\theta) = \frac{I_0}{\cos^3\theta} \tag{4.41}$$

which might be called the $\cos^{-3}$ law.

Simulation 4.4: A $\cos^{-3}$ source

We simulate a $\cos^{-3}$ intensity distribution set up using user-defined angular apodization in LightTools (Figure 4.7).

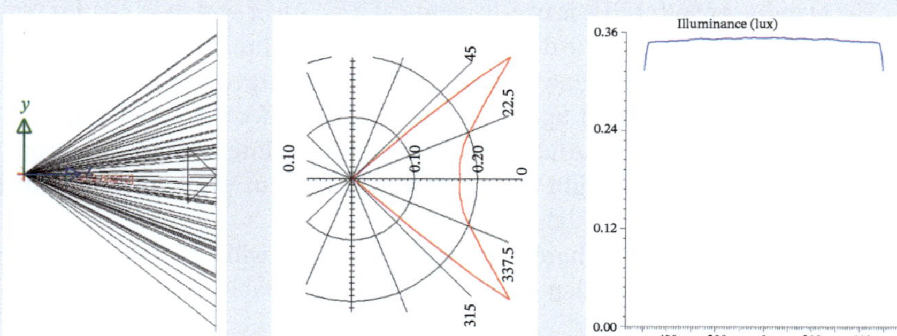

Figure 4.7 Left: point source with a $\cos^{-3}$ intensity distribution (shown in the center). Right: Constant illuminance distribution on the receiver obtained.

Simulation file is provided.

Exercise 4.12: Illuminance maximum at a side wall

A Lambertian source S is positioned at a ceiling at a distance $D_\perp = 1$ m from a wall with its axis pointing downwards (Figure 4.8).

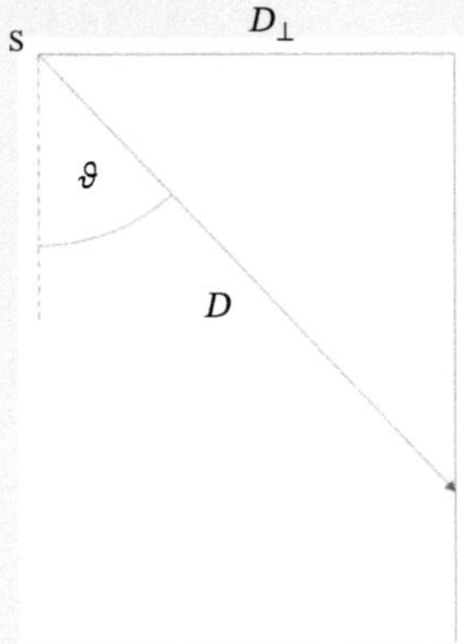

Figure 4.8 Source shining to a side wall.

Find the angle ϑ for which the illuminance at the wall will be at a maximum. Is this different for an isotropic source (defined by $I = I_0 = \text{const}$)?

Solution 4.12

We use the inverse square law (Eq. 3.142) with oblique incidence and some geometry

$$E(\vartheta) = \frac{I(\vartheta)}{D^2}\text{sr} \cdot \sin\vartheta = \frac{I(\vartheta)}{D_\perp^2}\text{sr} \cdot \sin^3\vartheta \tag{4.42}$$

For Lambertian emission (Eq. 3.117), we have

$$E(\vartheta) = \frac{I_0}{D_\perp^2}\text{sr} \cdot \sin^3\vartheta\cos\vartheta \tag{4.43}$$

The illuminance maximum is found via $\text{d}E/\text{d}\vartheta = 0$:

$$3\sin^2\vartheta\cos\vartheta - \sin^4\vartheta = 0 \tag{4.44}$$

and with Eq. 2.2, this becomes

$$\cos^2\vartheta + 3\cos\vartheta - 1 = 0 \tag{4.45}$$

with the result

$$\vartheta_{\text{max}} = 72.4° \tag{4.46}$$

(Continued)

Solution 4.12 (Continued)

Alternatively, we may calculate the etendue defined by area elements A and A' at the source and at the wall. As A' is oriented perpendicularly to A, we have

$$U(\vartheta) = \frac{AA'}{D^2} \cos\vartheta \sin\vartheta = \frac{AA'}{D_\perp^2} \cos\vartheta \sin^3\vartheta \tag{4.47}$$

with the same etendue maximum as above. For a Lambertian emitter (L = const) this corresponds to the point of highest illuminance.

In the isotropic case of an intensity,

$$E(\vartheta) = \frac{I_0}{D_\perp^2} \mathrm{sr} \cdot \sin^3\vartheta \tag{4.48}$$

and the maximum is found at $\vartheta = 90°$ (in the corner), as expected.

Simulation 4.5: Illuminance at the side wall

The simulation confirms the maximum (red color region) as calculated analytically, see Figure 4.9.

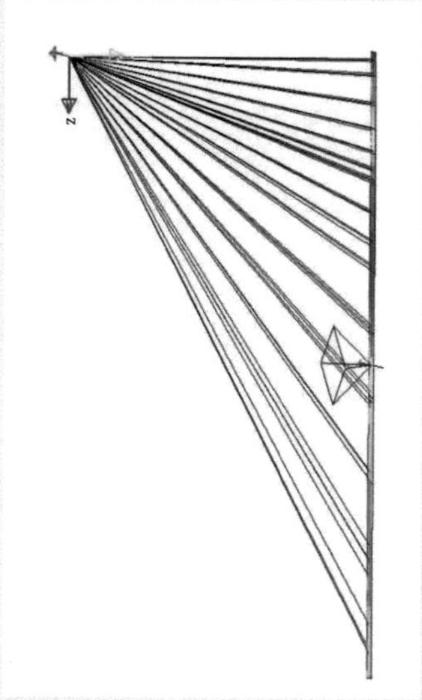

Figure 4.9 Lambertian source on top (at the coordinate origin) radiating at a side wall. The receiver there (see the glyph) shows an illuminance maximum (indicated by red color) where it was calculated.

Simulation file.

Exercise 4.13: Sources at the ceiling

Compare the effect of the following sources at the ceiling of a room of height D regarding the illumination on the floor:

a. An isotropic point source.
b. A small Lambertian element (semi diameter r_S).
c. A Lambertian disk ($r_S \lesssim D$).
d. A large Lambertian disk ($r_S > D$).
e. A nearly infinitely extended Lambertian emitter (whole ceiling area).

Assume same flux for (a–d), and same luminance for (d) and (e).
In each case, calculate the illuminance peak value and the illuminance distribution on the floor.

Solution 4.13

a. This point source is isotropic with an intensity of

$$I = \frac{\Phi_S}{4\pi\ \text{sr}} \tag{4.49}$$

The inverse square law implies a center illuminance on the floor of

$$E_0 = \frac{I}{D^2}\text{sr} = \frac{\Phi_S}{4\pi D^2} \tag{4.50}$$

For the *illuminance distribution*, we have to include two cosine factors for the distance and one for the inclined area element on the floor (cf. Eq. 3.190 and Exercise 4.11). Thus,

$$E(\vartheta) = \frac{\Phi_S}{4\pi D^2}\cos^3\vartheta \tag{4.51}$$

b. With the etendue U_S of the Lambertian emitter (Eq. 3.9), its luminance is obtained from

$$L_S = \frac{\Phi_S}{U_S} = \frac{\Phi_S}{(\pi r_S)^2\text{sr}} \tag{4.52}$$

Together with the tiny source, a small receiver of radius r_T on the floor center (right below the source) forms a light tube with the etendue (Eq. 3.42)

$$U_0 = \left(\frac{\pi\, r_S r_T}{D}\right)^2 \text{sr} \tag{4.53}$$

We can apply this approximation because both sides are very small.

The flux from the source into the light tube is

$$\Phi_T = U_0 L_S = \Phi_S \frac{U_0}{U_S} \tag{4.54}$$

(Continued)

Solution 4.13 (Continued)

(flux times the configuration factor, Eq. 3.25) and is delivering a receiver illuminance

$$E_0 = \frac{\Phi_T}{\pi\, r_T^2} = \frac{\Phi_S}{\pi\, r_T^2}\frac{U_0}{U_S} = \frac{\Phi_S}{\pi\, D^2} \tag{4.55}$$

Compared to Eq. 4.50, this illuminance value is larger by a factor of 4 because in (b) the source is hemispheric and Lambertian instead of isotropic as in (a), which causes a kind of concentration toward the target on the floor.

As in Eq. 3.190, we have

$$U(\vartheta) = U_0 \cos^4 \vartheta \tag{4.56}$$

and

$$E(\vartheta) = E_0 \cos^4 \vartheta \tag{4.57}$$

This result is valid as long as the approximation Eq. 3.42 actually holds (i.e. as long as $r_S \ll D$).

c. For a source of non-negligible size ($r_S \approx D$, Figure 4.10), we can no longer apply the far-field approximation. But we may still apply a weaker far-field approximation based on $r_T \ll D$ and look from the floor element to the source. This way, center etendue (r_T and r_S being co-axial) can be written as

$$U_0 = A_T \Omega_p = \pi\, r_T^2 \cdot \pi\, \mathrm{sr} \frac{r_S^2}{r_S^2 + D^2} \tag{4.58}$$

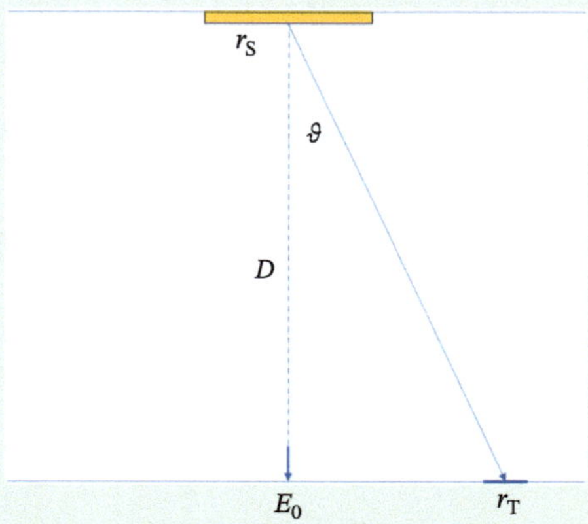

Figure 4.10 Extended source at the ceiling.

Hence,

$$E_0 = \frac{\Phi_S}{\pi\, r_T^2}\frac{U_0}{U_S} = \frac{\Phi_S}{\pi\,(r_S^2 + D^2)} \tag{4.59}$$

As the given source flux is coming from a larger area, the center illuminance is reduced with respect to (b). Again, the combination with the $\cos^4$ term gives the spatial distribution (approximately).

d. Here, no approximations of any kind lend themselves possible, but we can still calculate the illuminance for a *ring* of (small) width $2r_T$ and center distance R and calculate the etendue difference between the source and the lower and upper limits of the target using Hottel's formula (Eq. 3.42), see Figure 4.11.

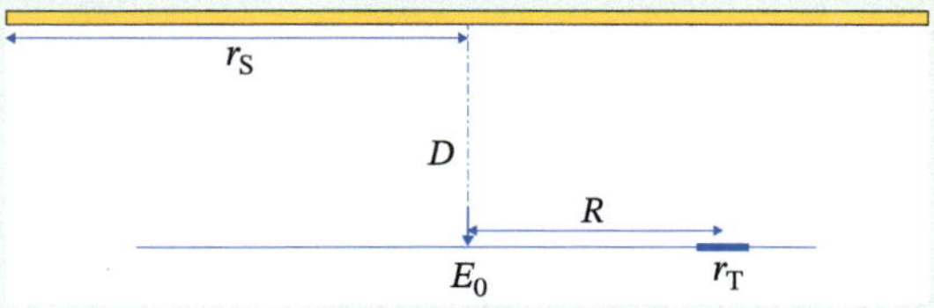

Figure 4.11 Extended source in a close distance.

$$U_{\text{ring}}(R) = \frac{\pi^2}{4}\left\{\left(\sqrt{(r_S + (R + r_T))^2 + D^2} - \sqrt{(r_S - (R + r_T))^2 + D^2}\right)^2 - \left(\sqrt{(r_S + (R - r_T))^2 + D^2} - \sqrt{(r_S - (R - r_T))^2 + D^2}\right)^2\right\} \tag{4.60}$$

and we need a spreadsheet for evaluation (see the following Simulation).

Next, we have to multiply the obtained ring etendue with the source luminance

$$L_S = \frac{\Phi_S}{(\pi\, r_S)^2 \text{sr}} \tag{4.61}$$

and obtain the illuminance according to

$$E(r_S, D, R) = \frac{L_S U_{\text{ring}}(R)}{2\pi\, R \cdot r_T} \tag{4.62}$$

e. A nearly infinitely extended source of luminance L_S will provide constant illuminance everywhere: every target point sees light from a whole hemisphere

$$\Omega_p = \pi \text{ sr} \tag{4.63}$$

hence

$$E_T = \frac{\Phi_v T}{\pi\, r_T^2} = \frac{L_S U_T}{\pi\, r_T^2} = L_S \pi \text{ sr} \tag{4.64}$$

which corresponds to $r_S \to \infty$ in Eq. 4.59.

Simulation 4.6: Ceiling sources

To prove results such as Eq. 4.59, we model the setup for $D = 100$ mm and $r_S = 1\ldots1000$ mm. The source is set to $\Phi = 1000$ lm (see Figure 4.12).

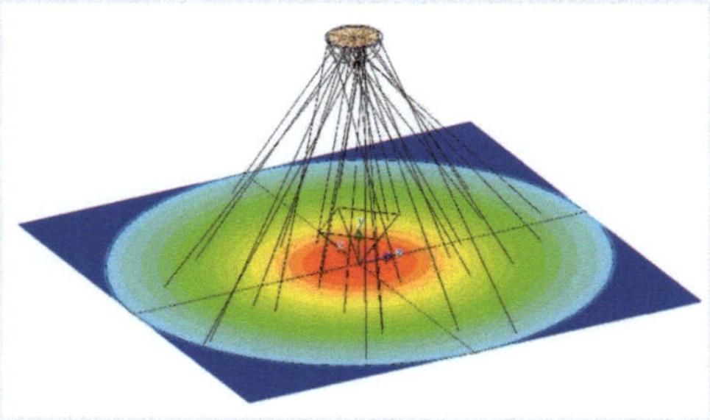

Figure 4.12 Extended ceiling source shining on the floor receiver.

The comparison shows that the center illuminance formulas work correctly. For details, see the provided file disk and ring.xlsx.

Simulations were performed using this simulation file.

Figure 4.13 shows illuminance distributions obtained in different ways.

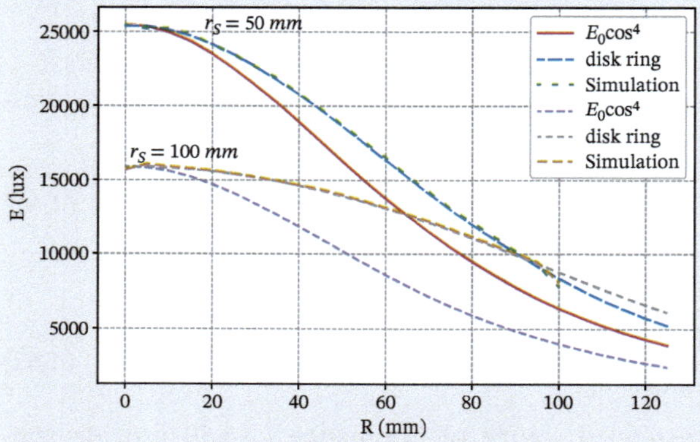

Figure 4.13 Illuminance distributions for two source sizes obtained from formula and simulation compared to simple $\cos^4 \vartheta$ distributions.

For small sources we can apply all approximations and get correct results. For large sources, the simple approach $E \approx E_0 \cdot \cos^4 \theta$ begins to fail. The ring etendue approach can help us to properly calculate such distributions.

We can now apply our newly acquired skill regarding the analysis of ring targets to solve a different task.

Exercise 4.14: Lambertian ring

A Lambertian source ($\Phi_S = 100\,\text{lm}$) with a ring-shaped exitance distribution (ring from $r_1 = 180\,\text{mm}$ to $r_2 = 220\,\text{mm}$) illuminates a target plane (parallel to the disk plane) at a distance $D = 100\,\text{mm}$ (see Figure 4.14).

Figure 4.14 Ring light.

Calculate the illuminance distribution.

Solution 4.14

We solve this task by generating illuminance distributions produced by an "outer" disk at $r_2 = 220\,\text{mm}$ and an "inner" disk at $r_1 = 180\,\text{mm}$ and subtracting them.

First, we have to ensure a proper flux scaling. The etendue of the ring source is

$$U_S = r_2^2\pi \cdot \pi\,\text{sr} - r_1^2\pi \cdot \pi\,\text{sr} = \pi^2(220^2 - 180^2)\,\text{mm}^2\text{sr} = 158\,000\,\text{mm}^2\text{sr} \tag{4.65}$$

The source luminance is

$$L_S = \frac{\Phi_S}{U_S} = 633\,\text{nit} \tag{4.66}$$

Thus, an outer disk has to be set up with a flux of

$$\Phi = L_S U_{\text{outer}} = 633\,\text{nit}(220\,\text{mm})^2\pi \cdot \pi\,\text{sr} = 302\,\text{lm} \tag{4.67}$$

and for the inner disk we need

$$\Phi = L_S U_{\text{inner}} = 633\,\text{nit}(180\,\text{mm})^2\pi \cdot \pi\,\text{sr} = 202\,\text{lm} \tag{4.68}$$

We calculate the target distributions for both disks as in Eq. 4.62 and subtract both. The results thus obtained are shown in Figure 4.15.

(Continued)

Solution 4.14 (Continued)

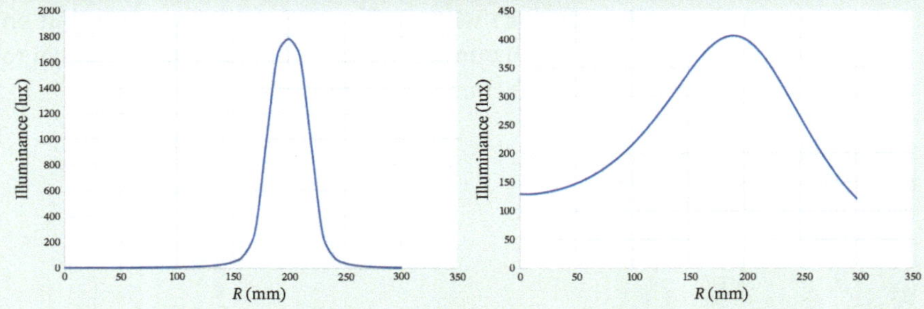

Figure 4.15 Target screen illuminance created by the ring source from Exercise 4.14 for target distances of 10 mm (left) and 100 mm (right).

Provided file Disk_and_ring.xlsx, Worksheet "Ringring".

Is this result correct? Let's simulate!

Simulation 4.7: Lambertian ring

The setup is quickly done in illumination software; the simulation result is displayed in Figure 4.16.

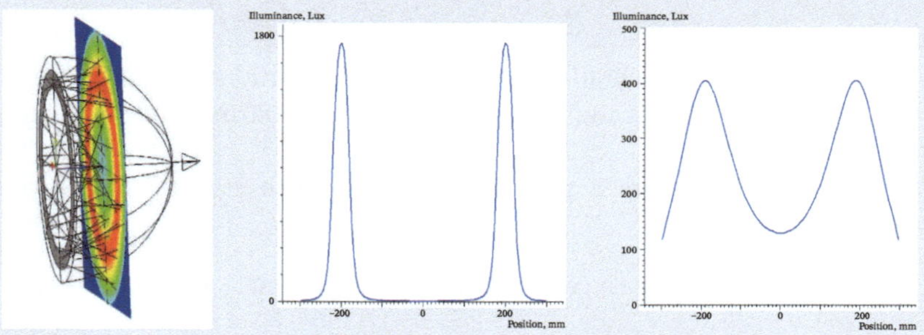

Figure 4.16 Lambertian ring source shining on a receiver at a distance D. Illuminance distributions for $D = 10$ mm and $D = 100$ mm.

Provided file Ring_Lambert.lts.

Simulated distributions as well as peak values agree well with the analytic solutions.

Keep this approach in mind: When facing a complex situation, see if you can analyze it by separating it into simpler situations that can be added or subtracted to re-create the original problem. Quite often, it is the subtracting approach that is not obvious at first sight.

Exercise 4.15: Office lighting

Office lighting standards such as DIN 5035 require 500 lx on the desk (see Figure 4.17). Your task is to select and to install a luminaire at the ceiling, at a height of $D = 2$ m from the desk surface.

Photo by Siteco

Figure 4.17 Modern office lighting.

Calculate the necessary luminous intensity of the fixture. How much of the source flux arrives at a 1 m · 2 m desk centered below the fixture?

Solution 4.15

Since the luminaire diameter is probably small compared to the distance between it and the desk, we can apply the inverse square law (Eq. 3.142) for an estimate, knowing that we will slightly underestimate. For the given illuminance and distance, you need an on-axis luminous intensity of more than

$$I = \frac{E_v D^2}{\text{sr}} = 2000\,\text{cd} \tag{4.69}$$

which is basically fulfilled, for example, by the products linked right below, although they are specified in different ways.

The flux into the given $2\,\text{m}^2$ target, on the other hand, is undefined, as the on-axis intensity carries no information on the beam shape. Even if we had a FWHM beam angle, we would still not know much about the beam shape outside this cone. What we need here is a light distribution curve (cf. Exercise 4.16 and the following Knowledge Corner Topic 4.2).

Theory and practice: While the calculation is correct, it is not exactly what we really need. This is a good example of a theory that sounds good just because it leaves out important facts. First, a typical office comprises more than one luminaire, and their contributions to the overall lighting add. Second, most office walls are not painted black: Multiple reflections, causing diffuse contributions from walls and ceiling, increase the illuminance levels, sometimes quite drastically. The job of a light planner is much more complex than implied in this simple exercise.

Knowledge Corner Topic 4.2: Light distribution curves (1)

For proper planning and design of office lighting solutions, you need to know the beam shape, known as the *light distribution curve* (LDC) of the fixtures you plan to use. (And obviously, you need to know much about the room itself as well.) You will find LDCs (as shown in Figure 3.33) from most luminaire suppliers.

The scale of the curve may come in terms of peak intensity (in cd) or *collimation strength* (Eq. 3.126, in cd/klm). In both cases, a total luminous flux will be specified in the datasheet.

We arbitrarily chose three products from different suppliers:

Supplier A - LDC scaled by peak intensity I_{max}
Supplier B - LDC scaled by collimation strength in cd/klm
Supplier C - just luminous flux given

Instead of the graphs, you may (and should!) use LDC data files provided for simulations offered as downloads by many lighting suppliers (Figure 4.18). This is true for stage lighting as well (cf. Knowledge Corner Topic 4.10). These data are provided in two flavors with more or less identical functions: IES and Eulumdat files. Typically, they can be recognized by their file extension: "`.ies`", properly standardized by the Illuminating Engineering Society [9], and "`.ldt`" (Eulumdat), an industry standard that has never been formalized to the best of our knowledge, but you can find unofficial descriptions on the internet, e.g. with the aid of the Wayback Machine.

There are free viewers for those files:

IESviewer
3D Viewer

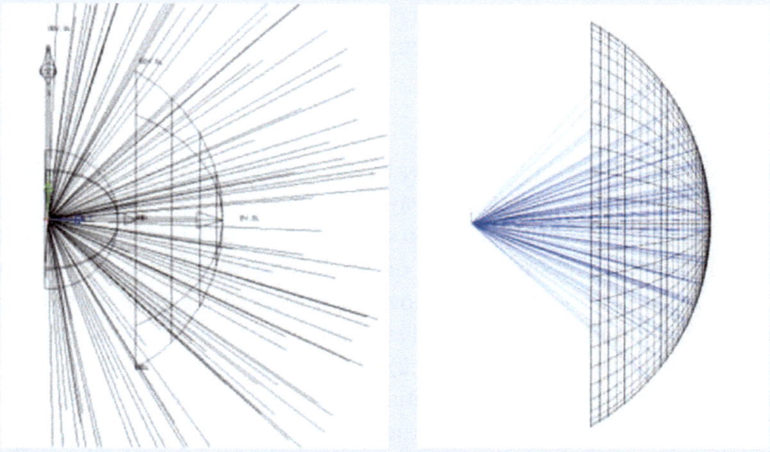

Figure 4.18 Import of Eulumdat files into illumination software: In LightTools, use the LDT/IES reader, ask it to set up a point source with an intensity distribution according to the LDT file, and let it shine on a far-field receiver. In Zemax Opticstudio, use the "Source LDT" and a polar detector.

In the following, we show how to use LDC files in illumination software.

LDC files work well in lighting design software such as Dialux and Relux, but most non-sequential software illumination optics packages support them as well.

LED optics suppliers provide light distribution files for the function of their lenses with a variety of LEDs, too.

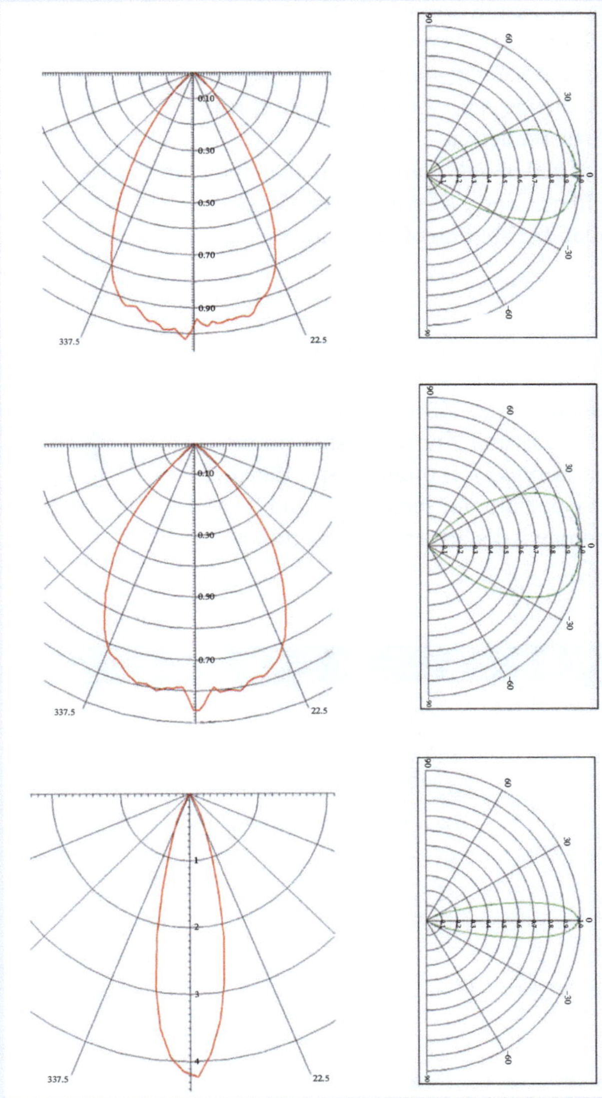

Figure 4.19 Far-field distributions generated on the basis of LDT files (top down: supplier A through C. LightTools (left column) shows absolute intensity values, Zemax (right column) shows normalized curves and the peak intensity. Both show almost the same shapes and peak values.

(Continued)

Knowledge Corner Topic 4.2: (Continued)

Dr. Wisenheimer:

The purpose of LDCs is to model the *far field of a luminaire*, i.e. it is implicitly assumed that the spatial distribution of luminance across the aperture of the device can be neglected (Figure 4.19). However, the length and width of the aperture are included in the IES and LDT formats. This information is used in light planning software by assuming that the luminance is spatially homogeneous, both for illuminance calculations and for photorealistic rendering.

There are also LDCs of actual *light sources* available, including LEDs, but we tend to use them rarely. In our work, we typically need to model near-field effects, i.e. the spatial and angular distribution of luminance. Thus, we need *ray files* for source simulations.

Dr. Wisenheimer:

For the far field of stage lights, check the photometry files such as those of the Sharpy.

Look out for tools such as the ARRI Photometrics App (Figure 4.20, available for iOS and Android, other luminaire makers may offer similar services). It provides specifications, light distributions, and a calculator for screen illuminance and beam diameter as a function of distance for a large number of lamp heads.

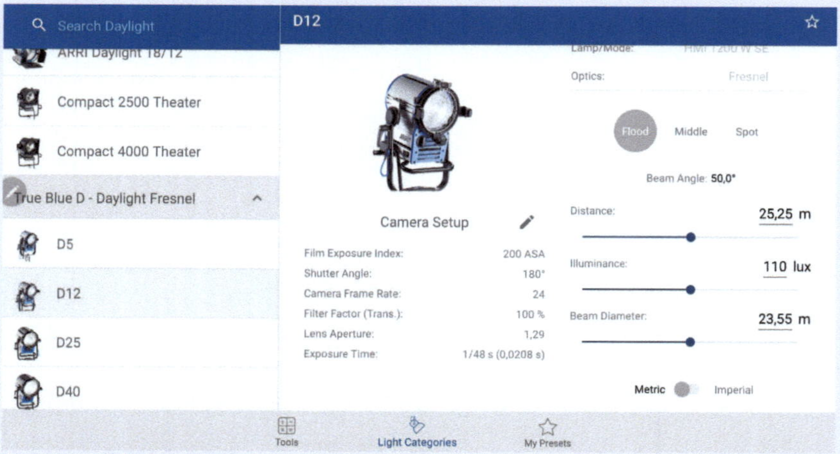

Screenshot from ARRI's photometry app

Figure 4.20 ARRI's photometry app.

Figure 4.21 Downlight. *Source*: Courtesy of Siteco.

We continue with predesign work for an actual light source. A *downlight* is a class of directional ceiling fixtures with a rather high luminance source that enables an exit area of a limited size (Figure 4.21). Most parts of the luminaire are hidden in the ceiling; from below, only some optical parts such as a reflector or a lens surface are visible. The light distribution is directional with a soft edge, illuminating the floor below without glare.

Exercise 4.16: Downlight

Imagine a downlight for operation with a chip-on-board (CoB) LED with a light emitting surface (LES) diameter of $2r_S = 6\,\text{mm}$ and a luminous flux of $\Phi_S = 1000\,\text{lm}$ such as this CoB LED.

Which beam angle corresponds to $I_0 = 2000\,\text{cd}$ as required for office lighting? Which downlight exit area diameter is necessary? Assume an ideal luminaire efficiency.

Solution 4.16

We analyze the specification with the help of the $\cos^n$ beam model (Eq. 3.133), which gives just the desirable kind of soft edge without undesired spillage into large angles. An exponent of

$$n = 2\pi\text{sr}\frac{I_0}{\Phi} - 1 = 4\pi - 1 \approx 12 \tag{4.70}$$

describes a beam with the given flux and intensity values.

The *beam angle* or full width at half maximum (FWHM) angle is a measure for the width of a rotationally symmetric collimated beam with a central peak. As the FWHM name implies, in a 40° beam, the intensity will drop to half its central peak value at an angle of 20° away from its axis of symmetry.

(Continued)

Solution 4.16 (Continued)

For our $\cos^{12}$ beam, the corresponding angle is calculated according to

$$\cos^{12}\theta_{0.5} = 0.5 \tag{4.71}$$

which yields $\theta_{0.5} \approx 20°$, corresponding to a beam angle of 40°.

The luminance of the CoB LED source is

$$L_S = \frac{\Phi_S}{r_S^2 \pi \cdot \pi \,\text{sr}} = \frac{1000\,\text{lm}}{89\,\text{mm}^2\text{sr}} = 11.25\,\text{Mnit} \tag{4.72}$$

Here, we should follow the advice we gave in Knowledge Corner Topic 1.3: 11.25 has too many digits for a radiometric end result, where a few percent precision is generally a good rule of thumb (due to measurement accuracy, LED binning, LED temperature etc.). Still, we should keep a few more digits in intermediate results to avoid truncation errors in the final result. Most likely, we would use software like Excel or Python, where intermediate results are stored with 15 digits, and then round the end result.

To achieve the specified target intensity, we need to increase the exit area,

$$A = \frac{I_0}{L_S} = \frac{2000\,\text{cd}}{11\,\text{Mnit}} = 177.7\,\text{mm}^2 \tag{4.73}$$

corresponding to a (now properly rounded) minimum diameter of 15 mm.

There are several ways to implement the necessary optics between the source and the exit areas – for example, deep-drawn, CPC-like aluminum reflectors, aspheric or Fresnel lenses and more will do the job.

Dr. Wisenheimer:

This diameter is too small for a typical downlight design where the diameter typically is around 100 mm. However, we've seen such tiny zero-glare near-invisible downlights in a suspended-ceiling project realized by Bartenbach. The fundamentals obviously leave room for other design requirements such as the avoidance of glare, industrial design, or ease of installation in suspended ceilings.

Exercise 4.17: Glare

You are sitting at your computer desk (with a screen luminance $L_B = 300$ nit) and receive illumination from a Lambertian emitter ($L = 1$ Mnit, seen under an angle of 0.001 sr) hanging from the ceiling in front of you (Figure 4.22). Would this be an acceptable working condition in terms of glare (DIN EN 12464-1)? Discuss (and take into account Figure 4.23)!

Figure 4.22 Glare in the office.

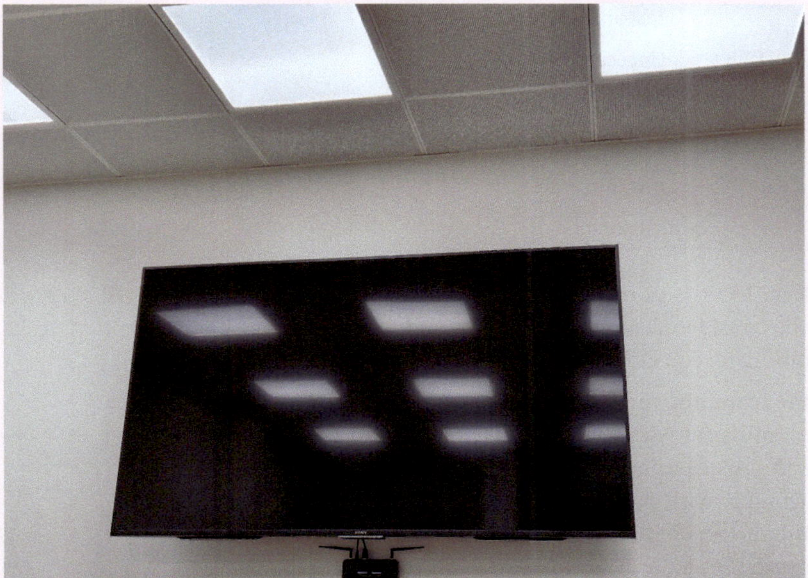

Figure 4.23 Glare from a class room screen. Remember the good old times when slides were *projected* at a white wall.

Knowledge Corner Topic 4.3: Unified glare rating

The good news is that the glare level can be calculated. The most common glare measure is the unified glare rating (UGR) defined by [10],

$$\mathrm{UGR} = 8\,\lg\left(\frac{0.25}{L_\mathrm{B}}\sum_i \frac{L_i^2\Omega_i}{p_i^2}\right) \tag{4.74}$$

with the background luminance L_B, the luminances L_i of the fixtures under consideration, and the solid angles Ω_i of the fixtures as seen by the observer [11]. (According to ISO 80000, $\lg = \log_{10}$ is the symbol for the decadic logarithm; ln for the natural logarithm).

The p_i are the *position factors* [12] or *Guth indices*, which take into account how far away from your line of sight a specific source is. On the line of sight, p assumes its minimum value of 1, decreasing the UGR contribution from light sources away from the light of sight.

UGR values are classified as follows [13, 14]:

UGR	Classification
10	Imperceptible
13	Just perceptible
16	Perceptible
19	Just acceptable
22	Unacceptable
25	Just uncomfortable
28	Uncomfortable
31	Just intolerable
34	Intolerable

Here, UGR < 19 relates to an office lighting situation, while UGR < 22 will still be considered acceptable for, say, industry workplaces and reception desks, according to, e.g. DIN EN 12464-1.

Order from above:
Our luminaires with UGR = 18.9 are too large! The customers love the distribution, but they want smaller luminaires that still comply with the specs. I promised them a new luminaire with half the diameter, creating the same light distribution, and with the same UGR value. Now do your job!

The UGR formula is a little complicated, making it hard to understand how luminaire design changes affect the result. We offer one remark: The nature of the logarithm effectively *adds* 8 lg(c) to the UGR when the argument of the lg is *multiplied* by c. Now, as a lighting designer, you will normally not just change the luminance. You typically choose luminaires with horizontal exit apertures and a certain intensity distribution $I(\theta)$ to achieve your lighting goals, where θ is the angle w. r. t. the vertical. As a designer in luminaire product development, that given $I(\theta)$ is a product specification your design must meet. Now, when you reduce the luminaire aperture area by some factor c, you reduce Ω_i by c. Thus, L_i must increase by the same factor c to keep $I(\theta)$ constant, so you change $L_i^2\Omega_i$ to $(c\,L_i)^2\Omega_i/c = c\,L_i^2\Omega_i$. Decreasing the luminance aperture area by a factor of c, i.e. decreasing aperture diameter by $\sqrt{c}$ as a luminaire design decision, adds 8 lg(c) to the UGR result. Half the luminaire diameter translates to adding 4.8 to the UGR. In conclusion: What Ike demands will not work – a luminaire with a UGR of $18.9 + 4.8 = 23.7$ is not a viable office lighting product.

Several software tools are available to compute UGR values. We are aware of Dialux®(a freeware version including UGR is available), AGI32®, and LightTools®, but there are certainly more options available. Shortcomings of the UGR standard have been recognized, especially regarding luminaires with non-uniform luminance [11]. CIE has published a new standard, CIE 232:2019 [15] which addresses these issues; a free software program that implements this standard is available from TechnoTeam.

Solution 4.17

With the specified values and assuming $p = 1$, Eq. 4.74 yields a value of 47, completely unacceptable and much larger than the value recommended for computer desks (19).

Provided file Glare.xlsx.

Dr. Wisenheimer:

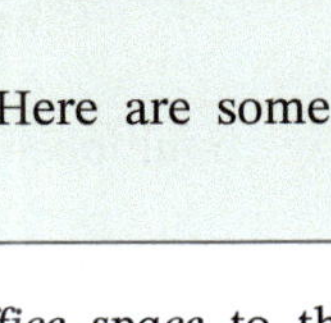

A Lambertian source should not be installed on the ceiling without further measures because of the glare it induces. Instead, it has to be shielded by *deglaring plates* or other means that block rays at shallow angles.

We need to reduce the luminance of the ceiling source in the direction of the eye. With mere reduction to $L = 10$ knit, e.g. by a de-glaring foil, we obtain UGR = 15 which is acceptable in terms of the standard.

The state of the art for classrooms is described in [16]. Here are some de-Glaring plates.

Next, we move to a completely different field: from the *office space* to the *outer space*.

4.3 Solar Collection, Astronomy and Space

Exercise 4.18: Solar collection

a. Calculate the irradiance of sunlight obtained with an ideal lens of $f = 100$ mm and a diameter of $2r_L = 50$ mm, mounted on a satellite.
b. What is the maximum possible irradiance with the best optics you can imagine?
c. Replace the lens from (a) by a disk source with the same radiance as the Sun. Calculate the irradiance on the target.

Solar quantities are presented in Knowledge Corner Topic 3.19 and in the provided file Sun.xlsx.

Solution 4.18

a. The image radius is

$$r_T = f \tan \theta_S = 100\,\text{mm} \tan 0.266° = 0.47\,\text{mm} \tag{4.75}$$

All light that hits the lens aperture will end up in the image area. Thus, the concentration with such a lens is

$$C = \frac{E_T}{E_S} = \frac{r_L^2}{r_T^2} = 2829 \tag{4.76}$$

Hence, the irradiance in the image of the Sun is

$$E_T = 1360\,\frac{\text{W}}{\text{m}^2} \cdot 2829 = 3.8\,\frac{\text{MW}}{\text{m}^2} \tag{4.77}$$

b. We combine Eqs. (4.75) and (4.76) and get HANDLE WITH CARE

$$C \approx \left(\frac{r_L}{f \sin \theta_S}\right)^2 = \left(\frac{1}{2F \sin \theta_S}\right)^2 \tag{4.78}$$

with the $F/\#$ value

$$F = \frac{f}{2r_L} \tag{4.79}$$

The lowest possible $F/\#$ is $F = 0.5$, which yields

$$C_{\text{max}} = \frac{1}{\sin^2 \theta_S} \approx 46\,400 \tag{4.80}$$

multiplied by the standard solar constant,

$$E_{\text{max}} = C_{\text{max}} \cdot E = 46\,400 \cdot 1.36\,\frac{\text{kW}}{\text{m}^2} = 63\,\frac{\text{MW}}{\text{m}^2} \tag{4.81}$$

(cf. Exercise 5.3), identical to the Sun's radiant exitance, and in accordance with the *brightness theorem* [17].

c. Question (c) is attributed to Lambert (cf. [18, § 23]).

Our way to a solution is the following: the system etendue is given by the lens–image light tube (and we are using the far-field etendue expression from Eq. 3.42)

$$U_T = (\pi\, r_T \sin\theta_L)^2 \approx \left(\pi\, r_T \frac{r_L}{f}\right)^2 = \left(\pi \frac{r_T}{f} r_L\right)^2 \approx (\pi\, r_L \sin\theta_S)^2 \quad (4.82)$$

and with a source of illuminance L_S at the lens position, we obtain the irradiance

$$E_T = \frac{\Phi}{A_T} = \frac{L_S U_T}{A_T} = E_S\left(\frac{r_L}{r_T}\right)^2 = E_S C \quad (4.83)$$

which is the same result as with the Sun itself and the lens.

Exercise 4.19: Large telescope

Imagine you could point the ELT (*Extremely Large Telescope*, first light scheduled for 2028, Figure 4.24) at the full Moon. Calculate the illuminance in the primary image located in the so-called "Nasmyth" focus formed by the first three mirrors!

Figure 4.24 Artist's impression of the Extremely Large Telescope. *Source*: Courtesy of ESO.

(Continued)

Exercise 4.19: (Continued)

ESO provides a good description of the ELT. Some data you may need:

Quantity	Value
Illuminance of sunlight	$E_M = 137$ klx (Eq. 3.224)
Moon surface scattering	Lambertian
Moon albedo	$\varrho = 10\%$ [19]
Diameter of primary mirror	$d_{ELT} = 39.3$ m
$F/\#$ of M1–M3 system	$F/17.4$
Angular extent of the Moon from Earth	$\theta_M = 0.27°$

The latter angle is much larger than the FoV of the telescope. The fact that the angular subtense of the Sun and of the Moon are almost identical is a pure coincidence.

(ELT optical design data taken from ESO.)

Solution 4.19

For the calculation of the image illuminance, we need the luminance of the Moon as the source:

$$L_M = \frac{E_M}{\pi \text{ sr}}\varrho = \frac{137 \text{ klx}}{\pi \text{ sr}} \cdot 10\,\% = 4360 \text{ nit} \tag{4.84}$$

The etendue formed by the Moon and the telescope is (Eq. 3.42)

$$U_{M,ELT} = AX = (19.6 \text{ m})^2 \pi \cdot (\sin 0.27°)^2 \pi \text{ sr} = 0.084 \text{ m}^2\text{sr} \tag{4.85}$$

so the flux of the Moon collected by the telescope is

$$\Phi_{M,ELT} = L_M U_{M,ELT} = 4360 \text{ nit} \cdot 0.084 \text{ m}^2\text{sr} \approx 366 \text{ lm} \tag{4.86}$$

The focal length of the telescope is

$$f = d_{ELT} \cdot F/\# = 684 \text{ m} \tag{4.87}$$

so the geometric radius of the Moon's image is

$$r = f \tan \theta_M = 3.2 \text{ m} \tag{4.88}$$

which is very likely larger than the ELT's actual sensor size or the entrance aperture of subsequent instruments. The illuminance in the image is then

$$E_{\text{image}} = \frac{\Phi_{M,ELT}}{\pi r^2} \approx 11 \text{ lx} \tag{4.89}$$

which is valid even if the telescope shows only a portion of the Moon.

Knowledge Corner Topic 4.4: Star photometry

Among astronomers, the radiant flux of a star is often called its *luminosity*. This may be acceptable, but beyond that, many of the basic astronomy books present a perfect confusion of radiometric and photometric terms and units. For example, the ubiquitous "brightness" is used for various quantities such as illuminance and luminance.

So let's start from scratch.

The luminosity Φ_S of a star can be calculated from its temperature and surface area (Eq. 3.196). The inverse square law (Eq. 3.142) states that the illuminance on Earth is (neglecting atmospheric absorption)

$$E = \frac{\Phi_S}{4\pi D^2} \tag{4.90}$$

with the star's distance D.

The *apparent brightness* of stars, a dimensionless relative value, is measured in *magnitudes*, using a logarithmic scale [20, Ch. 6.3],

$$m_1 - m_2 = -2.5 \log_{10} \frac{E_1}{E_2} \tag{4.91}$$

which means

$$\frac{E_1}{E_2} = 10^{-0.4(m_1-m_2)} \tag{4.92}$$

For once, we use the term "brightness" here because it is primarily related to an observation with the naked eye.

The illuminance (as observed on Earth) ratio of two stars of magnitudes m and $m+1$ is then [21]

$$\frac{E(m)}{E(m+1)} = 10^{0.4} = 2.512 \tag{4.93}$$

which was first suggested by Halley [22] and finally introduced by Pogson [23]. An illuminance ratio of 100 corresponds to a difference of five magnitudes.

Eq. 4.91 can be written as

$$m = -2.5 \log_{10} \frac{E}{\text{lx}} + m_0 \tag{4.94}$$

or

$$E = 10^{-0.4(m-m_0)}\text{lx} \tag{4.95}$$

with the constant m_0 chosen for historical reasons such that an *apparent magnitude* of $m = -14.2$ will correspond to $E = 1$ lx [24]. Hence,

$$E(m) = 10^{-0.4(m+14.2)} \text{ lx} \tag{4.96}$$

If we observed a star from two different distances, we would find

$$m_1 - m_2 = -2.5 \log_{10} \left(\frac{D_2}{D_1}\right)^2 \tag{4.97}$$

(Continued)

Knowledge Corner Topic 4.4: (Continued)

The *absolute magnitude* M of a star (of apparent magnitude m and actual distance D) is related to a distance of $D_0 = 10\,\text{pc} = 32.6$ light years

$$M = m - 2.5\ \log_{10}\left(\frac{D}{D_0}\right)^2 \tag{4.98}$$

to be able to compare different stars regardless of their distance.

Dr. Wisenheimer:

How does the magnitude of a star relate to its luminance or radiance? Did we not learn that luminance is conserved from the star's surface to our eyes (neglecting atmospheric absorption, Knowledge Corner Topic 3.11), and that a star's luminance depends only on its temperature (Figure 3.52)? Should we therefore not assume that all stars with the same temperature will look equally bright (which they obviously don't)?

The answer is *resolution*. The theoretical luminance definition assumes infinitesimal area, solid angle and flux. However, all luminance measurements, and all excitations of a single receptor in our retina, require finite power: They average over a finite area, solid angle and flux. A distant star is way too small to let surface features be resolved by even the strongest telescopes, let alone luminance meters or our eyes. There are two limits to observe: The receptor or pixel size, and the resolution (point spread function size) of the optical system, which is – at best – diffraction-limited. Let us denote the larger of both areas (receptor or pixel) A_{min}. If the image of the feature we want to observe is much larger than A_{min}, luminance conservation applies as in geometric optics; it is much smaller, then we effectively distribute the flux from the star incident on the optic's entrance over the etendue of the light tube from the exit aperture to A_{min}. This latter etendue cannot be smaller than λ^2 sr, the etendue of a single mode [25]. In between, there is a smooth transition, which raises interesting theoretical questions about instrument functions; see e.g. Chapter 12 of Winston's Nonimaging Optics book [26].

Exercise 4.20: Double star

The binary star system Alkor and Mizar in the Ursa major constellation have apparent magnitudes of 4.0 and 2.06, respectively (Figure 4.25). What is the common apparent magnitude of the two if unresolved?

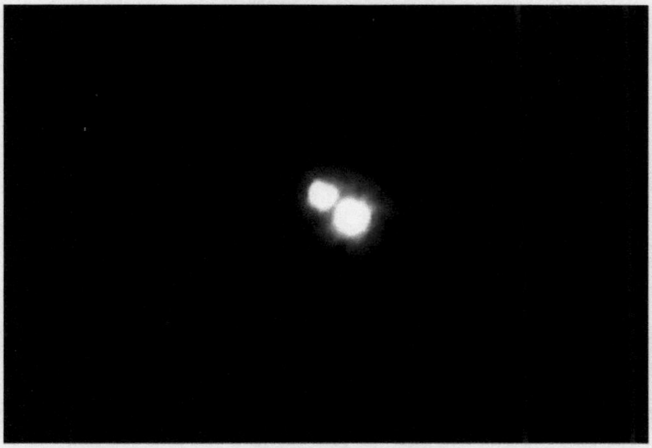

Figure 4.25 Mizar and Alkor. *Source*: Thomas Bresson/Wikimedia Commons/CC BY 2.0.

Solution 4.20

We may not add magnitudes – adding logarithms means multiplication of the underlying quantities. We need to use Eq. 4.96 and add the individual illuminance values. Alkor and Mizar contribute

$$E_{\mathrm{A}} = 10^{-0.4(4+14.2)}\,\mathrm{lx} = 5.25 \cdot 10^{-8}\,\mathrm{lx} \tag{4.99}$$

and

$$E_{\mathrm{M}} = 10^{-0.4(2.06+14.2)}\,\mathrm{lx} = 3.13 \cdot 10^{-7}\,\mathrm{lx} \tag{4.100}$$

Combined we have

$$E = E_{\mathrm{M}} + E_{\mathrm{A}} = 3.66 \cdot 10^{-7}\,\mathrm{lx} \tag{4.101}$$

Therefore, our result is

$$m_{\mathrm{M+A}} = -2.5 \log_{10}\left(\frac{E}{\mathrm{lx}}\right) - 14.2 = 1.9 \tag{4.102}$$

Exercise 4.21: Sirius (1)

What is the irradiance and illuminance of Sirius' light on Earth? Calculate its apparent magnitude.

(Continued)

Exercise 4.21: (Continued)

Some data about Sirius:

Quantity	Value
Surface temperature	9900 K [27]
Full angular extent	$5.7 \cdot 10^{-3}$ arcsec [28]

Solution 4.21

Equation 3.196 yields a radiance $L_{\mathrm{Si}} = 170\,\mathrm{MW/m^2sr}$ for Sirius. The half angle of Sirius is $\theta_{\mathrm{Si}} = 1.38 \cdot 10^{-8}$ rad, and its solid angle is $\Omega_{\mathrm{Si}} = 6.0 \cdot 10^{-16}$ sr. Its irradiance on Earth is therefore

$$E_{\mathrm{E}} = L_{\mathrm{Si}} \cdot \Omega_{\mathrm{Si}} = 1.04 \cdot 10^{-7}\,\frac{\mathrm{W}}{\mathrm{m}^2} \tag{4.103}$$

We take the luminous efficacy of the radiation to be $K = 74\,\mathrm{lm/W}$ at the given temperature from the file CIE_Planck.xlsx and obtain an illuminance

$$E_{\mathrm{v}} = KE_{\mathrm{E}} = 7.7\,\mu\mathrm{lx} \tag{4.104}$$

The related magnitude is found from Eq. 4.96:

$$m = -2.5\log_{10}\frac{E}{\mathrm{lx}} - 14.2 = -1.42 \tag{4.105}$$

which compares well with values from literature or Wikipedia.

Provided file: Sirius.xlsx.

Exercise 4.22: Sirius (2)

What is the maximum concentration ratio for light from Sirius? Derive the key properties of a fictitious optical system to implement the collection of Sirius light. Calculate the necessary optics diameter for a circular solar cell of $r_{\mathrm{T}} = 1\,\mathrm{mm}$.

Order from above:

Since a CPC is thermodynamically perfect, it is mandatory to use it here as well!

Solution 4.22

With the angular subtense of Sirius of $\theta_{Si} = 5.7 \cdot 10^{-3}$ arcsec [28], we obtain

$$C_{max,Si} = \frac{1}{\sin^2 \theta_{Si}} \approx 5.2 \cdot 10^{15} \tag{4.106}$$

For the lens system, we need a lens of a focal length

$$f = \frac{r_T}{\sin \theta_{Si}} = 72\,\text{km} \tag{4.107}$$

which means that an ideal $F/0.5$ lens would need to have a diameter

$$D = \frac{f}{F/\#} = 2f = \frac{2r_T}{\sin \theta_{Si}} \approx 144\,\text{km} \tag{4.108}$$

Ideally, this lens would be designed free from spherical aberrations.

Since this is not realistic, one may indeed try a CPC as suggested. Its dimensions are r_T on the receiver side, R on the light entrance (i.e. Sirius) side, and the same ratio as for the lens applies, so

$$R = \frac{r_T}{\sin \theta_{Si}} = 72\,\text{km} \tag{4.109}$$

The length of such a CPC (we assume a hollow one, with $n = 1$) would be

$$L \approx \frac{R}{\tan \theta_{Si}} \approx \frac{r_T}{\theta_{Si}^2} \approx 5 \cdot 10^9\,\text{km} \tag{4.110}$$

which is roughly the radius of the Neptune orbit. Perhaps we might be able to bring it down to the radius of Jupiter' orbit without too much loss of performance. Anyway, it was a good decision to make it hollow ...

In comparison, the above lens system appears almost feasible.

Let's better get back indoors.

4.4 Daylighting

Exercise 4.23: Light from the opposite house wall

A desk is illuminated with light that is scattered diffusely from the opposite building wall (distance $D = 10$ m, reflectance $\eta = 71\%$). Sunlight is shining ($E_S = 118$ klx) at the wall under 45° (Figure 4.26).

Can we reach the standard $E = 500$ lx goal of desk lighting (cf. Exercise 4.2)?

(Continued)

Exercise 4.23: (Continued)

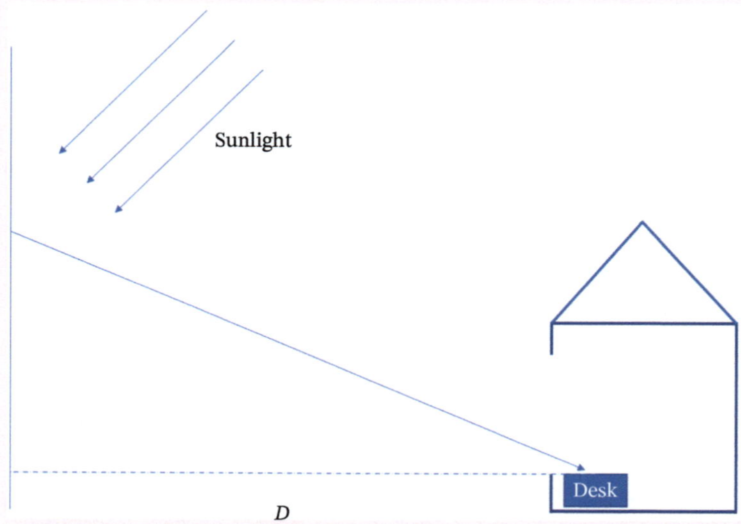

Figure 4.26 Daylighting example.

Solution 4.23

We consider an area element dA on the desk. Our plan is to estimate the illuminance dE_T on the desk that is caused by light scattered from an area element dA at the wall.

The luminance of the wall is

$$L_W = \frac{E}{\Omega_p} = \eta \frac{E_S \cos 45°}{\pi \text{ sr}} \approx 0.71 \frac{118 \text{ klx} \cdot 0.71}{\pi \text{ sr}} \approx 18 \text{ knits} \tag{4.111}$$

We have already estimated the etendue defined by dA and a half hemisphere in Exercise 3.1. With Eq. 3.34, we obtain for the desk illuminance

$$E_T = \frac{L_W U}{\text{d}A} = L_W \Omega_p = L_W \cdot \frac{\pi}{2} \text{sr} \approx 28 \text{ klx} \tag{4.112}$$

for an infinite-size scattering area.

That would be far more than enough for doing paperwork, in terms of the 500 lx office lighting standard. Of course, glare will be an issue here.

Realistic wall sizes and shapes are difficult to analyze analytically. We need a simulation for this.

Simulation 4.8: Daylighting simulation

We place a more realistic $10 \cdot 10\,\text{m}^2$ source at a distance of 10 m. The emitted flux is set to $\Phi_S = 6 \cdot 10^6$ lm to realize $L_v = 18$ knit as calculated above.

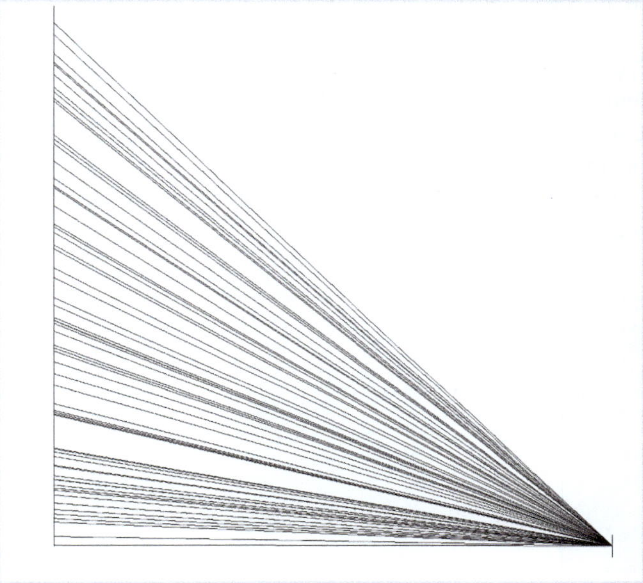

Figure 4.27 Daylighting exercise solved by a simulation: The Lambertian wall source (set to to achieve the calculated wall luminance of 18 knit) is on the left. A backward simulation was done from the receiver (right) to the source (left).

Provided file Daylighting.lts.

Dr. Wisenheimer:

This problem literally begs to be solved by *backward ray tracing*, also known as *reverse ray tracing*. If the receiver looks small as seen from a source that radiates into a large solid angle (like here), the problem is that nearly all rays will miss the receiver (Figure 4.27). Without special measures, we need a huge number of rays to obtain reasonably low noise information about the light distribution on the receiver.

If the receiver location in the angular space of the source is known, however, we can use one such special measure: Some software packages support applying an *aim area* to the source, which ensures that only those rays are actually traced that have a chance to reach the aim area and thus the receiver.

(Continued)

Simulation 4.8: (Continued)

In more complicated situations, however, it is more promising to look at the problem from the receiver side: If the source looks large, as seen from the receiver, a better approach is *backward ray tracing*. Here, the roles of receiver and source are reversed. Essentially, we shoot "vision rays" back into the optical system and find which luminance we see coming from there to determine the integrand in $E = \int L\,\mathrm{d}X = \int L(\theta, \varphi) n^2 \sin(\theta)\,\mathrm{d}\theta\,\mathrm{d}\varphi$. The interpretation is somewhat similar to etendue measurement via ray tracing, introduced in Simulation 3.2.

Backward raytracing is discussed extensively in computer graphics [29, 30]. It can be a valuable tool in stray light analysis, too [31].

With only 10*k* backward rays, a result equivalent to the forward simulation with 10 million rays can be obtained (Figure 4.28).

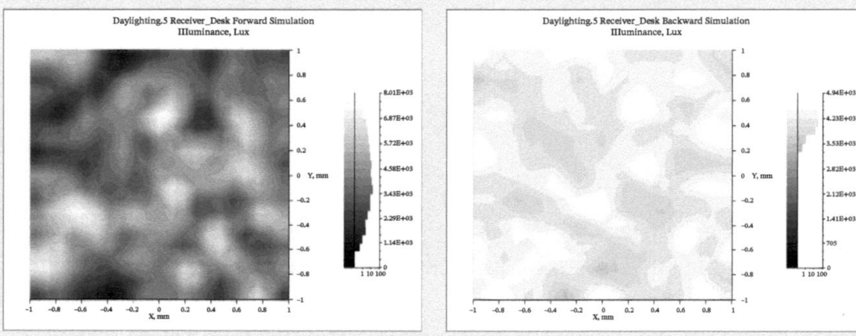

Figure 4.28 Simulation results of forward (left) and backward (right) simulations with 10 million and 10 000 rays, respectively. Both show roughly the same illuminance level, but the backward result has less peak-to-valley variations.

The result of 5 klx is smaller than the above estimate of 18 klx because of the limited wall size, but still way exceeding the requirements for office lighting illuminance. A hundredfold increase of the source area would already double the desk illuminance value.

Exercise 4.24: Translucent roof element

Consider a translucent and scattering roof window (horizontally arranged, area 1 m^2). Half of the incident light is scattered downwards. Calculate the number of 40 W bulbs from Exercise 4.1 to produce the same luminous flux as the window with full solar illuminance under 45° on a clear day.

Solution 4.24

The flux is

$$\Phi = E_S \cdot \cos 45° \cdot 0.5 \cdot A \approx 42\,\mathrm{klm} \tag{4.113}$$

Thus, we need approximately 104 of the 40 W light bulbs for the same luminous flux. Amazingly little daylighting would suffice to replace artificial light during the day.

Theory and practice: Just for the record – we ignored window transmission losses in this analysis. However, there is a factor that is more crucial: When there is even a single cloud covering the Sun, the illuminance drops by a factor of about ten – and this may happen suddenly. Artificial lighting combined with daylighting therefore requires automated control. Moreover, the color temperature of daylight varies over the day, and when the supplementing artificial lighting does not follow that pattern, people living and working in the building will be affected. On the other hand, the dynamic changes in daylight have a positive influence on the well-being of inhabitants. Just think how you feel in a windowless cubicle room with dull, constant artificial lighting, as opposed to working in a room not too far from large windows. Even when you rarely look outside (after all, you should be working!), you feel the rhythm of nature outside. Good artificial lighting should only reduce the dynamic to a comfortable level, not completely compensate it.

Exercise 4.25: Dome lighting

Sun from the top! You are dreaming of a large dome source (concave hemisphere, radius r_S, distance D from sphere center to the floor, flux Φ_S) in the ceiling of your living room. The source is Lambertian and ideally backscattering (Figure 4.29).

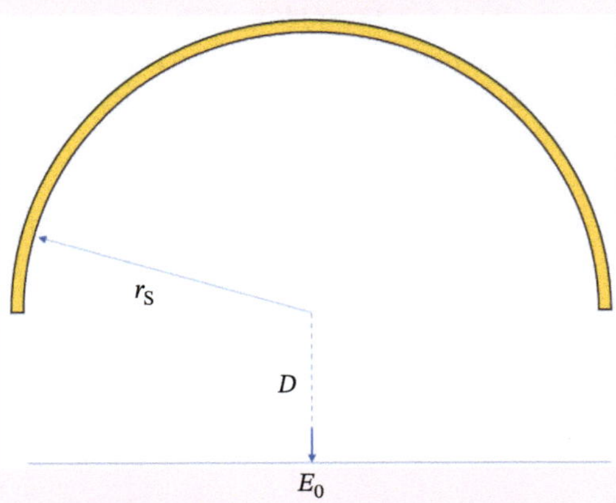

Figure 4.29 Light emitting dome.

Calculate the source luminance. What is the illuminance on the floor? Calculate the center value E_0 (see Figure 4.29) and make an educated guess for the distribution.

Solution 4.25

At a first glance, the luminance should be defined via the source flux Φ_S divided by hemispheric source area A_S and the angular size, i.e. the projected solid angle π sr, of all of its area elements,

$$L_S = \frac{\Phi_S}{\pi\,\text{sr} \cdot A_S} = \frac{\Phi_S}{\pi\,\text{sr} \cdot 2\pi\, r_S^2} \tag{4.114}$$

On the other hand, the complete flux is leaving the dome through the base area A_T

$$L_T = \frac{\Phi_S}{\pi\,\text{sr} \cdot A_T} = \frac{\Phi_S}{\pi\,\text{sr} \cdot \pi\, r_S^2} \tag{4.115}$$

leading to a higher effective luminance of such a *secondary source*. So what is really going on here? We will check this by means of a simulation.

Dr. Wisenheimer:

This higher luminance value is actually "implemented by physics" on the dome by *light recycling* [32]: rays hitting another part of the source and being reflected there. This is a legitimate means to increase the luminance of a source and does not contradict the Second Law, but the system has to be implemented properly (other than in Exercise 3.19 or 4.33; see also our discussion surrounding Kirchhoff's law in Knowledge Corner Topic 3.21).

Without any optical components such as retro-reflectors [33, 34], the effect is described by the *self configuration factor* (cf. Eq. 3.25 and [35–37]).

For $D > 0$, the floor disk will catch only a part of the emitted flux that can be calculated via the light tube etendue (Eq. 3.47) multiplied by L_T. For the center illuminance, we need the projected solid angle of the dome base,

$$\Omega_p = \pi\,\text{sr} \cdot \sin^2\theta = \pi\,\text{sr} \cdot \frac{r_S^2}{r_S^2 + D^2} \tag{4.116}$$

and

$$E_0 = L_T \cdot \Omega_p = \frac{\Phi_S}{\pi} \frac{1}{r_S^2 + D^2} \tag{4.117}$$

which includes the case $D = 0$.

Area elements on the edge of the floor will see the dome under a smaller projected solid angle than those of the center. We expect a lower illuminance at the rim, an effect we discussed already in Exercise 4.13. For $D = 0$, we expect a homogeneous distribution, because every area element on the detector gets light from the the full hemisphere. In practice, the illuminance will be even higher because of backscattered light from the floor itself.

Simulation 4.9: Dome lighting

We set up the dome as an object source (a large concave lens surface) with $r_S = 1000\,\text{mm}$ and $\Phi_S = 100\,\text{lm}$ (Figure 4.30).

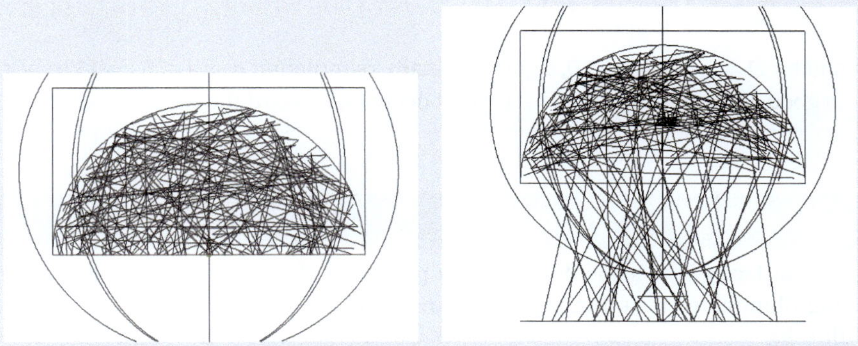

Figure 4.30 Dome shining light on the floor. Left: $D = 0$ (floor receiver at the dome center). Right: a distance $D = r_S$ between them.

Provided file: Dome.lts.

The effective luminance with these parameters is

$$L_T = \frac{100\,\text{lm}}{\pi\,\text{m}^2 \cdot \pi\,\text{sr}} \approx 10\,\text{nit} \tag{4.118}$$

and the simulation (using an additional small receiver close to the radiating surface measuring the luminance including recycling) shows a "measured" value of $L_T \approx 9.6\,\text{nit}$.

The obtained distributions on the target are shown in Figure 4.31.

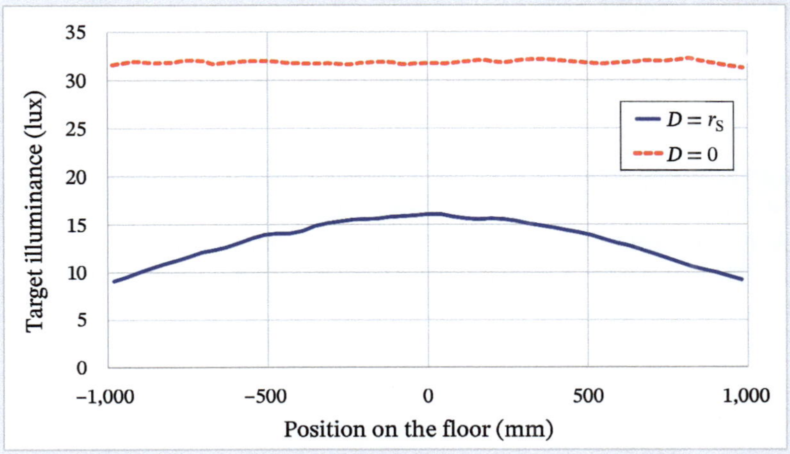

Figure 4.31 Illuminance distributions on the floor.

(Continued)

Simulation 4.9: (Continued)

For $D = 0$ (Figure 4.30, left), simulation yields $E_0 = 32\,\text{lx}$ as obtained from Eq. 4.117,

$$E_0 = \frac{100\,\text{lm}}{\pi}\frac{1}{m^2} = 32\,\text{lx} \tag{4.119}$$

and for $D = r_S$ (Figure 4.30, right) we obtain a simulation result of $E_0 = 15$ lx, also in good agreement with Eq. 4.117, that decreases toward the edge as predicted.

Exercise 4.26: Cove lighting

The surface of the same dome is now painted with an ideally scattering white and illuminated by a thin toroidal source from the rim (see Figure 4.32) with a flux Φ_S.

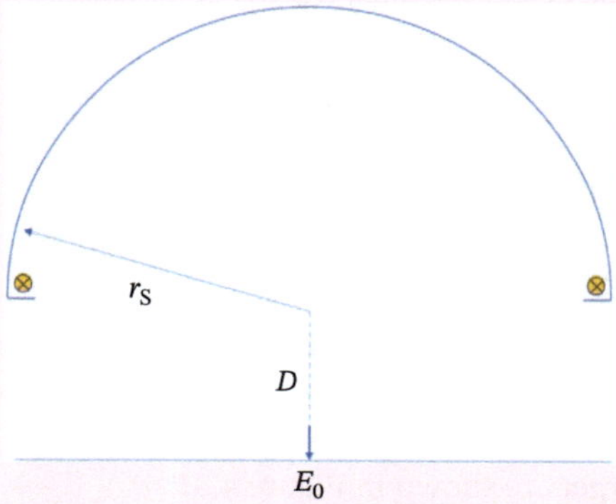

Figure 4.32 Illuminated dome or *cove*.

Is the luminance still constant over the surface of the dome?

Solution 4.26

A look in the world of interior lighting shows a widespread usage of cove lighting. There is always a bright region at the rim. Can we reproduce this by simulation?

Simulation 4.10: Cove lighting

We set up the arrangement with a toroidal light source radiating upwards only (Figure 4.33). In a real luminaire, one would probably hide an LED strip behind a baffle, also for interior architecture design reasons.

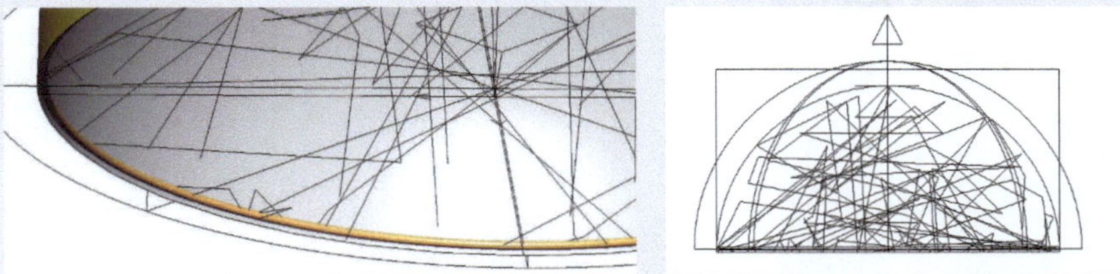

Figure 4.33 Toroidal source illuminating a cove and illuminance at the bottom of the cove.

Apart from the direct neighborhood of the source (which is not really visible from below anyway), the cove appears homogeneously illuminated (Figure 4.34).

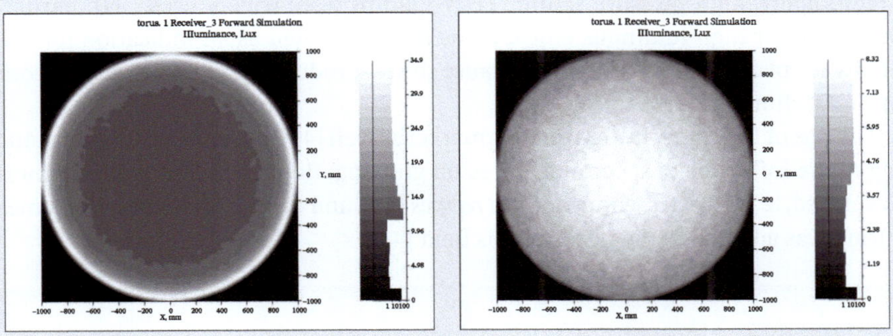

Figure 4.34 Illuminance distributions on the floor for $D = 0$ (floor receiver at the dome center) and a distance $D = r_S$ between them.

There are similar applications in architectural lighting that focus on a uniform illumination of the cove. In 2014, the Sistine Chapel in Vatican City was equipped with LED lighting (Figure 4.35) [38]. LED luminaires producing high-quality light (excellent color rendering and reduced blue, in order to not affect the medieval paint) were placed on ledges and thus hidden from visitors.

After our excursion into space and subsequent return to indoors, we will now proceed to visit a greenhouse.

Figure 4.35 Sistine Chapel illuminated by LED lights placed on the ledges on both sides *Source*: Antoine Taveneaux/Wikimedia Commons/CC BY SA 3.0.

4.5 Horticulture Lighting

Greenhouse lighting is not a new phenomenon, especially in the polar regions. Traditionally, high-pressure sodium (HPS) lamps were widely used [39], partially because their high luminous efficacy was misinterpreted as an indication for suitability for plant lighting, partially because the by-product heat was indeed welcome in a cold climate.

The rise of LEDs revolutionized this market as well [40]. An increased understanding of the influence of spectral features on growing, flowering, and photosynthesis developed, leading to research on *light recipes* [41], and even the science of light measuring was adapted to the needs of this field [42].

Exercise 4.27: Energy of a photon

How many photons per second correspond to 1 lm of monochromatic blue light at $\lambda = 460$ nm?

Solution 4.27

The $V(\lambda)$ curve has a value of 0.06 at 460 nm. Including the $K_{\mathrm{cd}} = 683$ lm/W factor, we obtain a luminous efficacy of the radiation (Eq. 3.243) of $K = 41$ lm/W.

Thus, 1 lm of blue light at 460 nm has a radiant flux of $\Phi = 24$ mW.

The energy of a photon at 460 nm is

$$Q_{\mathrm{P}} = \frac{hc}{\lambda} = 4.3 \cdot 10^{-19}\,\mathrm{Ws} \tag{4.120}$$

The photon flux is therefore

$$N_P = \frac{\Phi}{Q_P} = \frac{24\,\text{mW}}{4.3 \cdot 10^{-19}\,\text{W s}} = 5.6 \cdot 10^{16}\,\text{s}^{-1} \tag{4.121}$$

which would be usually expressed using Avogadro's constant

$$N_A = 6 \cdot 10^{23}\,\text{mol}^{-1} \tag{4.122}$$

as

$$n_P = \frac{N_P}{N_A} \approx 0.1\,\frac{\mu\text{mol}}{\text{s}} \tag{4.123}$$

Figure 4.36 Growlight. Is this a real photo or an AI fake? Discuss!

Knowledge Corner Topic 4.5: Horticulture light metrics

The portion of light that is actually used by plants is called *photosynthetically active radiation* (PAR). Originally, the range of 400–700 nm was specified for PAR, but not all wavelengths in this range are equally efficient and important, and light outside it may play a role as well [43, 44].

Photosynthesis is not a single chemical reaction, but a whole group [45]. The average photosynthetic efficiency is described by the McCree curve [46].

In horticulture, light is characterized as the input required for a specific chemical reaction, hence its amount is measured in terms of the *number* of photons instead of their total energy. Thus, the most suitable quantity to describe PAR is the *photosynthetic photon flux* (PPF), denoted by the symbol n_P with the unit µmol/s [47].

If the PPF is incident onto an area, the equivalent to irradiance is the *photosynthetic photon flux density* (PPFD), denoted by the symbol E_P with the unit µmol/(m^2 s). This unit is sometimes omitted in journals and websites (as in "tomatoes need 300 PPFD"). As radiometric and photometric units do not

(Continued)

Knowledge Corner Topic 4.5: (Continued)

provide a proper weighting for photosynthesis, LEDs may be even *specified* in terms of PAR, and fixtures may be evaluated accordingly [48].

Take a look at some horticulture LEDs.

The following table summarizes the units of horticulture lighting [49]:

Name	Acronym	Description	Symbol	Unit
Photosynthetically active radiation	PAR	Radiant flux within 400…700 nm	Φ_P	W
Photosynthetic photon flux	PPF	Total amount of PAR light	n_P	μmol/s
Spectral quantum distribution	SQD	Composition of PPF by wavelength		μmol/nm s
Photosynthetic photon flux density	PPFD	PPF per area	E_P	μmol/m^2 s
Photosynthetic photon intensity	PPI	PPF per solid angle	I_P	μmol/sr s
Photon radiance [50]	PPR	PPF per etendue	L_P	μmol/s mm^2 sr
Photosynthetic photon efficacy	PPE	PPF per electrical power	η_P	μmol/J

Dr. Wisenheimer:

Obviously, this brief introduction covers only the basics of the effects of light on plants.

Take care ! There are "LED growlights" featuring only red and blue LEDs as well as "full-spectrum LED growlights" with a little additional green or yellow light, and maybe even "far - red" or near - infrared LEDs.

Have a look at the world of light recipes and (according to the supplier) matching luminaires.
There are useful Horticulture lighting webtools.

Exercise 4.28: Stefan–Boltzmann law by photon numbers

Express the Stefan–Boltzmann law (Eq. 3.197) in terms of photon numbers.

(Exercise suggested by Rich Pfisterer.)

Solution 4.28

We define the *spectral photon radiance* [50] as (cf. Eq. 4.121)

$$L_{P,\lambda} = \frac{dL_P}{d\lambda} = \frac{N_\lambda}{U} \tag{4.124}$$

with the photon flux per wavelength interval N_λ and the (total) photon radiance

$$L_P = \frac{N_P}{U} \tag{4.125}$$

We start with the spectral radiance (Eq. 3.194; the index "e" is short for "energetic" to avoid confusion with other quantities)

$$L_{e,\lambda} = \frac{\Phi_\lambda}{U} = \frac{N_\lambda Q_\lambda}{U} = L_{P,\lambda} Q_P(\lambda) \tag{4.126}$$

Hence,

$$L_{P,\lambda} = \frac{L_{e,\lambda}}{Q_P(\lambda)} = \frac{2c}{\lambda^4} \frac{1}{\exp\left(\frac{hc}{\lambda k_B T}\right) - 1} \text{sr}^{-1} \tag{4.127}$$

With the same substitution $x = hc/(\lambda k_B T)$ as above, we get

$$L_P = \frac{2(k_B T)^3}{h^3 c^2 \text{ sr}} \int \frac{x^2 \, dx}{\exp(x) - 1} \tag{4.128}$$

with the result (cf. Eq. 3.196)

$$L_P(T) = \frac{\sigma_P T^3}{\pi \text{ sr}} \tag{4.129}$$

which is the Stefan–Boltzmann law (cf. [50]) for photon radiance (with the Riemann Zeta function ζ)

$$\sigma_P = \frac{2\pi k_B^3}{h^3 c^2} \cdot 2\zeta(3) \approx 1.52 \cdot 10^{15} \frac{1}{\text{s m}^2 \text{ sr K}^3} = 2.5 \frac{\text{nmol}}{\text{s m}^2 \text{ sr K}^3} \tag{4.130}$$

Accordingly, the dependence of photon radiance on the refractive index and on temperature is quite different.

We "prove" the numbers using the following estimate:

For the Sun with 5772 K we obtain a photon radiance of $L_P \approx 9.3 \cdot 10^{25}/(\text{s m}^2 \text{ sr})$. Let us consider 555 nm (with a photon energy of $Q_P = 3.6 \cdot 10^{-19}$ Ws) as the average wavelength of the Sun's radiation.

The product of both values is $L_S \approx 33$ MW/m² sr, which is larger than the actual value of $L_S = 20$ MW/m²sr (cf. Knowledge Corner Topic 3.19). While 555 nm is a good typical value for the visible part of sunlight, this estimate fails to account for the long infrared tail with low photon energy.

Exercise 4.29: HPS growlight

High-pressure sodium (HPS) lamps are the legacy light sources in horticulture. Estimate

- the wall-plug efficiency (WPE, Eq. 3.245),
- the luminous flux,
- and the PPF

of a 1000 W HPS lamp.

To do that, retrieve the spectrum and the datasheet of such a source.

Solution 4.29

We feed the retrieved spectra to a spreadsheet or to John Selverian's Color Calculator (Figure 4.37).

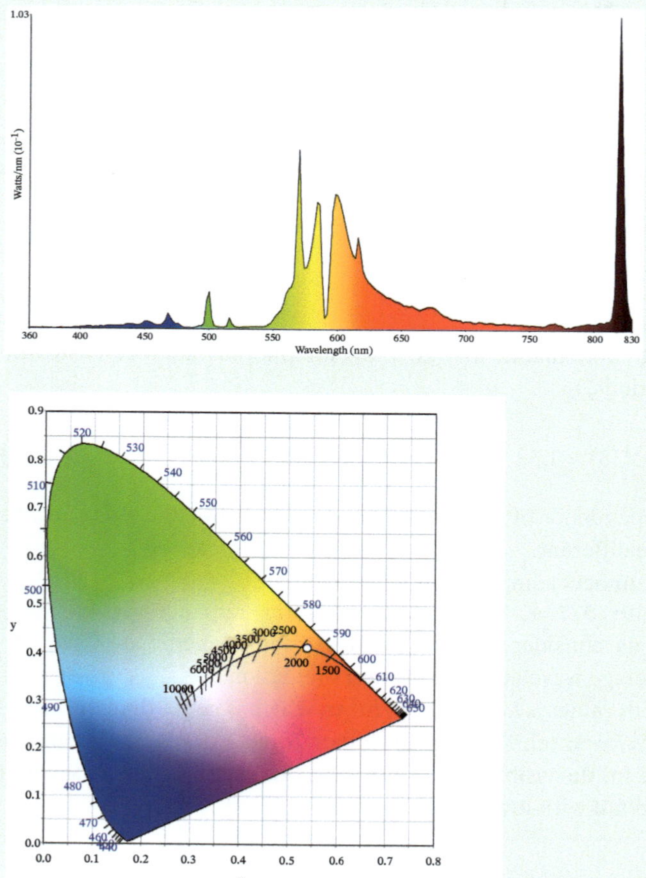

Figure 4.37 HPS spectrum and color coordinates.

We see broadened yellow sodium lines and some infrared content. From the spectrum, we obtain a luminous efficacy of radiation (Eq. 3.243) of $K = 289$ lm/W (including the IR peak). 77% of the radiant flux are within the PAR range.

The datasheets specify a luminous efficacy of $\eta_v \approx 155$ lm/W for the HPS lamp, so the radiant flux is about $\Phi = 536$ W, which means a WPE of 54%, or 46% of the electrical power being converted to heat.

Using a spreadsheet, we can scale the spectral radiant flux according to the total radiant flux of the source, calculate the number of photons per nm, and sum over the PAR range. The result is $n_P = 2030\,\mu$mol/s, close to the datasheet value of $n_P = 2120\,\mu$mol/s.

Provided file HPS.xlsx.

77% of the radiant flux being within the PAR range sounds better than it is, because in contrast to PAR's constant weighting, plants are not able to make much use of green and yellow radiation. We move on to LED devices.

Exercise 4.30: LED growlight

Growlights (Figure 4.36) benefit from LED technology, since a luminaire employing different LEDs can match green leaf absorption (Figure 3.54) much better than, for example, HPS lamps.

For the exercise, we arbitrarily choose an LED device operating at 1000 W and with a spectrum similar to that of chlorophyll absorption. The "PAR efficacy" of the device is about 3 μmol/J, and we assume Lambertian emission.

(a) Estimate the WPE (Eq. 3.245).

(b) Calculate the required distance for flowering, assuming a PPFD requirement of 1000 μmol/m^{2}s.

Solution 4.30

The PPF is

$$n_P = 3\frac{\mu\text{mol}}{\text{J}} \cdot 1000\,\text{W} = 3000\,\frac{\mu\text{mol}}{\text{s}} \tag{4.131}$$

Let us simplify the spectrum of the device by simply describing it as consisting of a blue and a red line, with half of the PPF in each channel. The photon energies (Eq.4.120) are then $Q_{P,b} = 4.5 \cdot 10^{-19}$ J and $Q_{P,r} = 3 \cdot 10^{-19}$ J for the blue and red photons, respectively.

(a) The combined radiant flux is then

$$\begin{aligned}\Phi &= Q_{P,b} \cdot n_{P,b} + Q_{P,r} \cdot n_{P,r} \\ &= 7.5 \cdot 10^{-19}\,\text{J} \cdot 1500\,\frac{\mu\text{mol}}{\text{s}} = 675\,\text{W}\end{aligned} \tag{4.132}$$

corresponding to a WPE of 67.5%.

(Continued)

Solution 4.30 (Continued)

(b) The share of the total flux which is reaching the target is

$$\Phi_{\mathrm{T}} = \frac{\Omega_{\mathrm{T}}}{2\pi\,\mathrm{sr}}\Phi \tag{4.133}$$

With Eq. 2.8 this becomes

$$E_{\mathrm{T}} = \frac{1}{2\pi\,D^2}\Phi \tag{4.134}$$

and thus we get (translated into the photon-flux picture)

$$D^2 = \frac{1}{2\pi}\frac{\Phi}{E_{\mathrm{T}}} = \frac{1}{2\pi}\frac{3000\,\mu\mathrm{mol/s}}{1000\,\mu\mathrm{mol/(m^2\,s)}} = 0.48\,\mathrm{m}^2 \tag{4.135}$$

with the result $D = 0.7$ m.

Dr. Wisenheimer:

The HPS technology provides a PAR efficacy of approximately 40%, but even this radiation is not well-matched to the needs of plants with their red and blue absorption peaks. Only for greenhouses in polar regions, there may be still some HPS lamps involved because their heat may be welcome in this case.

In contrast, the spectrum of LED growlights can be tailored to the needs of plants. The above WPE of 67.5% is based on a red-blue-dominated spectrum. Such high values are serendipitous: Due to sheer luck, InGaN LEDs have their highest efficiency in the deep blue, and InGaAlP LEDs have theirs in the deep red, thus offering an almost perfect match to the chlorophyll absorption. Consequently, LED-operated greenhouses will be much more efficient, especially economically, and this is what we observe on the market. In many cases, one would add some green or white and some far red (730 nm) to serve some secondary needs of plants.

4.6 Light Detection

Exercise 4.31: Screen luminance

How can you use a photometer (an instrument that measures illuminance on a small sensor, typically a few millimeters in diameter) to estimate the luminance of a TV screen? How would you proceed?

Solution 4.31

If you press the sensor head directly on the screen, it shows the screen's *luminous exitance* E_S. Assuming that the TV screen is a Lambertian emitter, its luminance is

$$L_S = \frac{E_S}{\Omega_p} = \frac{E_S}{\pi \text{ sr}} \tag{4.136}$$

However, you might be (rightly!) worried about how hemispherical the photometer really is, and how Lambertian the screen is. An alternative is to make a cube out of (for example) black cardboard, with a side length about ten times the sensor diameter. Cut a hole somewhat larger than the photometer sensor diameter into one side (A) and another hole on the opposite side (B), such that the second hole hold subtends an angle of about $\alpha = \pm 20°$. When you now hold the photometer sensor at hole A and orient hole B toward the screen, the screen luminance is again given by

$$L_S = \frac{E_S}{\Omega_p} = \frac{E_S}{(\sin 20°)^2 \pi \text{ sr}} \tag{4.137}$$

Knowledge Corner Topic 4.6: Integrating spheres

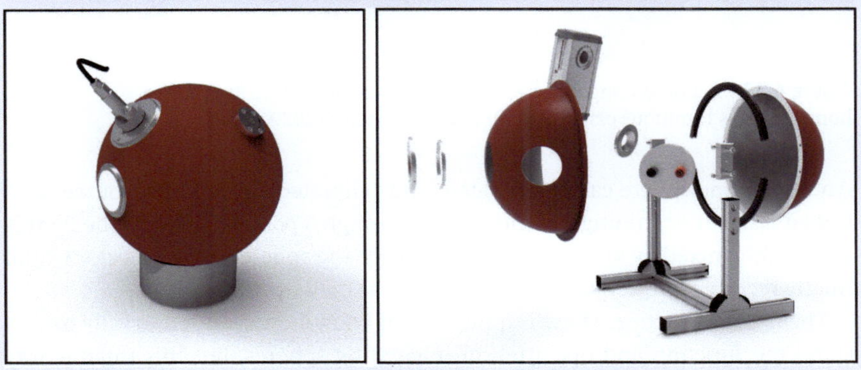

Figure 4.38 Left: An integrating sphere with two ports and a bayonet for a sensor head. Right: Blowup of a similar sphere. Integrating spheres are typically made of two halves for easy painting and maintenance.

Integrating spheres are helpful detector accessories to enable reliable flux measurements, independent of the spatial and angular luminance distribution of the incident light (Figure 4.38). The light that is generated by a source in the sphere or entering through an aperture, will be integrated, and a small sensor at the wall will show a signal that is proportional to the total flux of the source.

(Continued)

Knowledge Corner Topic 4.6: (Continued)

Their invention is attributed to the German railway engineer Richard Ulbricht [51].

Other shapes may work as well, such as the shoebox shown in Figure 4.39.

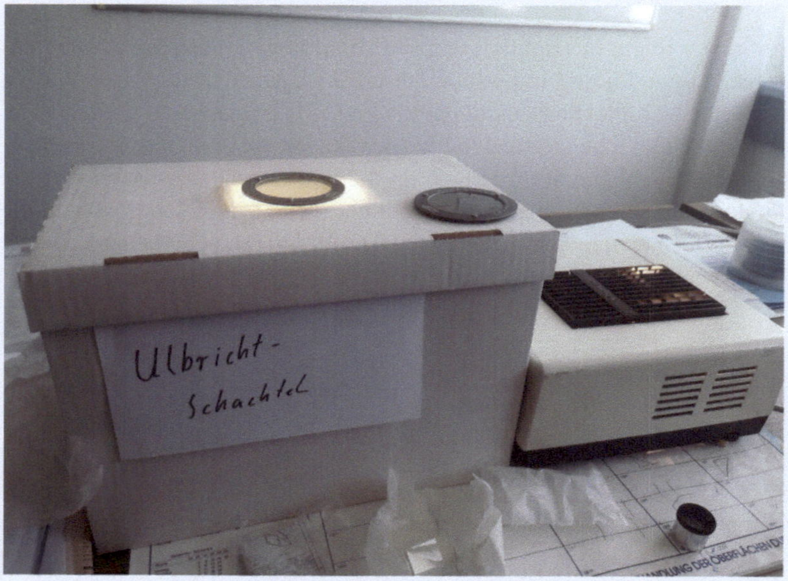

Figure 4.39 Eckhard Langenbach's integrating shoebox set up to produce homogeneous luminous exitance from a slide projector (2020 at Fisba AG).

An integrating sphere can be operated (and calibrated) with a light source positioned *inside* it or shining into the sphere through a *port*. Either way, the light is thoroughly mixed by multiple scattering inside the sphere, and a detector would finally receive a portion of the light through a small opening.

Theory and practice: There is quite a lot of know-how required to fully understand the function and operation of integrating spheres, but this topic is way too extensive for this book. A starting point is provided by Lighting Global [52].

Exercise 4.32: Integrating sphere (1)

A Lambertian emitter is placed at a port and radiates into the sphere (Figure 4.40).

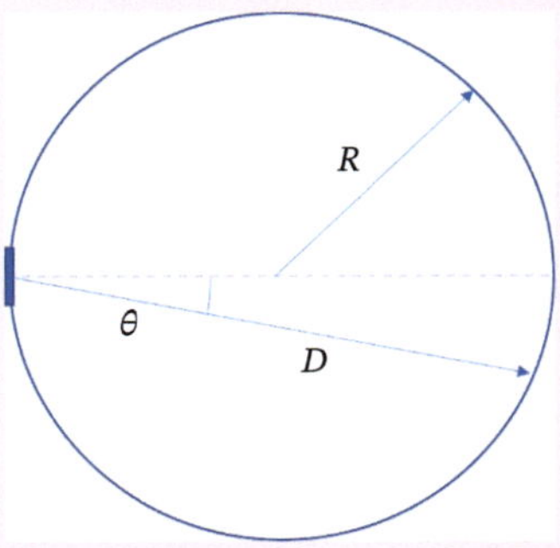

Figure 4.40 Lambertian emitter radiating into a sphere (radius R).

Calculate the illuminance distribution at the opposite wall of the sphere (considering only a single pass).

Solution 4.32

Basic trigonometry yields

$$D = 2R\cos\theta \tag{4.138}$$

hence illuminance is found using the inverse square law (Eq. 3.142), the intensity of a Lambertian emitter (Eq. 3.111), and the factor $\cos\theta$ for obliquity,

$$E(\theta) = \frac{I(\theta)}{D^2}\text{sr}\cos\theta = \frac{I_0\cos\theta}{4R^2\cos^2\theta}\text{sr}\cos\theta = \frac{I_0}{4R^2}\text{sr} = E_0 \tag{4.139}$$

We obtain a constant illuminance at the opposite surface of the sphere.

Now, each surface element of the sphere reflects like a Lambertian source. Even if the actual source is not at all Lambertian, the reflected light (second pass) will illuminate the sphere homogeneously – this is why integrating spheres work so well. After the second reflection, the light has completely "forgotten" where it came from.

Dr. Wisenheimer:
This example features the same geometry as the *Rowland Circle* [53] (a spectrometer arrangement with a curved grating).

Simulation 4.11: Integrating sphere

We place a receiver at a concave surface and a source at the opposite side. The concave surface is absorbing, to ensure that only the light directly from the source is considered. As shown in Figure 4.41, we obtain indeed constant illuminance.

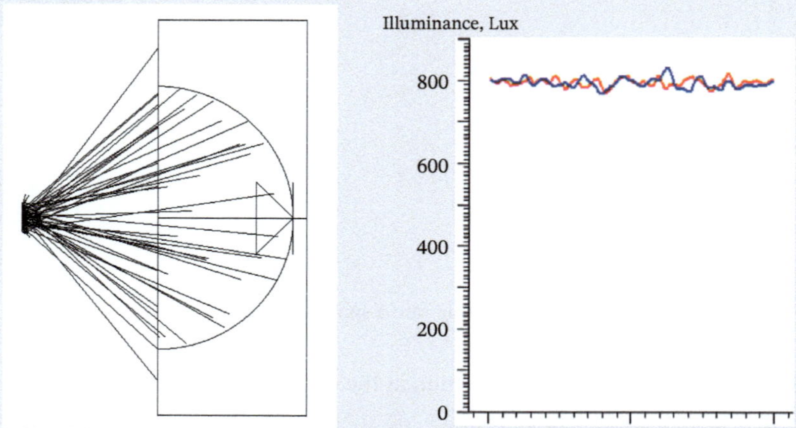

Figure 4.41 Left: Lambertian emitter illuminating a concave spherical surface. Right: Surface illuminance in the concave hemisphere (the diagram's x axis is the arc length along the surface of the sphere).

Exercise 4.33: Integrating sphere (2)

Let's invent something!

It is easy to produce high luminance: place a primary light source (with flux Φ_S, radius r_S) in the center of an integrating sphere (with radius R). The sphere surface is a Lambertian scatterer with $\eta_r = 90\%$ or more. We provide a small opening (radius r_T) to extract light of very high luminance from the sphere (Figure 4.42). The smaller the hole, the higher the available luminance!

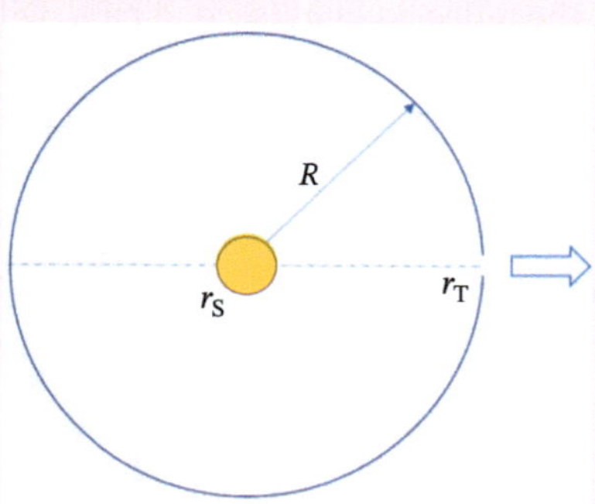

Figure 4.42 Light source inside a sphere with a small opening.

The argument is as follows: The first pass (before any scattering) produces an illuminance at the sphere's wall that can be calculated via the inverse square law (Eq. 3.142):

$$E_0 = \frac{I}{R^2}\text{sr} = \frac{\Phi_S}{4\pi R^2} \tag{4.140}$$

Then, the whole surface is radiating with a luminance of

$$L_1 = \eta_r \frac{E_0}{\pi \text{ sr}} \tag{4.141}$$

(for each surface element) into an hemisphere of $\Omega_p = \pi$ sr, which gives an *additional* illuminance of

$$E_1 = L_1 \cdot \pi \text{ sr} = \eta_r E_0 \tag{4.142}$$

and so on (we assume $r_T \ll R$). This yields a geometric series [54, Ch 4.1.6]

$$E = \sum E_i = (1 + \eta_r + \eta_r^2 + \ldots)\, E_0 = \frac{1}{1-\eta_r}\frac{\Phi_S}{4\pi R^2} \tag{4.143}$$

Therefore, we can extract light with a luminance of (Eq. 3.112)

$$L_T = \frac{E}{\pi \text{ sr}} \tag{4.144}$$

from the small hole. We compare with that of the bulb (Eq. 4.5)

$$L_S = \frac{\Phi_S}{(\pi\, r_S)^2 \text{sr}} \tag{4.145}$$

and obtain an increase in luminance by a factor of

$$\frac{L_T}{L_S} = \frac{1}{1-\eta_r} \cdot \frac{r_S^2}{4R^2} \tag{4.146}$$

The higher the reflectance, the higher the gain – just operate a source in an only slightly larger sphere of good reflectance.

Discuss this conclusion!

Solution 4.33

The total flux at the wall of the sphere is apparently

$$\Phi = E \cdot 4\pi R^2 = \frac{1}{1-\eta_r}\Phi_S > \Phi_S \tag{4.147}$$

which seems to violate the First Law.

There are two reasons for this conclusion. First, the model of the light source inside the integrating sphere is oversimplified [55]. The interaction of the radiation with the source itself was neglected, and a considerable portion of the flux would actually be absorbed by the light source itself, another portion leaving through the small port. Second, the First Law was misunderstood in the

(Continued)

Solution 4.33 (Continued)

argumentation. It is *energy* that's conserved, not *power*. Storage of energy at low power with the aim of releasing it with higher power is quite common and not forbidden by the First Law – think of capacitors. In our case, when the source is turned on, it fills the sphere with electromagnetic energy at constant power. Absorption at the walls, within the source itself, and losses through the port are all proportional to the *energy density* in the sphere. The lower the absorption rate, the higher the energy density will become until equilibrium is reached. From this moment, the energy emitted by the source in each time interval will be either absorbed and converted to heat or lost through the small port, in perfect compliance with the First Law.

The authors remember failed attempts in the nineties to operate an electrodeless lamp (a small sphere with a sulfur fill excited by microwaves [56, 57]) inside an integrating sphere and to extract light through a small hole to obtain a luminance value higher than that of the source.

Dr. Wisenheimer:

For thermal sources at a temperature T, the recycling effect in a perfect integrating sphere can reach spectral radiance values up to black-body radiance, but no more. When looking closely at a tungsten filament coil through a strong filter, we see that the inner parts of the wire look brighter because they do not only emit, they also reflect, whereas for the outer parts, there is no incoming light to reflect. In gas discharge lamps, the arc is partly transparent, decreasing luminance according to Kirchhoff's Law (Knowledge Corner Topic 3.21).

Parabolic mirror segments [58] or various types of retro-reflectors [59, 60] can be cleverly arranged to zigzag the rays back through the arc, thereby increasing luminance. However, such arrangements are of little practical relevance, since they are bulky, expensive and hard to adjust. Of great theoretical interest is a generalization of Kirchhoff's law from thermal sources to LEDs and, in fact, all the way to lasers [61].

Exercise 4.34: Integrating sphere (3)

Integrating spheres can be used for the measurement of diffuse reflectance. Here, a sample is placed at the rear side of an integrating sphere (Figure 4.43). It is now easy to hit the sample with an incident laser beam.

But lasers have their own practical problems, which is why an LED might be used instead. Image a Lambertian disk source of $r_S = 0.5\,\text{mm}$ to the rear of an integrating sphere (diameter $2R = 150$ mm, image semi-diameter $r_1 = 5$ mm, port semi-diameter $r_2 = 5$ mm).

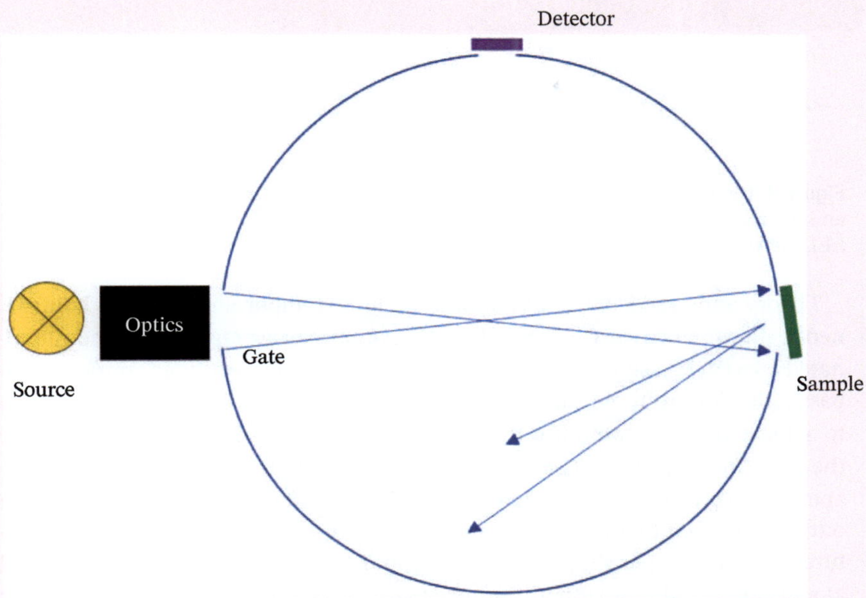

Figure 4.43 Reflectance measurement using an integrating sphere.

Solution 4.34

The etendue of the source has to be equal to that of the imaging part, which is defined by two disks in this case – the exit pupil (the *port* of the sphere) and the target image (the sample). We use the far-field approximation, Eq. 3.42, because of the small apertures.

$$U_{\mathrm{T}} = \left(\frac{\pi\, r_1 r_2}{2R}\right)^2 \mathrm{sr} = \left(\frac{\pi \cdot 5\,\mathrm{mm} \cdot 5\,\mathrm{mm}}{150\,\mathrm{mm}}\right)^2 \mathrm{sr} = 0.27\,\mathrm{mm}^2\mathrm{sr} \tag{4.148}$$

This value must be equal to the etendue of that fraction of the source's light that is collected within an appropriate angle. We find

$$U_{\mathrm{S}} = (\pi\, r_{\mathrm{S}} \sin\theta)^2 \tag{4.149}$$

which yields $\theta = 20°$ and allows to start the design of a lens system. Light outside this angle cannot be collected to the target etendue; the optical efficiency is given by the configuration factor

$$\eta = \frac{U_{\mathrm{T}}}{U_{\mathrm{S}}} = \sin^2 20° \approx 12\,\% \tag{4.150}$$

and can hardly be improved as there are not many *smaller* high-power LEDs. The remainder of the source radiation is lost or (in etendue lingo [62]) spilled.

The imaging design is carried out, for example, from the sample via the port (as the entrance pupil) to the LED.

(Continued)

Solution 4.34 (Continued)

Figure 4.44 Optical design to illuminate a sample for reflectance measurements inside an integrating sphere. The design is usually carried out from the sample (right) to the LED (left).

Theory and practice: In this configuration, it is important to have a homogeneous illuminance on the target (to achieve proper averaging over the illuminated area) with a sharp boundary that is within the port (to avoid spurious first-pass reflection from the rim of the port). A direct LED image is not a good way to achieve this. We could use Köhler illumination [62, 63] (Figure 4.44): Image the LED with a primary condenser with somewhat larger NA into a secondary aperture to create a well-defined circular cutoff baffle at the primary condenser output, and then to image that circular baffle with a lens at the secondary aperture onto the sample. Alternatively, we could place a hexagonal integrating rod at a distance to the LED that captures a slightly larger angle and image the exit of this integrating rod to the sample. Both approaches would allow us to discard the excess light from the source in a controlled way, thereby avoiding stray light. Did you note how the guidance we get from the simple theory allows us to foresee how Köhler or mixing rod systems will need to be set up?

Exercise 4.35: Smartphone brightness sensor

Contemporary smartphones feature a *light sensor* to enable an adaptive display brightness. Funny enough, some *light meter* apps are available that promise the use of this sensor as a true illuminance meter (Figure 4.45).

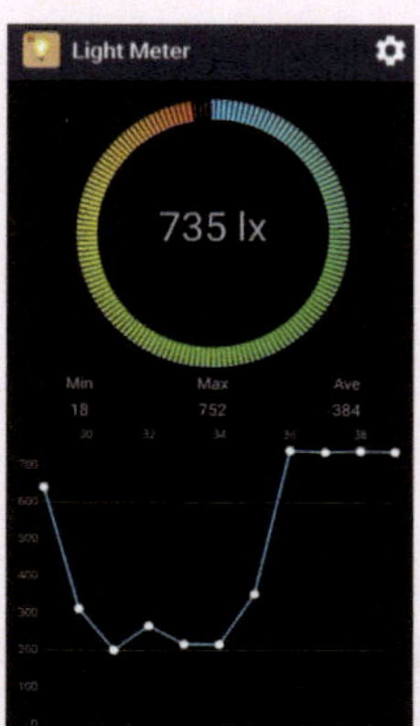

Screenshot from Light Meter app

Figure 4.45 Mobile phone Light Meter app.

Of course, the numbers are neither reliable nor are they necessarily a linear function of the incident illuminance. And there is probably neither a cosine correction nor a $V(\lambda)$ sensitivity [64]. Anyway – suggest a way to calibrate such a sensor *at home*!

Solution 4.35

You may use an omnidirectional LED retrofit bulb of the large spherical soft type, with, say $\Phi_S = 1000\,\text{lm}$ and $r_S = 30\,\text{mm}$.

As a first calibration point, we use the surface. Here, the luminous exitance is

$$E_S = \frac{\Phi_S}{A_S} = \frac{\Phi_S}{4\pi\, r_S^2} \tag{4.151}$$

which gives a value on the order of

$$E = \frac{1000\,\text{lm}}{12 \cdot 900\,\text{mm}^2} \approx 10^5\,\text{lx} \tag{4.152}$$

Second, we place the phone at a larger distance D from the bulb. Its luminous intensity is

$$I_S = \frac{\Phi_S}{\Omega} = \frac{\Phi_S}{4\pi\,\text{sr}} \tag{4.153}$$

and therefore, the illuminance is (using 3.142)

$$E = \frac{I_S}{D^2}\text{sr} = \frac{\Phi_S}{4\pi\, D^2} \tag{4.154}$$

We place the smartphone at a distance of $D = 1$ m from the bulb's center (bonus question: why from the center?), and obtain

$$E_S \approx \frac{1000\,\text{lm}}{12\,\text{m}^2} \approx 80\,\text{lx} \tag{4.155}$$

These two reference points will enable a rough calibration.

We will now turn to real light meters, i.e. instruments that are intended to give precise illuminance measurements.

A first issue that all light meters face is this: illuminance is defined on a small planar patch. Planarity is important, and orientation is important as well. Light meters usually have a primary sensor area with a circular boundary defining that planar circle over which illuminance is to be averaged (Figure 4.46). Now, collimated light shining onto this circular receiver from an angle θ would "see" a reduced receiver area that catches a corresponding portion of the illuminance:

$$E(\theta) = E_0 \cos\theta \tag{4.156}$$

where E_0 is the incidence normal to the surface. In reality, we face more or less severe deviations. The receiver area may be recessed with respect to the case to block stray light from the side and/or for mechanical protection and construction reasons.

Moreover, a scattering cover on top of the actual receiver may not work perfectly and suffer from angle-dependent Fresnel coefficients. Technical measures to overcome the problem are termed "cosine correction", whose quality is an important property of the light meter [65] To improve the situation, suppliers sometimes use scattering domes instead of flat scatterers.

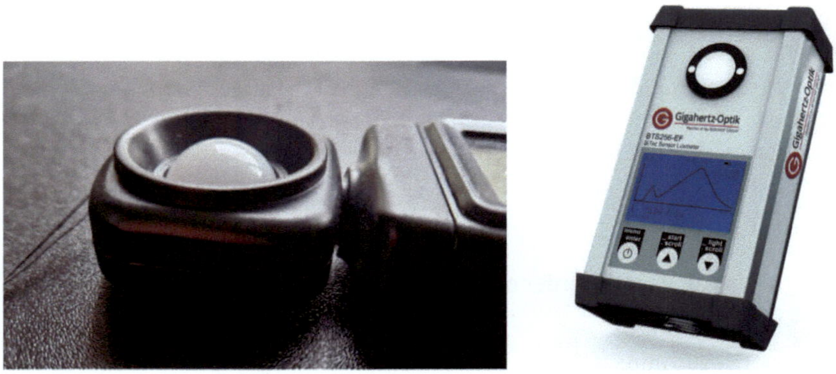

Figure 4.46 Left: Photometer head with scattering dome on top of the sensor. Right: A higher-end device with a spectrometer inside and a flat sensor surface. *Source*: Courtesy of Gigahertz Optik.

Exercise 4.36: Cosine correction

A disk receiver ($d = 10$ mm) is arranged in a black tube ($h = 10$ mm). Calculate the deviation from perfect cosine sensitivity for a collimated beam incident under an angle of $\theta = 26.5°$ with respect to the disk axis.

Solution 4.36

We switch to a simulation to investigate the topic.

Simulation 4.12: Angular sensitivity

To model the angular sensitivity of a receiver in several configurations (such as those shown in Figure 4.47), we set it up as Lambertian source and measure the corresponding far-field distribution. As long as the actual sensor surface at the bottom of the tube has perfect cosine correction, this is a valid approach; in a sense, it corresponds to backward ray tracing again. Results are shown in Figure 4.48.

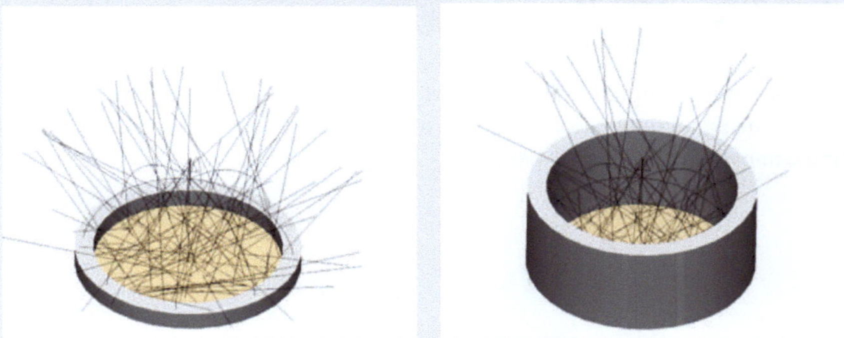

Figure 4.47 Receiver with cosine sensitivity (left) and recessed receiver (right).

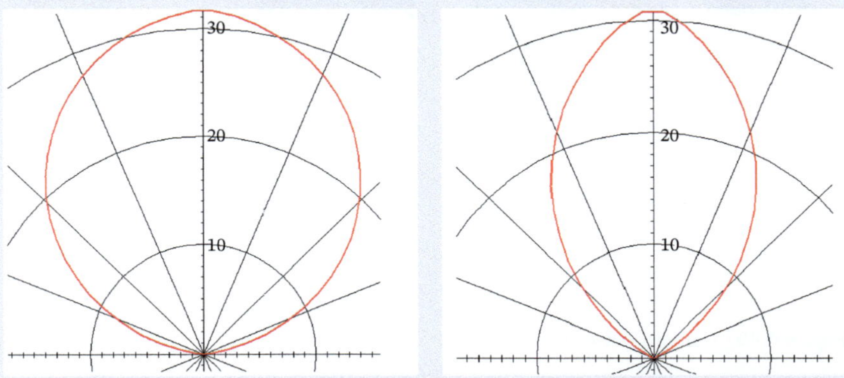

Figure 4.48 Angular sensitivity (polar diagrams with arbitrary but same scale) for both receivers. Left: nearly cosine shaped sensitivity. In the recessed case (right), the sensitivity is reduced for larger angles of incidence.

Provided file: Receiver_cos.lts.

Camera sensors

When developing optical systems for *camera sensors* such as machine vision systems, microscopes, or endoscopes, there is always a need to calculate the necessary amount of light at the object. This requires a basic understanding of the sensor's function and key properties.

We present just a brief introduction to the world of sensor science and refer the interested reader to the EMVA 1288 standard [66], some books [67] and web pages [68, 69].

Be prepared to find that proper data are rather sparse. Most suppliers tend to publish only a subset of the key parameters, and the names and acronyms may differ between vendors.

Knowledge Corner Topic 4.7: Camera sensors

Conversion

The basic sensor function is to convert incident photons (irradiance or illuminance during an integration time τ) to a digital output signal via charge carriers and camera voltage (Figure 4.49).

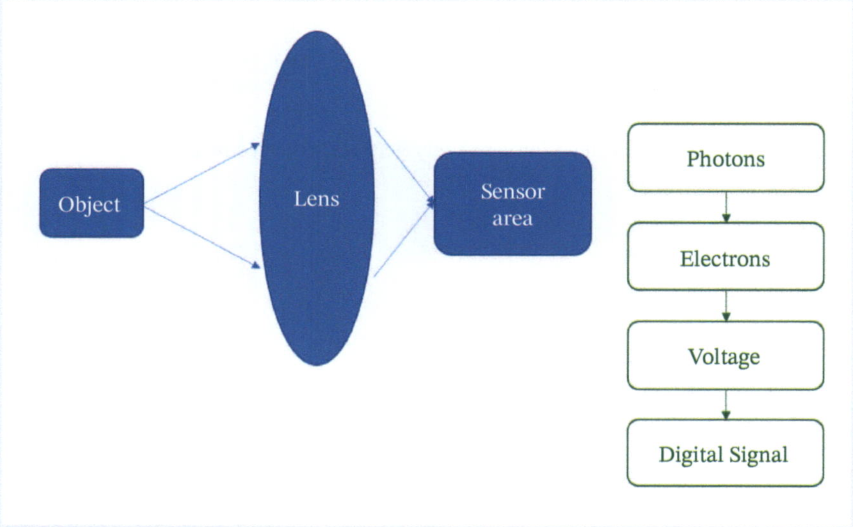

Figure 4.49 Camera sensor function.

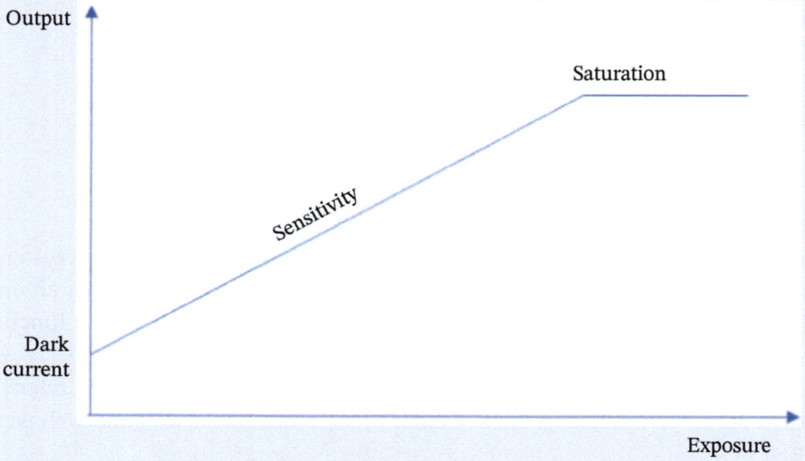

Figure 4.50 Basic characteristics of a camera sensor.

The conversion of N photons to electrons is described by the total quantum efficiency η (on the order of 50%)

$$N = \eta\, \tau N_{\mathrm{P}} \tag{4.157}$$

with the photon flux N_{P} that is related to irradiance through the photon energy Q_{P} (cf. Eq. 4.120) as

$$N_{\mathrm{P}} = \frac{\Phi}{Q_{\mathrm{P}}} = E\frac{A}{Q_{\mathrm{P}}} \tag{4.158}$$

with the pixel area A and an average photon energy Q_{P} (according to the spectral distribution). [66]

The ratio

$$s = \frac{N}{E\tau} \tag{4.159}$$

(the slope of the graph in Figure 4.50) is called the *sensitivity* or *responsivity* of the device. Typical values are in the range of a few thousands of electrons per lx s.

During pixel readout, the number of stored electrons is converted to camera voltage,

$$U = CN = CsE\tau \tag{4.160}$$

with the *conversion gain* or *output gain conversion* (OGC) C [67] (values around $10\ldots100\,\mu\mathrm{V/e^-}$).

Sometimes, the value Cs

$$\frac{U}{E\tau} = Cs \tag{4.161}$$

with the unit V/(lx s) is provided as well.

Still other sources specify sensitivity just as a voltage, which is intended to describe the voltage of a pixel for a reference setup (object luminance, lens $F/\#$, integration time). Here, one has to calculate system etendue based on $F/\#$ and sensor area and multiply it with the object luminance to get some kind of a "reference flux". A comparison of the sensitivity voltage to that of saturation yields a saturation illuminance.

The *digital output* is ideally

$$d(U) = 2^X \frac{U}{U_{\mathrm{max}}} \tag{4.162}$$

with resolution X (bits) and maximum voltage U_{max} possibly related to a full well [67, Ch. 6].

Conversion gain may also be defined as

$$\frac{d}{N} = \frac{C}{U_{\mathrm{max}}} \tag{4.163}$$

with the unit $\mathrm{LSB/e^-}$ (least significant bit per electron).

(Continued)

Knowledge Corner Topic 4.7: (Continued)

Sensitivity may be provided via

$$\frac{d}{E\tau} = \frac{Cs}{U_{\max}} \tag{4.164}$$

with the unit LSBs/(J/m^2) (least significant bit per fluence).

Dark signal
Without irradiation, there is still a thermally generated *dark current* producing a number of N_D charge carriers in the well per second. For CMOS pixels, this value is in the order of about ten electrons per second. Sometimes we see the dark signal with the unit LSBs/s and refer to Eq. 4.163. Alternatively, the *dark current density* may be provided, which also allows to calculate the number of dark electrons [67].

The *signal-to-noise ratio* (SNR)

$$\mathrm{SNR} = \frac{N}{N_{\mathrm{T}}} \tag{4.165}$$

(N_{T} containing all types of noise, including N_{D}) is normally reported for a given number of photons per pixel, for example

$$\mathrm{SNR}_{10} = \frac{10}{N_{\mathrm{T}}(10)} \tag{4.166}$$

and usually given in dB,

$$\mathrm{SNR} = 20\,\lg \frac{N}{N_{\mathrm{T}}} \tag{4.167}$$

For increasing electron numbers, SNR will increase as well, but with a declining slope.

Saturation
The capacity limit to store charge carriers in a pixel, N_{S}, is termed *well size* or *full well capacity* FWC (unit: electrons), typically in the order of a few ten thousand electrons. This is indicated as the *saturation level* in Figure 4.50.

The saturation irradiance level is

$$E_{\mathrm{S}} = \frac{N_{\mathrm{S}}}{A} \frac{Q_{\mathrm{P}}}{\eta\tau} \tag{4.168}$$

Saturation equivalent exposure (SEE) [67] is defined via

$$\mathrm{SEE} = E_{\mathrm{S}}\tau \tag{4.169}$$

The *dynamic range* [68] is

$$\mathrm{DR} = 20\,\lg \frac{N_{\mathrm{S}} - N_{\mathrm{D}}}{N_{\mathrm{T}}} \tag{4.170}$$

(the denominator referring to the total noise [67]).

DR is related to the saturation level, whereas SNR corresponds to the current signal level. However, there is a *maximum* SNR related to the full well capacity [70].

Dr. Wisenheimer:
Camera suppliers have now started reporting according to the EMVA1288 standard, which makes life easier for the user [66].

A typical illumination design task is to provide a system that can just about saturate the camera sensor for a given integration time. Let's take a look how this is achieved.

Exercise 4.37: Camera sensor sensitivity

Assume a camera system with the following specifications.

Quantity	Value
$F/\#$	5
Exposure time	20 ms
Wavelength	$\lambda = 550$ nm
Sensitivity	$4500\,\mathrm{e^-/lx\,s}$
Dynamic range	DR = 72 dB
Pixel size	1 μm
Quantum efficiency	$\eta = 50\%$
Lambertian object reflectance	$\eta_R = 25\%$

What is the necessary illuminance level on an object that is observed with this camera?

Solution 4.37

First, we need a rough estimate of the well capacity. We assume 10 dark electrons as the only noise source HANDLE WITH CARE and approximate with Eq. 4.170

$$N_S = 10^{\frac{DR}{20}} \cdot 10 \approx 40\,000 \tag{4.171}$$

(Continued)

Solution 4.37 (Continued)

We use Eq. 4.168 to approximate the saturation illuminance. The energy of a green photon is

$$Q_P = \frac{ch}{\lambda} = 3.6 \cdot 10^{-19}\,\mathrm{W\,s} \tag{4.172}$$

Hence, the saturation irradiance level based on our data is

$$E_{e,S} = \frac{N_S Q_P}{\eta \tau A_{pixel}} = 40\,000 \cdot \frac{3.6 \cdot 10^{-19}\,\mathrm{W\,s}}{0.5 \cdot 20\,\mathrm{ms} \cdot 1\,\mu\mathrm{m}^2} = 1.4\,\frac{\mathrm{W}}{\mathrm{m}^2} \tag{4.173}$$

and saturation illuminance is

$$E_{v,S} = K_{cd} E_{e,S} \approx 1\,\mathrm{klx} \tag{4.174}$$

The etendue of the camera system is

$$U = \frac{\pi A_S}{4F^2}\,\mathrm{sr} \tag{4.175}$$

with the sensor area A_S.

The luminance corresponding to the saturation level is

$$L = \frac{\Phi}{U} = \frac{E_{v,S} A_S}{U} = \frac{4}{\pi\,\mathrm{sr}} E_{v,S} F^2 = \frac{4}{\pi\,\mathrm{sr}} \cdot 1\,\mathrm{klx} \cdot 25 \approx 32\,\mathrm{knit} \tag{4.176}$$

The same luminance has to come from the observed object. If the object is illuminated by E_O and is scattering into the hemisphere (projected solid angle π), its luminance is

$$L = \eta_R \frac{\Phi}{U} = \eta_R \frac{E_O}{\Omega_p} = \eta_R \frac{E_O}{\pi\,\mathrm{sr}} \tag{4.177}$$

Consequently, the object has to be illuminated with

$$E_O = \frac{\pi\,\mathrm{sr} \cdot L}{\eta_R} = \frac{\pi\,\mathrm{sr} \cdot 32\,\mathrm{knit}}{\eta_R} = 400\,\mathrm{klx} \tag{4.178}$$

for sensor saturation. Note that object and sensor sizes are not important for this consideration.

Theory and practice: Each step of the calculation may be correct, but the whole value chain hinges on a single rough estimate: the 10 dark electrons. They lead to a full well capacity estimate which appears large for a 1 $\mu\mathrm{m}^2$ pixel. Even if the very rough noise estimate is off by only a factor of two, then the same is true for the required illuminance level. Half the number of LEDs and optics may well be the difference that makes a product viable. Always question your assumptions and try to corroborate. If you have access to a camera with the sensor under consideration, the simplest test is to remove the lens, expose the naked sensor to a defined illuminance and check where you get saturation.

Exercise 4.38: Lidar

Lidar (an acronym for "light detection and ranging") measures distances by sending out pulses and measuring time-of-flight. Again, illumination must provide sufficient light on the sensor.

How much irradiance E_O do we need to shine onto an object (distance 50 m, reflectance $\eta = 10\,\%$) to irradiate a camera sensor with 1 $\mu W/mm^2$?

Assume an $F/2$ camera lens and the object as a Lambertian scatterer.

Solution 4.38

We derive a relation between object and image irradiance on the basis of etendue conservation. The etendue U_C defined by the object and image areas A_O, A_I and the camera lens is

$$U_C = \frac{\pi A_O}{4F_O^2}\,\text{sr} = \frac{\pi A_I}{4F_I^2}\,\text{sr} \tag{4.179}$$

The radiance of the reflecting object (a Lambertian scatterer) is

$$L = \eta\frac{\Phi}{U} = \eta\frac{E_O}{\pi\,\text{sr}} \tag{4.180}$$

The flux from A_O through the lens is

$$\Phi_C = LU_C = \eta\frac{E_O}{\pi\,\text{sr}}\frac{\pi\,A_I}{4F_I^2}\,\text{sr} = \eta E_O\frac{A_I}{4F_I^2} \tag{4.181}$$

Hence, the necessary irradiance on the object is

$$E_O = \frac{4F_I^2}{\eta}E_I = \frac{4\cdot 2^2}{10\,\%}\cdot 1\,\frac{\mu\text{W}}{\text{mm}^2} = 160\,\frac{\mu\text{W}}{\text{mm}^2} \tag{4.182}$$

For this estimate we need neither the object distance nor the magnification, only the fact that object and image are conjugates.

4.7 Projectors and Stage Lights

Projectors and stage lights are based on the same principle: a slide, gobo (*graphic optical blackout* or *goes before optics*), or imager is imaged onto a screen and illuminated appropriately for this purpose.

Knowledge Corner Topic 4.8: ANSI lumens

Measurement of the luminous flux of a data projector is governed by the ANSI/NAPM IT7.228-1997 standard [71]. The document defines 9 points on the illuminated rectangle (Figure 4.51). This can be done manually with a photometer head at the wall, with a detector array or with the help of a luminance camera.

(Continued)

Knowledge Corner Topic 4.8: (Continued)

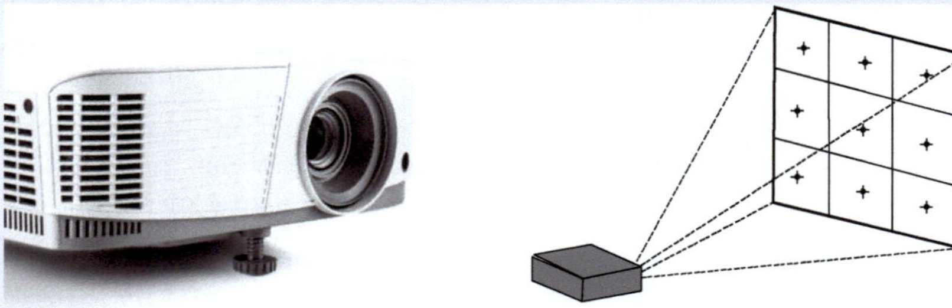

bborriss/Adobe Stock Photo

Figure 4.51 Conference room projector. Luminous flux is specified and measured in *ANSI lumen* with an corresponding measurement setup.

The specification of a data/video projector would ideally comprise the following data: display technology, native resolution, ANSI lumen "Brightness", full on/off contrast and checkerboard (ANSI) contrast, throw ratio, zoom factor, projection offset, and light source data.

Exercise 4.39: Projector luminance

A projector shines on the $3 \cdot 4\,\text{m}^2$ screen from Exercise 4.53 from a distance of $D = 3.4$ m. We assume a projection lens with an exit pupil diameter of $2\,r_\text{P} = 4\,\text{mm}$ and a total efficiency of the projector of $\eta = 12\%$. The measured flux at the screen is 2000 lm. Can this projector be an LED device? If not: Guess which type of light source might be inside.

Solution 4.39

We calculate the etendue using the pupil area and the solid angle

$$\Omega_\text{p} \approx \Omega \approx \frac{A_\text{screen}}{D^2}\text{sr} \approx 1\,\text{sr} \tag{4.183}$$

which yields

$$U = \pi\, r_\text{P}^2 \cdot 1\,\text{sr} = 12\,\text{mm}^2\text{sr} \tag{4.184}$$

suggesting a 0.55″ diagonal (standard pixel) DMD, e.g. the DLP5500 from TI. How do we know? DMD size and mirror tilt are in the data sheet, and the maximum numerical aperture of the projection lens is given by the mirror tilt. This lets us estimate the etendue of the DMD/projection lens subsystem – a bonus exercise for you!

We can calculate the necessary source luminance

$$L_\text{S} = \frac{\Phi}{U\eta} = \frac{2000\,\text{lm}}{12\,\text{mm}^2\,\text{sr} \cdot 12\%} = 1400\,\text{Mnit} \tag{4.185}$$

which is clearly out of range for LEDs, at least at the time when this text was written (in 2024).

The required source flux is about

$$\Phi_S = \frac{\Phi}{\eta} = \frac{2000\,\text{lm}}{12\%} = 16\,500\,\text{lm} \tag{4.186}$$

which suggests a 200 W short-arc discharge source [72].

The development of devices that are expected to deliver a certain luminous flux is always characterized by a "lack of lumens". During the development of the first data projectors in the nineties, the pull by the market (there was a differential price of \$7/lm) was bigger than the available knowledge about illumination design. The authors remember a situation where a certain projector prototype delivered 250 lm instead of the design target of 600 lm. The reason was a condenser design based on imaging optics without any non-sequential fine-tuning by optimization (which was not available at that time). So a solution had to be developed by experiment.

Knowledge Corner Topic 4.9: Luminance diagram of a DLP projector

The luminance diagram (see [62] for a detailed explanation of this very useful high-level concept) of Figure 4.52 explains the photometric conditions inside this projector. It allows us to trace flux and luminance through the projector and to illustrate loss mechanisms (geometrical or physical).

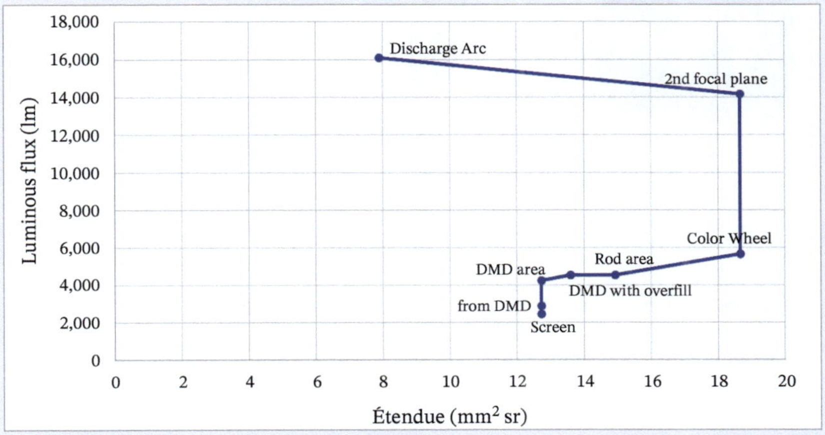

Figure 4.52 Luminance diagram of a discharge-lamp-based DLP projector: The light of the arc is reflected to the secondary focal plane of an elliptical reflector. It is distributed to a larger etendue due to variable magnification (and some light is lost by leaving the reflector without hitting its surface). The area of the mixing rod cuts out a certain share of this distribution. The subsequent color wheel has an average transmission of about 40%. The integrator exit area is imaged to the surrounding of the DMD, and the actual DMD area cuts out a certain share. The DMD reflectance of about 70% and the lens transmission leave some light that is finally reaching the screen.

Exercise 4.40: Projector brightness

Order from above:

Ike tells the team,[a] much to the team's dismay: "I have hired an external consultant to qualify you to increase the 'brightness' of our recently developed LED DLP projector [73]. His name is Rube."

Let's invent something!

I am thrilled by those projectors! I will show you where to go and have prepared the following suggestions:

a. Increase the area of the LEDs so they will deliver a higher flux.

b. Place a reflector (a corner cube or an array of those) into the DMD's off-light cone for back reflection (Figure 4.53, cf. [74, Figure 2]).

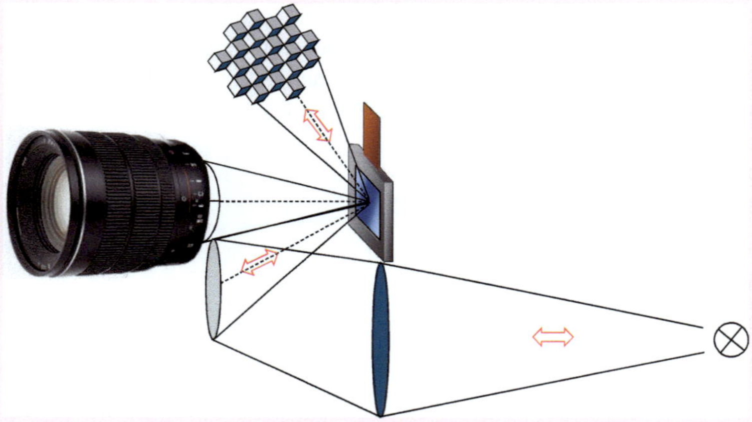

Figure 4.53 Corner cube reflector in the dark-state beam.

[a]These things happen all the time.

c. Recycle some of the off- state light by using a loop (Figure 4.54) [75].

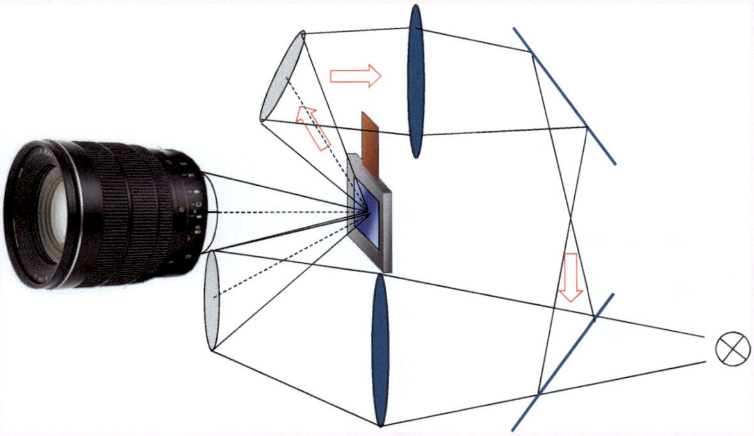

Figure 4.54 Recycling of the dark-state beam.

d. Improve the loop by means of a polarizing beam splitter (PBS) and rotating the polarization of the recycled light by 90° (Figure 4.55).

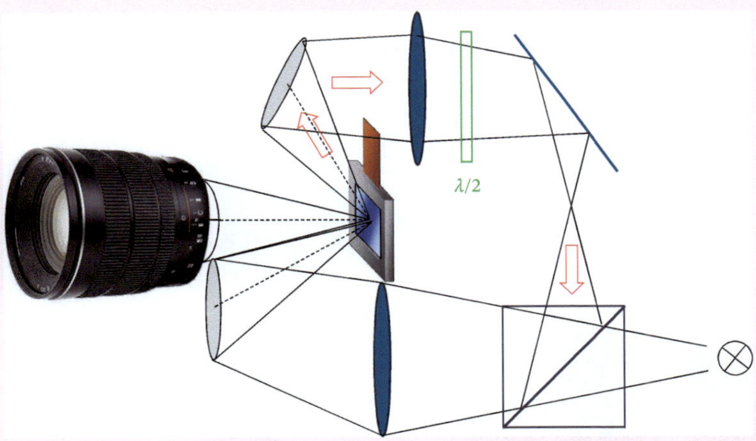

Figure 4.55 Recycling loop with a polarizing beam splitter.

(Continued)

Exercise 4.40: (Continued)

e. Use a source of higher luminance that fills only half of the etendue. The other half could be used to add recycled light (Figure 4.56).

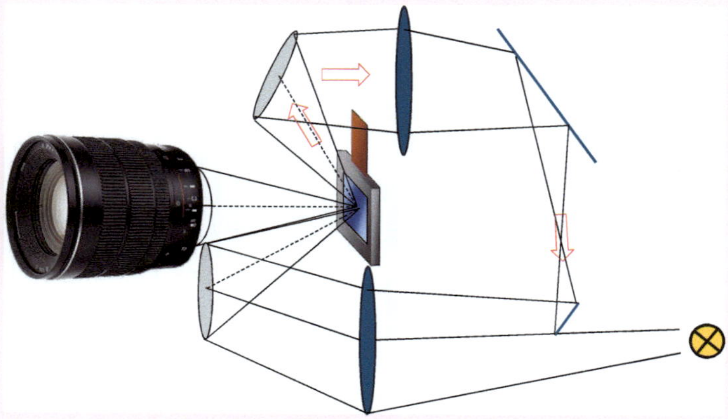

Figure 4.56 Recycling using a higher-luminance source.

f. Add a second illumination system shining light from the original off-state direction onto the DMD (using a negative image for projection from this side), see Figure 4.57). Operate both systems alternating [76].

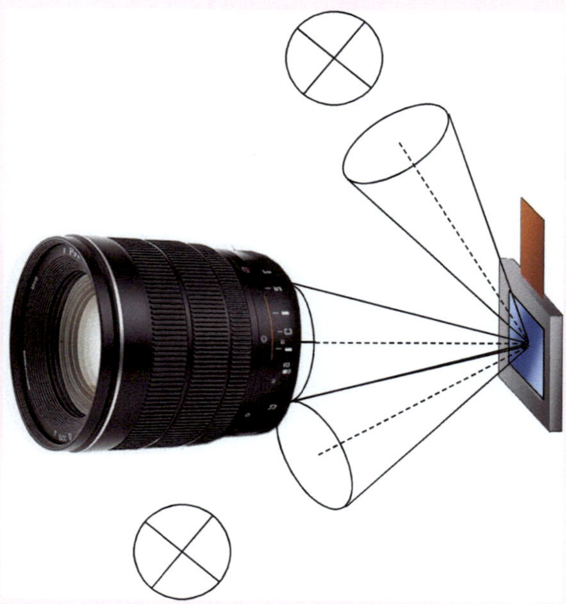

Figure 4.57 DMD operation with a second source.

Dr. Wisenheimer:

For one-chip-lamp projectors, a couple of such hacks have been tried and even implemented, the first having been the color wheel's white segment, which adds luminous flux to the white image at the cost of less flux in the primary colors. There were also attempts to implement recycling schemes such as *scrolling color recapture* [77] and manipulations of the light pipes [78]. Some were successful and became state of the art, others were discarded due to secondary problems.

Solution 4.40

Assumption (1): For our discussion, we assume that the original projector design was properly set up, realizing equal etendue of LED and DMD.

a. Because of (1), the additional flux is just wasted (spilled).

b. The reflected photons will have to travel back into the light source and be recycled there [79], which is hard to achieve efficiently. You should not assume a potential gain of more than 20%.

c. Never try to cheat a beam splitter or the Second Law. A beam splitter always works according to the implemented ratio. Thus, if recycling works, you lose some of the initial light. In other words: if you trace a ray back from the exit, it can end up only in one destination.

d. A PBS and rotation of the polarization plane in the loop would work. However, if (1) holds, half of the original flux would be wasted at the initial polarization. The only way to avoid this would be to use an intrinsically polarized LED. You could try to recycle the other half of the polarization; however, the potential gain will not be large.

e. This may work if properly designed.

f. This may work as well, as one can operate an LED at a higher current for a lower duty cycle.

Dr. Wisenheimer:

The flux of a projector is usually measured by presenting a *white image*. All ideas except for (e) and (f) would just fail.

For static projection purposes such as for hospitality and signage (such as a "Welcome" message in a hotel or the projected floor arrows in the IKEA shops), gobo projectors are still quite popular (Figure 4.58). They deliver a flux range of about 100 lm up to several 10 klm. Stage lights such as moving heads feature a similar optical system.

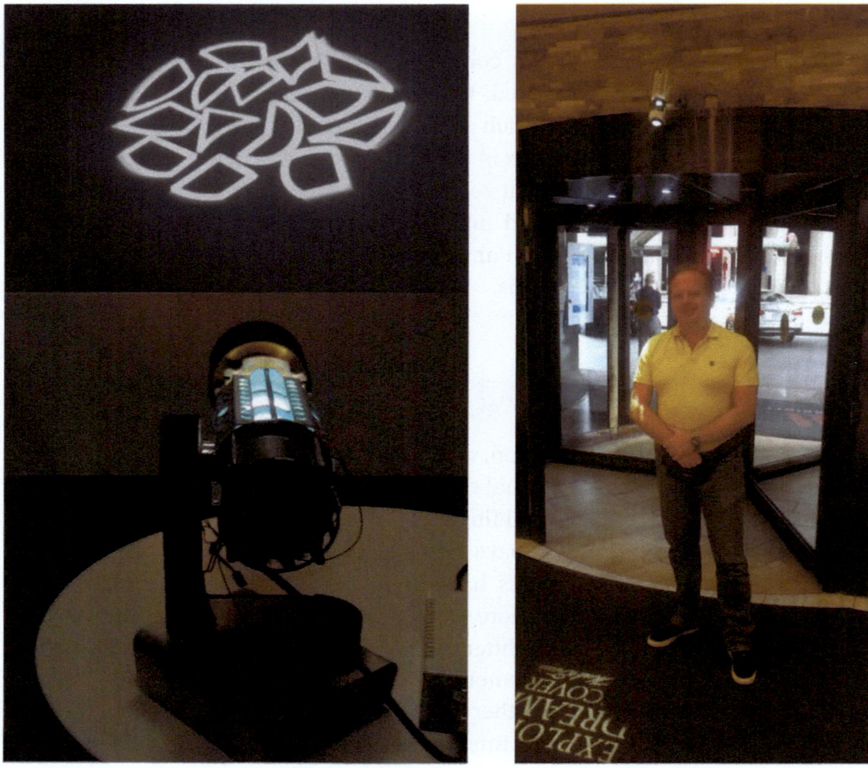

Figure 4.58 Left: Gobo projector. Right: Henning in a hotel entrance just after spotting a device he designed.

Exercise 4.41: Gobo size

Order from above: Develop a new 20 klm hospitality and discothèque gobo projector adhering to a strict cost budget. Use a "bright and cheap" CoB LED (LES $d = 33$ mm, $\Phi_S = 25$ klm). A 30° field of view (FoV) is required. For cost reasons, a $F/4$ lens needs to be used.

How much flux can you deliver? Which gobo diameter r_G is necessary? What size of the front lens is required?

Solution 4.41

The etendue of the LED is

$$U_S = A_S X_S = r_S^2 \pi \cdot \pi \,\text{sr} = (\pi \cdot 16.5\,\text{mm})^2 \text{sr} \approx 2700\,\text{mm}^2\text{sr} \tag{4.187}$$

which needs to be equal to that of the projection system

$$U_G = A_G \cdot X_G = r_G^2 \pi NA^2 \pi \text{ sr} = \left(\frac{r_G \pi}{2F}\right)^2 \text{sr} \tag{4.188}$$

Hence, the gobo radius is given by

$$r_G = r_S 2F = 16.5 \text{ mm} \cdot 8 \approx 132 \text{ mm} \tag{4.189}$$

requiring impractically large lens elements with a semi diameter of at least twice that radius.

The same calculus leads us to the front (or better: exit) pupil radius r_F:

$$U_T = r_F^2 \pi (\sin 15°)^2 \pi \text{ sr} = U_S \tag{4.190}$$

and therefore

$$r_F = \frac{r_S}{\sin 15°} \approx 64 \text{ mm} \tag{4.191}$$

which is smaller than that of the entrance side.

Assuming one could really build a lens of this size, the flux will depend on the efficiency of the illumination system that transports the light from the source to the gobo and further to the lens entrance pupil.

Dr. Wisenheimer:

Projection is a high-luminance business. The LED flux in the example sounds impressive, but the luminance is only on the order of

$$L_S = \frac{\Phi_S}{U_S} = \frac{25 \text{ klm}}{2700 \text{ mm}^2 \text{sr}} \approx 10 \text{ Mnit} \tag{4.192}$$

which indicates that such an LED is rather suitable for general lighting or maybe high-bay lighting. So we have either to accept the large lens elements or to look for higher-luminance LEDs.

Unfortunately, these usually do not come with such a high-lumen package. On the other hand, many smaller high-luminance LEDs may be combined for a projector of such a flux but much smaller size.

Exercise 4.42: Stage lighting

The Source Four [80] is probably the most popular ellipsoidal reflector spotlight for stage and theater, especially in North America (Figure 4.59).

(Continued)

Exercise 4.42: (Continued)

It was originally equipped with halogen lamps of various powers, but nowadays, there are LED versions. The device comes with rotating shutters and a choice of optical systems of different field angles and performance levels.

Product Video.

Figure 4.59 Source Four ellipsoidal reflector spotlight. *Source*: Courtesy of ETC.

The device is traditionally operated with a 750 W, 3200 K tungsten halogen lamp (such as shown in Figure 4.60, called "HPL", *high performance lamp*). Key parameters are

Figure 4.60 HPL 750 W halogen lamp. *Source*: Courtesy of ams-OSRAM.

- bulb diameter 19 mm,
- $\Phi \approx 20$ klm out of 750 W,
- four filaments with 7.5 mm · 11.5 mm total size (giving the device its name).

The setup of the Source Four is displayed schematically in Figure 4.61.

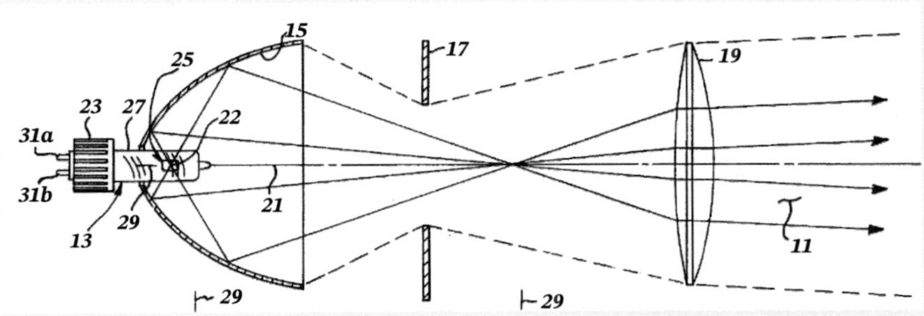

Figure 4.61 Source Four illumination system ([81]/David W. Cunningham/Public Domain): a burner in a faceted elliptical reflector, a gate or gobo, and a projection lens.

These numbers justify an educated guess for some key parameters:

- Gate diameter $2r_G = 75$ mm (size B gobo).
- Approximately $F/2$ optics.
- Beam angle $2\theta = 26°$ (the initial version, later were actually many variants).
- Reflector: elliptical with radial facets, Par38 size.

Estimate the fixture's total flux and its luminous intensity on-axis.

Solution 4.42

A closer look on the lamp picture (Figure 4.60) shows that it comes with four coils of approximately $l_S = 11.5$ mm times $r_S = 1.5$ mm, in a distance of 5 mm from the axis, each of them producing 5 klm.

The etendue of each such cylinder (mantle) is

$$U_S = A\Omega_p = 2\pi\, r_S l_S \cdot \pi \text{ sr} \approx 340\,\text{mm}^2\text{sr} \tag{4.193}$$

or perhaps a little more when including the end faces of the coils.

Effective luminance will be

$$L_{\text{eff}} = \frac{\Phi_S}{U_S} = \frac{5\,\text{klm}}{340\,\text{mm}^2\text{sr}} \approx 15\,\text{Mnit} \tag{4.194}$$

which is a little larger than a source luminance for general lighting but smaller than the actual 3200 K filament surface luminance of $L = 50$ Mnit to Eqs. 3.196 and 3.199 (shown in Figure 3.52).

(Continued)

Solution 4.42 (Continued)

The etendue of the Source Four is approximately

$$U_T = \left(\frac{\pi\, r_G}{2F/\#}\right)^2 \text{sr} = \left(\frac{\pi \cdot 75\,\text{mm}}{4 \cdot 2}\right)^2 \text{sr} = 870\,\text{mm}^2\text{sr} \tag{4.195}$$

which is well within the typical range for ellipsoidal stage lights.

Four filaments with an etendue of 340 mm^2 each and an effective luminance of 15 Mnit are (over)filling the projection system etendue with light. The maximum flux (cf. Eq. 3.95) is limited by their effective luminance

$$\Phi = L_{\text{eff}} U_T = 15\,\text{Mnit} \cdot 870\,\text{mm}^2\text{sr} \approx 13\,050\,\text{klm} \tag{4.196}$$

which is close to the specified Source Four 26° datasheet value of 13 690 lm.

Based of the description in the datasheet, we assume a cosine-shaped illuminance distribution in the gate area:

$$E(r) = E_0 \cos\left(\frac{\pi}{2}\frac{r}{r_G}\right) \tag{4.197}$$

ensuring a decrease to zero at $r = r_G$, the edge of the gate area.

Integration yields a luminous flux

$$\Phi = \int_0^{r_G} E(r)\,\text{d}A = E_0 \int_0^{r_G} \cos\left(\frac{\pi}{2r_G} r\right) 2\pi\, r\,\text{d}r \tag{4.198}$$

We substitute x according to

$$r = \frac{2r_G}{\pi} x \tag{4.199}$$

to obtain

$$\Phi = E_0 \pi\, r_G^2 \frac{8}{\pi^2} \int_0^{\pi/2} \cos x \cdot x\,\text{d}x = E_0 \pi\, r_G^2 \cdot \frac{8}{\pi^2}\Big[\cos x + x \sin x\Big]_0^{\pi/2} \tag{4.200}$$

The center illuminance is then

$$E_0 = \frac{\Phi}{\pi\, r_G^2} \frac{\pi^2}{8} \frac{1}{(\pi/2) - 1} = \frac{13\,050\,\text{lm}}{37.5\,\text{mm}^2} \frac{\pi}{8} \frac{1}{(\pi/2) - 1} \approx 6.4\,\text{Mlx} \tag{4.201}$$

This value translates to the far field via the focal length of the projection lens, which is about

$$f = \frac{r_G}{\tan\theta} = \frac{37.5\,\text{mm}}{\tan 13°} = 162\,\text{mm} \tag{4.202}$$

This is only slightly smaller than the specified value (in the above video) for the original single asphere.

A radial line element dr close to the center of the gate is imaged into an angle of

$$d\varphi = \frac{dr}{f} \tag{4.203}$$

and the flux of a disk of radius dr must be equal to that of a cone of angle dφ,

$$E_0\,\pi(dr)^2 = I_0\,\pi sr(d\varphi)^2 \tag{4.204}$$

Hence,

$$I_0 = E_0 f^2 \frac{1}{sr} = 7\,\text{Mlx} \cdot (115\,\text{mm})^2 \frac{1}{sr} = 168\,\text{kcd} \tag{4.205}$$

pretty close to the datasheet value shown in Figure 4.62. This estimate does entirely neglect the actual $F/\#$ of the lens.

We have seen that based on just a few datasheet values and the laws of optics, we can reverse-engineer quite a few more details of the lamp's optical design.

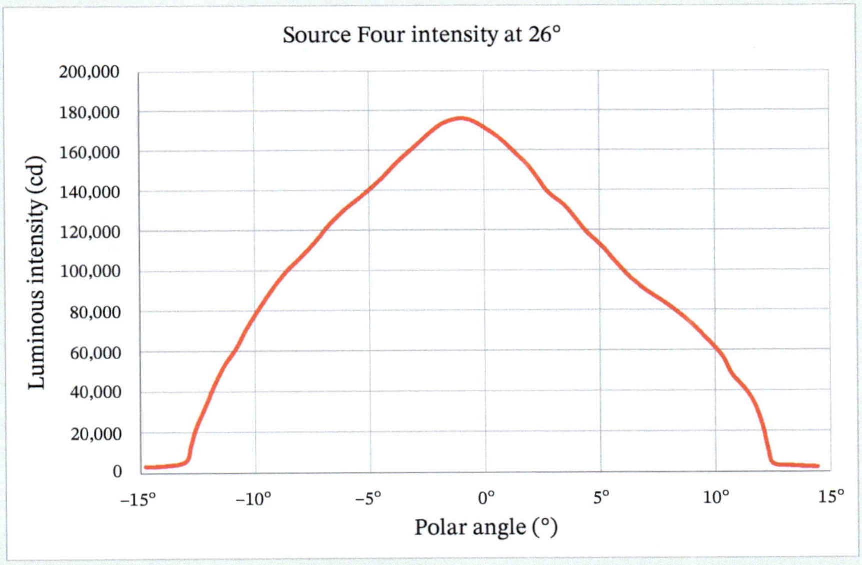

Figure 4.62 Luminous intensity distribution of the Source Four 26° operated with the HPL 750W (after the datasheet).

Simulation 4.13: Source Four simulation

We set up a simulation with the four filaments of the HPL lamp, a reflector based on data from the above video, and a 26° aspheric lens.

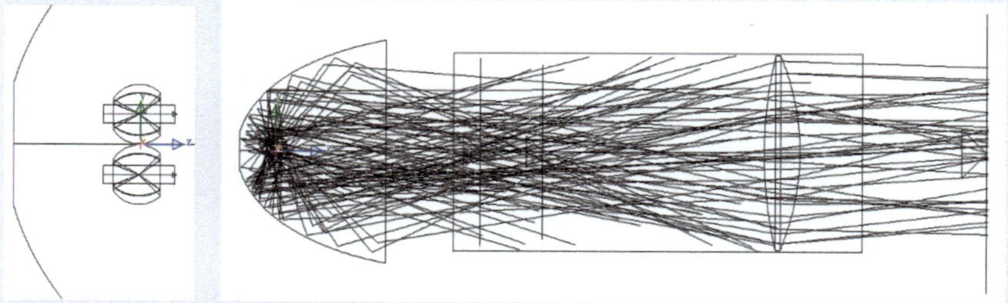

Figure 4.63 Source Four simulation. Left: source models. Right: reflector, gate, and lens.

The coils were set to 5 klm each and the correct far-field distribution. The reflector is a best guess matching the drawings, which strongly influence the photometric results. As it needs some radial grooves for light mixing (otherwise the fourfold symmetry of the filaments would show), we applied a 4° Gaussian scattering. The illuminance in the gobo plane (6 Mlx) does match the above estimate, but the far-field intensity on axis is only one third of both the estimate and the actual value because the *F*/# of the lens is too large and the match between reflector and lens is not really good (in the simulation, see Figure 4.63). The simulation is helpful to illustrate the function and the design choices of the device, but is far from perfect in terms of the far-field intensity. It is not that easy to design a Source Four!

Provided file Source_four.lts.

A Fresnel light or Fresnel fixture is simpler, creating less focused light (Figure 4.64, see also Simulation 4.14). It consists of a light source, some (but not much) optics, and a Fresnel front lens made of glass or plastic. The lens can be moved along the axis to alter the beam angle.

Exercise 4.43: Fresnel light

Your task is to measure the luminous intensity of a Fresnel light. Unfortunately, the space in your laboratory is limited and does not allow far-field measurements at a large distance. Therefore, you intend to measure the illuminance with a photometer at various distances and use the inverse square law (Eq. 3.142) to extrapolate. However, as we have seen in Knowledge Corner Topic 3.16, it is important to determine where the "center of light" or "virtual focus" lies. How can you identify the location related to zero distance?

AI rendering by the authors

Figure 4.64 Fresnel fixture.

Solution 4.43

You will remember our discussion in Knowledge Corner Topic 3.16, where we discussed how to use a plot of $1/\sqrt{E}$ vs. distance D to find the minimum distance (the LPD) where the plot starts to be linear and then fitted a straight line via linear regression. The horizontal intercept D_0 of that straight line is the "zero distance location", also known as the "light center of mass" or "virtual focus".

We stop here: this is the solution! However, in the following Simulation, we create a model of a legacy tungsten Fresnel and analyze how this approach would work in practice. After all, Knowledge Corner Topic 3.16 discussed only theoretical light sources.

Simulation 4.14: Fresnel light

We simulate a Fresnel fixture and "measure" illuminance as a function of the distance from the front lens (Figure 4.65). Provided file: Fresnel.lts.

(Continued)

Simulation 4.14: (Continued)

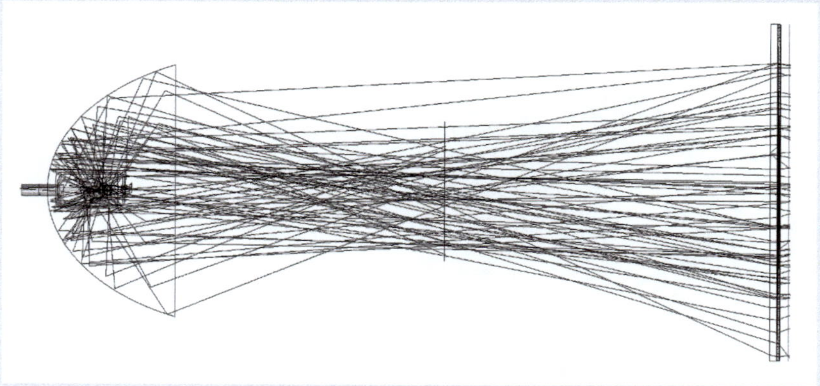

Figure 4.65 Model of a legacy Fresnel fixture: 1kW tungsten halogen source inside an elliptical reflector, secondary focal plane, Fresnel lens, and receiver (left to right).

If we plot the illuminance as a function of distance from the Fresnel lens (as usual), we cannot easily identify the zero-point of the distance scale (Figure 4.66).

From Knowledge Corner Topic 3.16, we know that it is better to plot $1/\sqrt{E_0}$ vs. distance and to apply a linear fit (outside the near-field region) that provides the desired answer.

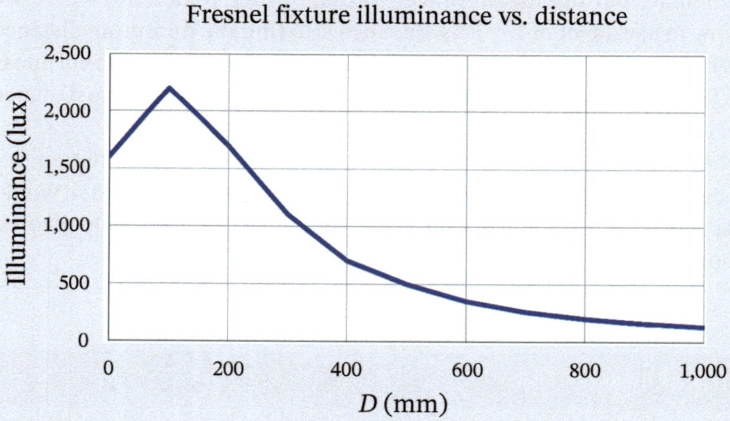

Figure 4.66 Receiver center illuminance E_0 vs. distance D.

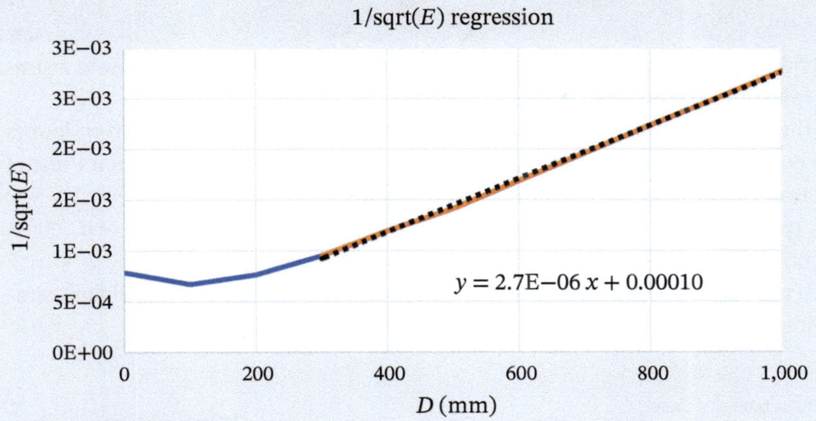

Screenshot from ARRI's photometry app

Figure 4.67 Top: Plot of $1/\sqrt{E_0}$ (in $1/\sqrt{\text{lx}}$) vs. distance D. The far-field region – the linear part of the curve – is indicated in orange. The linear fit is restricted to this region; the near field is ignored. Bottom: The ARRI Photometrics app predicts illuminance values from intensity measurements for many of their fixtures. Here, we use the T1 in spot position, corresponding to our simulation, to compare. The predicted illuminance of 1459 lux at 10 m corresponds to 146 kcd.

The linear fit yields a true zero at $D_0 = -39$ mm (inside the device, which is ok). The slope

$$\frac{1}{\sqrt{E_0}} = \frac{2.7 \cdot 10^{-6}}{\text{mm}\sqrt{\text{lx}}} D = \frac{2.7 \cdot 10^{-3}}{\text{m}\sqrt{\text{lx}}} D \tag{4.206}$$

tells us that

$$E_0 = \frac{137\,\text{klm}}{D^2} = \frac{137\,\text{kcd}}{D^2}\text{sr} \quad \Rightarrow \quad I_0 = 137\,\text{kcd} \tag{4.207}$$

The simulated result of $I_0 = 137$ kcd agrees very well with the prediction from the ARRI Photometrics app for their T1 1 kW Tungsten Fresnel fixture (146 kcd, see Figure 4.67).

Provided file: Fresnel_Illu.xlsx.

Dr. Wisenheimer:

Tungsten lamps are so 1990 ... today's Fresnel fixtures often use high-power RGBW LEDs. They save energy, but their key advantages are: they are tunable, requiring no "gels" (the color filter foils common on film sets), can be dimmed without color change, and their brightness and color can be remotely controlled by a software.

Of course, their optics have to be more complex due to the additional color mixing requirements. Some use a dense square RGBW array and a tapered integrating rod, which provides color mixing and sufficient collimation to avoid spillage beyond the (plastic) Fresnel lens [82]. Others use many individual LEDs with

(Continued)

Simulation 4.14: (Continued)

individual primary TIR lenses and a fly's eye array for color mixing before a similar (plastic) Fresnel lens creates the beam.

Other architectures with a *magic number* [83] of spherical plastic front lenses also became popular, e.g. this LED stage light, bridging the gap between Fresnel and traditional stage lights (ellipsoidal projectors).

A typical RGBW LED package is the High-power RGBW OSRAM OSTAR Stage.

We now double-check the above intensity estimate using our Fresnel light simulation (Figure 4.65) and find a good correspondence (Figure 4.68).

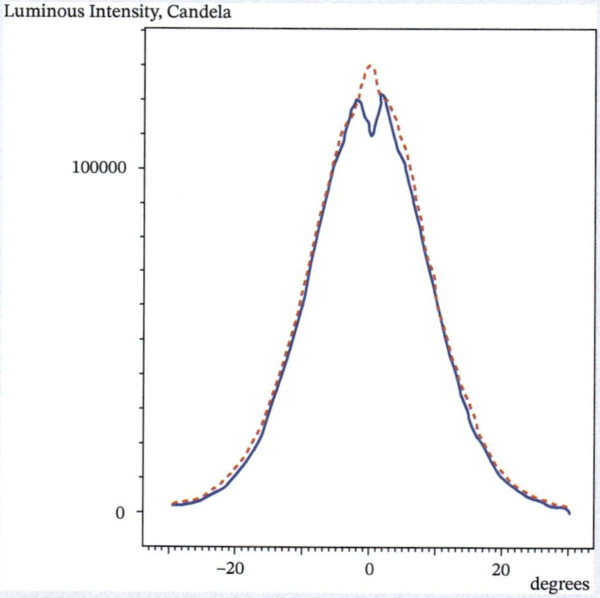

Figure 4.68 Luminous intensity of the simulated Fresnel fixture (red) showing a peak slightly above 130 kcd.

Knowledge Corner Topic 4.10: Light distribution curves (2)

Illuminance and luminous intensity distributions produced by stage lights are specified either in a "photometry" report such as in Figure 4.69 or by LDT / IES files that contain the far-field information (same as for office lights, cf. Knowledge Corner Topic 4.2) and can be imported into illumination software (Figure 4.70).

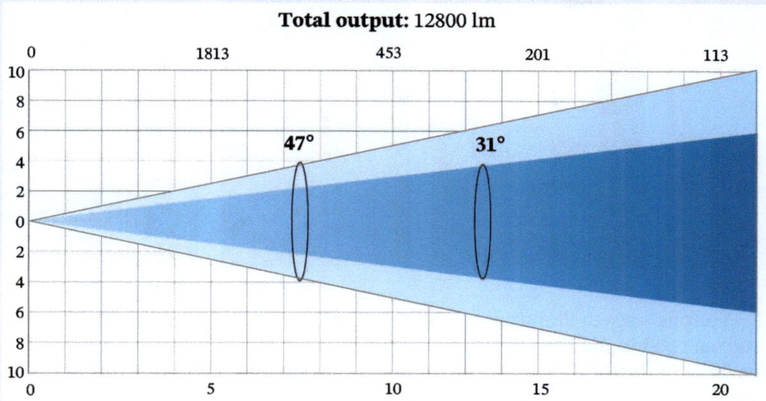

Figure 4.69 Part of a photometry report for a wash light showing the *field angle* (FW @ 10% of peak) and the *beam angle* (FWHM).

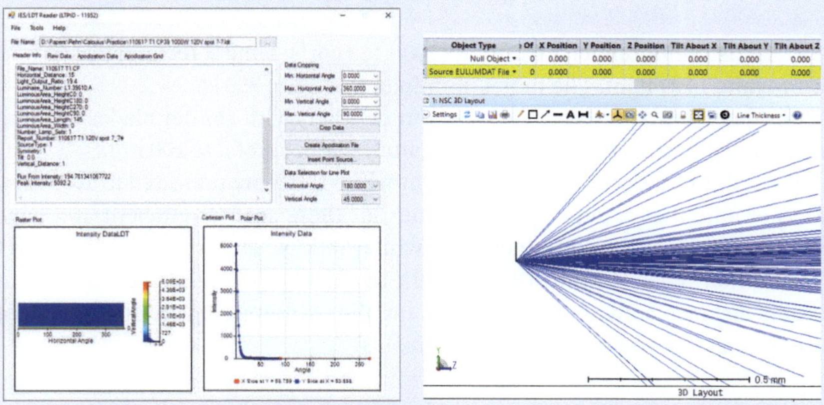

Figure 4.70 Import of an Eulumdat file (ARRI T1 Tungsten Fresnel) to LightTools (left) and Zemax (right). The way of importing the files may be slightly different, the outcome is the same. *Source*: Screenshot from ARRI's photometry app.

Small versions of such luminaires can be applied to illuminate paintings in a museum (Figure 4.71).

AI rendering by the authors

Figure 4.71 Various ways to illuminate paintings. Light homogeneity on the painting is important (not so good in the first picture from the left), as is the illumination *around* the painting (too large in the third picture). On the right, the light distribution perfectly matches the painting – whether this is good or bad would have to be be decided by the curator (AI renderings by the authors).

Exercise 4.44: Museum light

Your task is to illuminate a painting ($100 \cdot 133\,\mathrm{cm}^2$) using a rectangular spot of warm-white LED light. The target illuminance is $E_\mathrm{T} = 750\,\mathrm{lx}$.

Consider using a projector (Source Four or other) with shutter blades forming a rectangular gate and a projector (exit pupil) diameter of $d = 100$ mm.

For the illumination of paintings, warm-white LEDs are preferred to avoid any excess of blue that could harm the paint. But these are less efficient and have less luminance compared to their cold-white siblings, and we can probably get no more than $L_\mathrm{S} = 40\,\mathrm{Mnit}$. Does it work at all? If so, calculate the maximum possible distance between the projector and the painting. Neglect any overfill of the painting area you would provide in reality.

Solution 4.44

The required flux on the painting is

$$\Phi_\mathrm{T} = E_\mathrm{T} A_\mathrm{T} = 1000\,\mathrm{lm} \tag{4.208}$$

With the given luminance, you can illuminate the painting under a projected solid angle of

$$\Omega_\mathrm{p} = \frac{E_\mathrm{T}}{L_\mathrm{S}} = \frac{750\,\mathrm{lx}}{40\,\mathrm{Mnit}} = 2 \cdot 10^{-5}\,\mathrm{sr} \tag{4.209}$$

So it can clearly be done; we would be in fundamental trouble only in the case $\Omega_\mathrm{p} > \pi$. The actual value corresponds to a cone angle of $\theta = 0.14°$ and (with the suggested projector diameter) a distance of

$$D = \frac{50\ \mathrm{mm}}{\tan\theta} \approx 20\ \mathrm{m} \tag{4.210}$$

We could calculate this also via the system etendue and Hottel's formula (Eq. 3.42),

$$U_T \approx \left(\frac{\pi\, r_O r_P}{D}\right)^2 \text{sr} \tag{4.211}$$

If we can fill this etendue, we can later always apply shutter blades and make the beam rectangular.

At the same time, we have

$$\Phi_T = L_S U_T \tag{4.212}$$

which when combined yield

$$\begin{aligned} D &= \pi \cdot r_O r_P \sqrt{\frac{L_S}{\Phi_T}\,\text{sr}} \\ &= \pi \cdot 833\,\text{mm} \cdot 50\,\text{mm} \cdot \sqrt{\frac{40\,\text{Mnit}}{1000\,\text{lm}}\,\text{sr}} \approx 26\,\text{m} \end{aligned} \tag{4.213}$$

This is slightly larger than our previous result, but we still have to reduce our illuminated area using the blades.

Dr. Wisenheimer:

You could of course design a free-form element and shape the rectangle by tailoring, but as a staff member in a museum you would most likely have only a limited number of lights available and also no money for customized solutions. And since the supply of point sources lags behind demand, you would need an *extended source algorithm* for the free-form optics – not yet available commercially, but the subject of current research.

A flying-spot projector works by scanning a laser beam along a screen by means of changing the tilt of a tiny mirror, for example. If the three lasers are driven according to the color information of the current spot position, we can obtain a full-color image at the screen (Figure 4.72).

Exercise 4.45: Resolution of a flying-spot laser projector

Estimate the horizontal resolution of a flying-spot projector on the basis of its parameters.

Consider a laser with $\lambda = 500$ nm and a beam parameter product of 1 mm · mrad, whose scanning mirror has a diameter of 1 mm and can be tilted by $\pm 10°$. Calculate the maximum possible pixel resolution.

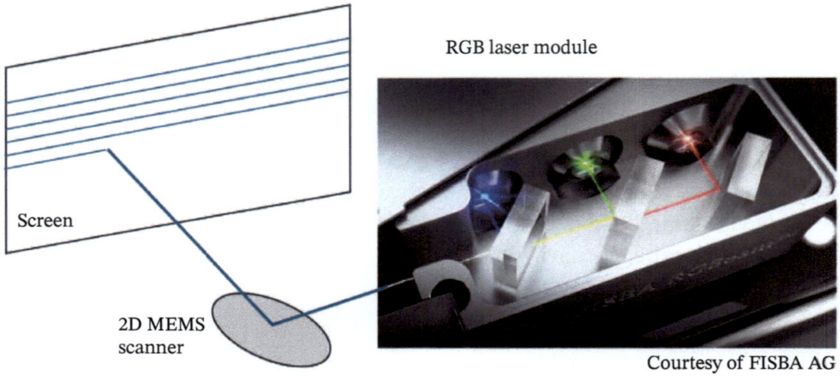

Figure 4.72 Flying-spot laser projection scheme.

Solution 4.45

Why should the resolution be restricted at all? Lasers are pretty close to point sources, aren't they?

We check this assumption and calculate the etendue of the laser using Eq. 3.21,

$$U_{\mathrm{L}} = (\pi\, w_{\mathrm{L}}\theta_{\mathrm{L}})^2 = (\pi \cdot 10^{-3})^2\, \mathrm{mm^2 sr} \approx 10^{-5}\, \mathrm{mm^2 sr} \tag{4.214}$$

The target etendue is defined by the size of the scanning mirror (r_{m}) and the pixels to be illuminated (r_{p}) on the screen (at an unknown distance D) and can be calculated from Eq. 3.42,

$$U_{\mathrm{T}} = \left(\frac{\pi\, r_{\mathrm{p}} r_{\mathrm{m}}}{D}\right)^2 \tag{4.215}$$

We request $U_{\mathrm{L}} = U_{\mathrm{T}}$ and calculate the (half) angular subtense of a pixel

$$\theta_{\mathrm{p}} \approx \frac{r_{\mathrm{p}}}{D} = \frac{\sqrt{U_{\mathrm{L}}}}{\pi\, r_{\mathrm{m}}} = \frac{w_{\mathrm{L}}\theta_{\mathrm{L}}}{r_{\mathrm{m}}} = \frac{1\,\mathrm{mm}\cdot\mathrm{mrad}}{0.5\,\mathrm{mm}} = 2\,\mathrm{mrad} \approx 0.1^\circ \tag{4.216}$$

With the $\pm 10^\circ$ mechanical movement of the mirror, the spot moves within approximately $\pm 20^\circ$, which means that we can have up to 400 pixels in a horizontal line – this is less than VGA resolution.

Dr. Wisenheimer:

This limitation can, in principle, be completely overcome by an optical system (an f-theta lens) to image the scanning mirror onto the wall. This would convert the etendue problem into a cost problem.

Exercise 4.46: Smartphone projector

Order from above: Why are projectors always so expensive? Why don't we use halogen lamps instead of those ridiculous discharge lamps? They only get hot and explode – and projectors use far too much optics inside!

Ike is excited about a really cheap projector he found on the internet (Figure 4.73) and requests the team "to do their homework".

Figure 4.73 Mini "smartphone projector" made from cardboard.

Mini smartphone projector – no cable, no noise!

Analyze this projector and estimate the flux delivered by it. How can you improve it?

Solution 4.46

You quickly find out that this projector is just cardboard with a plano-convex glass lens to image a phone display onto the wall.

A contemporary mobile-phone OLED display produces $L \approx 600$ nit, the display area is on the order of $A \approx 9000\ \text{mm}^2$ HANDLE WITH CARE. The total flux provided is then

$$\Phi = LA\pi\ \text{sr} \approx 17\ \text{lm} \tag{4.217}$$

From the picture, we guess $f \approx 160$ mm and a lens diameter $d \approx 40$ mm, hence

$$F/\# = \frac{d}{f} \approx 4 \tag{4.218}$$

The geometric efficiency (also known as configuration factor, Eq. 3.25) is then

$$\eta = \frac{U_\text{T}}{U_\text{S}} = \frac{1}{4F^2} \approx 1.6\,\% \tag{4.219}$$

(Continued)

Solution 4.46 (Continued)

(of the 17 lm) which means we get about 0.3 lm. This is the reason why this device needs a white screen in a very dark room for the projection to be seen at all.

An improvement would be to collect more light by using a lower $F/\#$ (i.e. a much larger lens). And you may dream of a Fresnel lens as a field lens on the display to guide the light into the lens pupil.

Another example that falls into the same category.

Exercise 4.47: OLED projector

An "OLED (organic light-emitting diode) projector" was advertised [84] with a display luminance of as much as $L_S = 10\,000\,\mathrm{cd/m^2}$.[c] Its OLED area was 6 mm · 8 mm and the lens had a NA = 0.3 on the OLED side. What luminous flux can be produced by such a projector? Estimate its efficiency.

[c]One of the authors remembers being on a business trip when his boss called at midnight, impressed by the article and expressing fear. Impressive numbers indeed!
The second law saved both of us.

Solution 4.47

The etendue of the OLED (a Lambertian emitter) is

$$U_S = \pi\,\mathrm{sr} \cdot A_S \approx 150\,\mathrm{mm^2 sr} \tag{4.220}$$

and the total flux from the OLED is thus

$$\Phi_S = U_S L = 1.5\,\mathrm{lm} \tag{4.221}$$

The projector etendue is

$$U_T = A_S \cdot \pi\,\mathrm{NA}^2 \mathrm{sr} \approx 14\,\mathrm{mm^2 sr} \tag{4.222}$$

Multiplication with the source luminance yields the output flux

$$\Phi = L_S U_T = 10\,000\,\mathrm{nit} \cdot 14\,\mathrm{mm^2 sr} = 0.14\,\mathrm{lm} \tag{4.223}$$

which is probably insufficient for any purpose.
Its geometric efficiency is at best (Eq. 3.19)

$$\eta = \frac{U_T}{U_S} = \frac{14\,\mathrm{mm^2 sr}}{150\,\mathrm{mm^2 sr}} \approx 10\% \tag{4.224}$$

From time to time, a hoax is popping up on the internet announcing the projection of images onto the Moon (Figure 4.74) [85].

Figure 4.74 Moonvertising. *Source*: Graphics by the authors.

Exercise 4.48: Moonvertising

Order from above: If you go the extra mile, we can be part of an exciting business. Prove that you are an extraordinary engineer and devise a projector for *moonvertising*!

You are granted access to a secretive "magic" LED that produces 10 000 lm from a circular chip area of 1 mm^2.

How many of those LEDs do we need? What is the necessary semi-diameter r_L of an optical system to project the LED area to the Moon?

You will need some data and requirements:

- The target is to achieve 1% of the solar irradiance on the Moon (so that the beam can be seen on the new Moon)
- The radius of the Moon is $r_M = 1700$ km
- The distance to the Moon is $D_M = 384\,000$ km
- The illuminance of sunlight at the Moon is $E = 137$ klx (cf. Knowledge Corner Topic 3.19)

Solution 4.48

The total luminous flux of the Sun that hits the Moon is

$$\Phi_v = E \cdot \pi r_M^2 = 137\,\text{klx} \cdot 9 \cdot 10^{12}\,\text{m}^2 \approx 1.2 \cdot 10^{18}\,\text{lm} \tag{4.225}$$

and we need 1 % of that value. So the required number of those magic LEDs is

$$\frac{1.2 \cdot 10^{18}\,\text{lm}}{100 \cdot 10\,000\,\text{lm}} = 1.2 \cdot 10^{12} \tag{4.226}$$

(remember to send your boss a note asking for a budget).

Projector pre-design: The etendue formed by the Moon area and front lens light tube is (cf. Eq. 3.42)

$$U_{LM} = \left(\frac{\pi r_L r_M}{D_M}\right)^2 \text{sr} \tag{4.227}$$

and this should be equal to that of the LED,

$$U_S = \pi\,\text{sr} \cdot A = 3.1\,\text{mm}^2\text{sr} \tag{4.228}$$

Hence, the semi-aperture of the front lens has to be

$$r_L = \frac{D_M\sqrt{U_S}}{\pi\, r_M} = \frac{384\,000\,\text{km}\sqrt{\pi}\,\text{mm}}{\pi\, 1700\,\text{km}} = 127\,\text{mm} \tag{4.229}$$

which indicates a lens diameter of about 25 cm per projector or 20 projectors per square meter. And you need many of them – they will occupy an area about twice as large as Switzerland.

And remember – this is based on the data of the “magic” LED.

See also Shining lasers on the Moon.

Now, back to Earth and to real devices employed by rescue teams such as the coast guard flying a helicopter and looking for shipwrecked sailors: search lights (Figure 4.75).

Figure 4.75 Search light. *Source*: AI renderings by the authors.

Exercise 4.49: Search light

A proud engineering team had finished a prototype of an LED search light with a front diameter of $2r = 200$ mm, more or less conforming to a customer's specification of $I_0 = 5$ Mcd and $\Phi = 12$ klm. It is based on the combination of four $\Phi_S = 5000$ lm LEDs, each with an area of $A_S = 9\,\text{mm}^2$. (For example, this is such an LED.)

The search light might perhaps use a light guide arrangement such as the one shown in Figure 4.76.

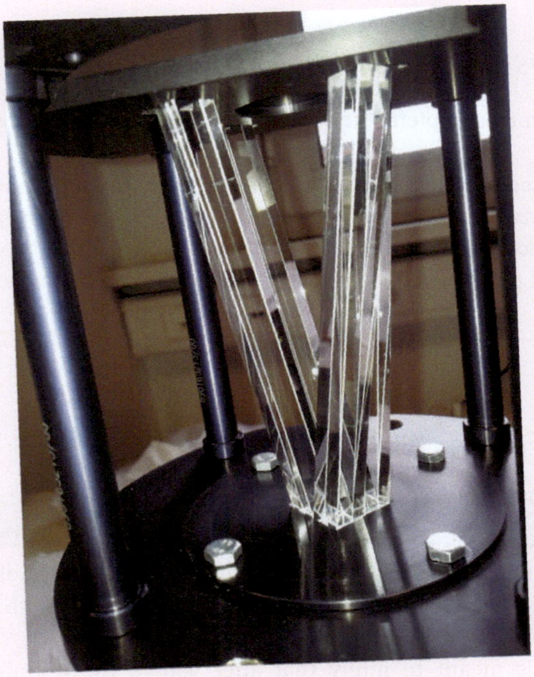

Figure 4.76 Light guides for LED light combination. Light is traveling top to bottom.

Order from above:

Ike went into negotiations with a customer (as always without participants from engineering) and promised them an increase of the central beam candle power (also known as on-axis intensity) to 6 Mcd while simultaneously *reducing the luminaire diameter* to 150 mm. To compensate for this, he obtained the customer's *authorization to use more LEDs*.

Discuss the specifications he promised.

Solution 4.49

The luminance of each of the four LEDs is

$$L = \frac{5000\,\text{lm}}{\pi\,\text{sr} \cdot 9\,\text{mm}^2} = 177\,\text{Mnit} \tag{4.230}$$

The upper limit for the luminous intensity of a beam with that luminance shining from the front area, disregarding transmission losses, is

$$I_0 = LA = 177\,\text{Mnit} \cdot \pi \cdot (0.1\,\text{m})^2 \approx 5.5\,\text{Mcd} \tag{4.231}$$

which was almost reached by the prototype. With the reduced front diameter, this changes to

$$I_0 = 177\,\text{Mnit} \cdot \pi \cdot (0.075\,\text{m})^2 \approx 3\,\text{Mcd} \tag{4.232}$$

which seems to imply that it can be implemented by using just twice the number of sources.

But for such etendue-limited systems, intensity is *limited by luminance*, not flux (cf. Eq. 3.95). This means that the result is *independent of the number of sources*. More flux would support a broader beam, whereas a small amount of flux at the same luminance could produce a needle-like distribution with the same peak intensity.

4.8 Headlights

Exercise 4.50: Hiking lamp

You are about to buy an outdoor headlamp for night hiking. On the package, it says "900 ft beam". Of course, its light will not stop traveling after 900 ft, though. You are puzzled.

a. You find out that this is meant to imply that after 900 ft, the illuminance *perpendicular* to the beam (at its central peak) has dropped to $E_0 = 0.25$ lx, according to ANSI FL1 [86]. What is the peak luminous intensity of the beam?

b. The specification also says "1000 lm", but tells nothing about the beam width. For a rough estimate, start with assuming a "top-hat" distribution: Constant intensity in a cone with a sharp cutoff and no intensity outside.

What is the angular width of this cone? Why should you have second thoughts about this "1000 lm" figure?

c. While hiking at night, you are most interested in the path up to two meters (or six feet) ahead. Is this lamp suitable for this purpose as-is? If not, what types of beam adjustment would you like to have, and why?

Solution 4.50

a. The inverse square law, Eq. 3.142, provides the answer (after implicit conversion from foot to meter):

$$I_0 = \frac{E_0 D^2}{\text{sr}} = \frac{0.25\,\text{lx} \cdot (274\,\text{m})^2}{\text{sr}} = 18.8\,\text{kcd} \tag{4.233}$$

b. With that intensity and the top-hat assumption $I = I_0$, we can easily calculate the solid angle that is filled with light:

$$\Omega = \frac{\Phi}{I} = \frac{1000\,\text{lm}}{18.8\,\text{kcd}} = 0.053\,\text{sr} \tag{4.234}$$

For such a small solid angle, we can compute the cone half-angle θ using the small-angle approximation,

$$\theta = \sqrt{\frac{\Omega}{\pi}} = 7.5° \tag{4.235}$$

We would not really be happy with such a lamp. To emit 1000 lm, it will need roughly 10 W of electrical power and will generate about 5 W of heat from it – hard to cool without frying your forehead and emptying your battery quickly.

c. With a beam angle of $\pm 7.5°$, the lamp will illuminate a circle with a diameter of 50 cm two meters ahead of you. This is by far too narrow to be practical – after all, you would want to see any stumbling blocks right in front of you. The beam would also be way too bright, almost 5000 lux (a full Moon in a clear night provides about 0.2 lx [87]); 20 lux would sound about right.

Most of the time, you need a wide beam with low illuminance in the near-field around you, and a narrow beam with high intensity for those few moments where you would like to look far ahead. In other words – you would want dimming and zooming, which is what good outdoor headlamps offer today.

Exercise 4.51: Automotive headlight

You are designing an automotive high-beam headlamp. According to ECE regulations, certain minimum/maximum illuminance values must be achieved on a screen perpendicular to the headlamp axis at a distance of $D = 25$ m. In this case, the key requirement is $E_0 = 100$ lx on-axis.

You are given an LED package, the OSRAM OSLON Black Flat, with a light output area of 5.5 mm · 1.1 mm in air (source area $A_S = 6\,\text{mm}^2$) and a luminous flux of $\Phi_S = 1700$ lm.

(Continued)

Exercise 4.51: (Continued)

For your initial estimates, you may think of the LED as Lambertian. You think you can capture 90% of the luminous flux from the LEDs with your optic, using the light from a $\theta = \pm 72°$ cone from the LEDs (the outermost 18° are hard to catch and they are probably yellowish anyway. Bonus question: Confirm that the outermost 18° contain 10% of the flux of a Lambertian source). You also assume that with mirrors, lenses and a front window, you will lose another 30% of the light due to transmission/reflection losses.

a. What luminous intensity does your headlamp have to emit on-axis to achieve 100 lx in 25 m?

b. Your customer defines (for design reasons) the exit aperture – a circle with a diameter of $2r_F = 30\,\text{mm}$. Is it possible to achieve the required 100 lx with this setup? In other words, what is the minimum circle diameter you need?

c. What would change so far if you could collect all the light, instead of just 90%?

d. What would change so far if you had better optics, thus achieving an overall transmission of 85%?

e. White LEDs get a little more efficient when they are *index-matched*, i.e. when the primary optics is attached to the LEDs using a silicone glue without an air gap. Optimistically (see Exercise 4.10), you can assume an efficiency gain of 20% from the LED into the silicone glue, if everything else remains the same. Which optical effect do you think causes this efficiency gain?

f. Assume a silicone glue with $n = 1.41$ and a 20% efficiency gain and that you still have 90% collection efficiency and only 30% transmission losses. How would your required headlamp aperture change if you would use index-matched optics?

Solution 4.51

a. The inverse square law (Eq. 3.142) delivers, one more time:

$$I_0 = \frac{E_0 D^2}{\text{sr}} = \frac{100\,\text{lx} \cdot (25\,\text{m})^2}{\text{sr}} = 62.5\,\text{kcd} \tag{4.236}$$

b. To answer this, we first need an estimate of the source luminance L_S. The LED's etendue U_S is

$$U_S = XA_S = \pi\,\text{sr} \cdot 6\,\text{mm}^2 \approx 19\,\text{mm}^2\text{sr} \tag{4.237}$$

Luminance is flux divided by etendue, so

$$L_S = \frac{\Phi_S}{U_S} = \frac{1700\,\text{lm}}{19\,\text{mm}^2\text{sr}} \approx 90\,\text{Mnit} \tag{4.238}$$

This value is reduced by the $\eta = 70\%$ transmission of the optics, but not by the collection from the LED, as this would just imply an (intermediate) target etendue smaller than that of the source, leaving an aperture luminance $L_A = \eta_S = 63\,\text{Mnit}$.

The maximum on-axis intensity that can be obtained after the front aperture with area $A_F = r_F^2\pi$ is (Eq. 3.107)

$$I_0 = 63\,\text{Mnit} \cdot 706\,\text{mm}^2 = 44.5\,\text{kcd} \tag{4.239}$$

which is clearly insufficient, compared to the specification.

The minimum front aperture size is

$$A_F = \frac{I_0}{L_A} = \frac{62.5\,\text{kcd}}{63\,\text{Mnit}} \approx 1000\,\text{mm}^2 \tag{4.240}$$

implying an increased diameter of about 35 mm.

Let's sit back and relax for a moment. We know from experience that these things happen all the time – a requirement is impossible to meet, but to most people the reasons are not obvious. In this case, however, you are able to state – based on just three facts (LED area, LED flux, optical transmission) – that for 100 lx at 25 m, a diameter of 35 mm is required and not 30 mm as the customer demands. Few people would love you for this, but in fact they should: you are saving yourself, your team and the customer many hours of endless grief. Your refusal is not due to a lack of competence – on the contrary –, but based on fundamental conservation laws. Herein lies the power of the theory of light for illumination.

c. We illustrate the situation with the help of a luminance diagram (Figure 4.77) [62].

We need to discuss one more aspect: The optics will fill the front aperture and a given cone (angle θ) with light from the LED. If the etendue defined by those (cf. Eq. 3.11),

$$U_T = (\pi\, r_F \sin\theta)^2\,\text{sr} \tag{4.241}$$

is smaller than the etendue of light collection from the LED, we will get an according portion of luminous flux, as the luminance diagram in Figure 4.77 illustrates.

(Continued)

Solution 4.51 (Continued)

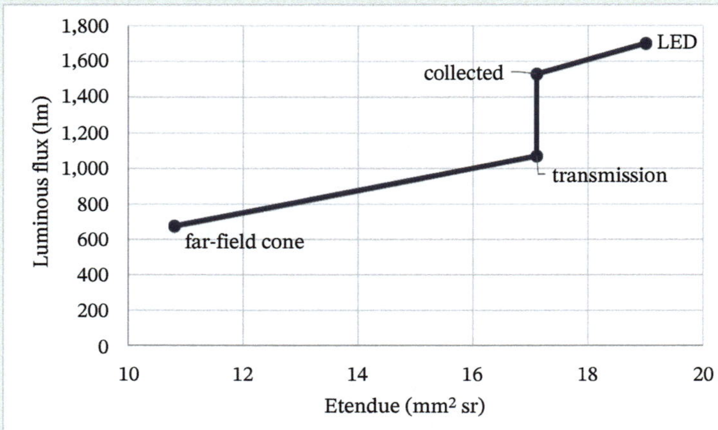

Figure 4.77 Flux and etendue of the discussed optical system.

If it is larger, the flux will be diluted, i.e. more or less homogeneously distributed over the larger etendue.

We provide the file headlamp.xlsx.

Play around with the LED collection efficiency, aperture radius, and the size of the cone angle. However, if all you can change is the collection efficiency, the 35 mm requirement will not change. But why?

Here I am to help you!

I'm hovering at the screen, looking back toward the aperture. But all I can see in the aperture is the same old luminance, 70% of what the LED emits – due to the conservation of luminance along a ray, except for transmission loss. Whatever you do in the optics to change collection efficiency: My eyes catch only the tiniest bit of light. Collection efficiency is utterly unimportant to achieve 100 lux where I am. What you need is a larger aperture so that I can see the luminance coming from a larger solid angle and thus larger angular extent: $E = LX = L\Omega_p \approx L\Omega$.

d. With an increase in transmission efficiency, the Lightman would see a higher luminance from the same aperture and we would get a higher far-field intensity. But even 85% transmission would not be sufficient to get down to 30%. Instead, we would need a transmission of $70\% \cdot (35/30)^2 \approx 96\%$. You will not achieve this transmission in practice except if you use unobtainium as your optical material (an inside joke for those who watched Avatar).

e. Here, we discuss a white LED, i.e. a blue-emitting LED chip on which a phosphor-grain-filled, thin silicone platelet is glued. Light scattered around in the silicone platelet will partly be refracted out, partly experience total internal reflection. Since the index of refraction of silicone is about 1.41, the refracted and TIR'ed fractions will be about equal (why?). When we embed the device in a silicone dome, there is perfect index matching and therefore no more TIR: It is now easier for the light to escape (Figure 4.78). Since a large part part of the TIR'ed light is recycled via scattering and back out, the effect is not that big. There will be some gain, but probably less than 20%. This mechanism has already been discussed in Exercise 4.10.

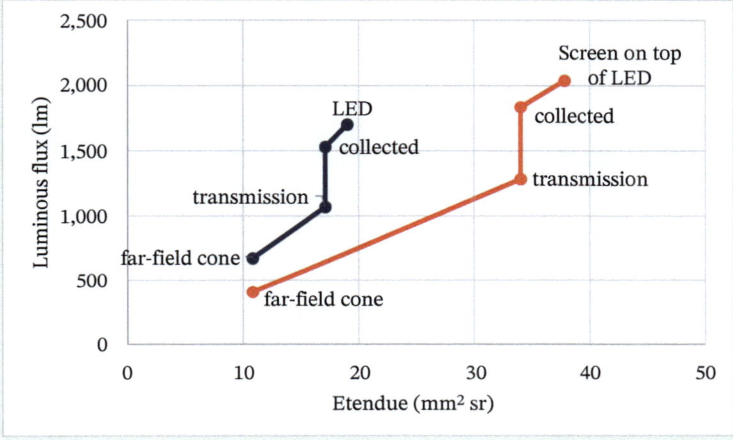

Figure 4.78 Luminance in the course of the optical system without (blue curve, left) and with (orange curve, right) immersion.

f. The immersion provides a collection of more light into a smaller etendue. As the defined gain (20%) is smaller than the maximum possible effect based on the refractive index (a factor of two), we end up with more light in a considerably larger etendue. As a result, the luminance is reduced and so is the far-field intensity with the same aperture. This is illustrated in Figure 4.78.

Feel free to investigate this case as well using the provided file. In the end, with the kind of immersion defined here you would need a larger aperture to get the same intensity.

If you are puzzled now because you feel it's hard to grasp why 20% more efficiency could possibly be bad, don't worry. For most people, it is not intuitively clear that adding a dome will decrease luminance. But the effect is real and fully consistent with the etendue, luminance and flux logic. In a sense, there is room for more rays in the silicone: Rays that would be TIR'ed at an air interface, because they cannot exist in air but can propagate into silicone. These rays get their fair share of the light, which thus is diluted over more rays so that each ray will carry less light.

Exercise 4.52: Headlamp diameter

After a long development process, the lighting division of a car manufacturer finished a LED headlamp with a luminous intensity of 225 kcd on-axis from an aperture of radius $r_F = 40$ mm.

a. Calculate the original source luminance. After prototype examination, it was decided to shrink the headlight for design reasons to a semi-diameter of $r_2 =$ 23 mm.

b. What luminous intensity can be expected?

c. Discuss!

Solution 4.52

a. Source luminance enables the original design with (cf. Eq. 4.239)

$$L = \frac{I_0}{\pi\, r_F^2} = \frac{225\,\text{kcd}}{\pi \cdot 1600\,\text{mm}^2} = 45\,\text{Mnit} \tag{4.242}$$

b. With the same source luminance but smaller aperture, we get

$$I_0 = L \cdot \pi\, r_2^2 = 75\,\text{kcd} \tag{4.243}$$

c. To achieve the original intensity with the reduced diameter we need a source with a luminance of

$$L_2 = \frac{I_1}{A_2} = \frac{225\,\text{kcd}}{1662\,\text{mm}^2} = 135\,\text{Mnit} \tag{4.244}$$

There is no way to achieve the same result with a larger number of lower-luminance LEDs.

Dr. Wisenheimer:

In the improbable case that the original design was done with a $L = 135$ Mnit LED and some *dilution* (spreading of the light over a larger etendue to achieve an *effective luminance* of $L =$ 45 Mnit), one could adapt the design to the new diameter without an accompanying reduction in intensity.

Exercise 4.53: Screen illuminance

We abuse a projection screen as shown in Figure 4.51 of 3 m · 4 m prepared for projector lumen measurement and shine an automotive high-beam from a distance of $D = 25$ m to the screen. Assume a $\cos^{100}$ distribution (cf. Knowledge Corner Topic 3.15) with a peak intensity of $I_0 = 62.5$ kcd (cf. Exercise 4.51). Estimate the luminous flux on the rectangular area.

Solution 4.53

The angle of a corner of the screen to the axis is

$$\theta = \arctan\left(\frac{\sqrt{(4\,\mathrm{m})^2 + (3\,\mathrm{m})^2}}{2 \cdot 25\,\mathrm{m}}\right) = 5.7° \tag{4.245}$$

and a relative intensity at the edge of $\cos^{100}\theta = 0.6$.

For simplicity, we assume a top-hat intensity distribution of $I = 0.8\,I_0$.

We calculate the solid angle of the screen as seen from the headlight (cf. Knowledge Corner Topic 2.4):

$$\Omega = \frac{A}{D^2}\,\mathrm{sr} = \frac{12\,\mathrm{m}^2}{625\,\mathrm{m}^2}\,\mathrm{sr} = 0.02\,\mathrm{sr} \tag{4.246}$$

Hence,

$$\Phi \approx 0.8 I_0 \Omega = 1000\,\mathrm{lm} \tag{4.247}$$

which is just a first estimate.

We check the result in the reverse direction. A cone of the same solid angle as the screen has a half angle $\theta = 4.6°$. The flux contained in such a cone can be estimated from Eqs. 3.133 and 3.130 as

$$\Phi = \frac{2\pi\mathrm{sr}}{101} I_0 \left[1 - \cos^{101} 4.6°\right] \approx 1080\,\mathrm{lm} \tag{4.248}$$

which is not too different from the above rough estimate.

We will cross-check this result using a simulation.

Simulation 4.15: Automotive headlight shining onto screen

We set up a small source, give it a $\cos^{100}$ intensity distribution (the maximum possible exponent of a $\cos^n$ source in Zemax), and scale the flux properly (the total flux of the *$\cos^n$* beam is given by Eq. 3.130):

$$\Phi = \frac{2\pi}{101}\mathrm{sr} \cdot 62.5\,\mathrm{kcd} = 3888\,\mathrm{lm} \tag{4.249}$$

With these source settings, we run a simulation and measure the total flux at the $3\,\mathrm{m} \cdot 4\,\mathrm{m}$ receiver. The simulation result is displayed in Figure 4.79.

(Continued)

Simulation 4.15: (Continued)

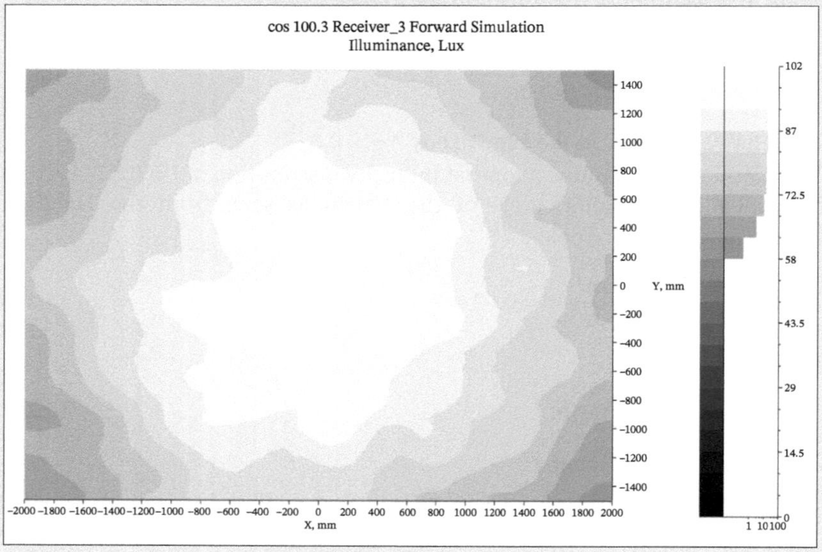

Figure 4.79 Screen illuminance. As expected, we find 100 lux in the center. The total screen flux in this simulation is 1007 lm.

We find that the simulated total flux is 1007 lm– almost too close to our estimate to be true.

Provided file: cos100.lts.

References

1. Bermel, P. (2012). High efficiency incandescent lighting. US Patent, US8823250B2.
2. Delmdahl, R., Kunzer, M., and Schwarz, U. (2011). Thin film LEDs gaining ground: excimer laser lift-off enables high brightness LED production. *Laser Technik Journal* 8 (3): 48–52. https://doi.org/10.1002/latj.201190027.
3. Hecht, E. (2010). *Optics*. San Francisco, CA: Addison-Wesley.
4. David, A.J., Aldaz, R.I., Butterworth, M. et al. (2012). LED with remote phosphor layer and reflective submount. US Patent, US8168998B2.
5. Ying, S.P., Fu, H.K., and Tu, H.Z. (2014). Curved remote phosphor structure for phosphor-converted white LEDs. *Applied Optics* 53 (29): H160–H164. https://doi.org/10.1364/AO.53.00H160.
6. Bierhuizen, S.J. and Harbers. G. (2009). Illumination system with optical concentrator and wavelength converting element. US Patent, US7543959B2. https://patents.google.com/patent/US7543959B2/en?oq=US7543959.

7 Harbers, G. and Keuper, M. (2005). Remote wavelength conversion in an illumination device. US Patent, US20050270775A1. https://patents.google.com/patent/US20050270775A1/en?oq=us2005270775a1.

8 William James, C. (2017). Using pinhole images to understand nonimaging optics. In: *Optical Design and Fabrication 2017 (Freeform, IODC, OFT), OSA Technical Digest (Online)*, ITh1A-1. Denver, CO: Optical Society of America.

9 IES (2019). ANSI/IES LM-63-19: IES standard file format for the electronic transfer of photometric data and related information. https://store.ies.org/product/approved-method-ies-standard-file-format-for-the-electronic-transfer-of-photometric-data-and-related-information.

10 Discomfort Glare in Interior Lighting (1995). Number 117 in Technical Report/International Commission on Illumination. CIE Central Bureau, Vienna.

11 Thorn, P. (2020). Discomfort caused by glare from luminaires with a non-uniform source luminance. *LED Professional* 78: 30–32.

12 Luckiesh, M. and Guth, S.K. (1949). Brightnesses in visual field at borderline Between Comfort and Discomfort (BCD). *Illuminating Engineering* 44 (11): 650–670.

13 Shuster, J. (2014). Addressing glare in solid-state lighting. https://www.cooperlighting.com/s/ephesus/assets/31-Bright-Paper-Adressing-Glare.pdf.

14 Carlucci, S., Causone, F., De Rosa, F. et al. (2015). A review of indices for assessing visual comfort with a view to their use in optimization processes to support building integrated design. *Renewable and Sustainable Energy Reviews* 47: 1016–1033. https://doi.org/10.1016/j.rser.2015.03.062.

15 CIE (2019). Discomfort caused by glare from luminaires with a non-uniform source luminance. https://cie.co.at/publications/discomfort-caused-glare-luminaires-non-uniform-source-luminance.

16 Giang, D., Pham, H., Van, Q.N. et al. (2024). A study on pendant and blackboard asymmetric lens LED luminaires for optimal illumination in classrooms. *Sustainability* 16 (22): 10015. https://doi.org/10.3390/su162210015.

17 Welford, W.T. and Winston, R. (1982). Nonconventional optical systems and the brightness theorem. *Applied Optics* 21 (9): 1531–1533. https://doi.org/10.1364/AO.21.001531.

18 Slevogt, H. (1974). *Technische Optik*. Berlin: de Gruyter.

19 Wu, Y., Wang, Z., Cai, W. et al. (2018). The absolute reflectance and new calibration site of the moon. *The Astronomical Journal* 155 (5): 213. https://doi.org/10.3847/1538-3881/aabaf5.

20 Unsöld, A. and Baschek, B. (2002). *The New Cosmos*. Berlin, Heidelberg: Springer.

21 Inglis, M. (2007). *Astrophysics Is Easy! A Complete Introduction for Amateur Astronomers*. New York, London: Springer.

22 Miles. R. (2007). A light history of photometry: from hippacarchus to the hubble space telescope. *Journal of the British Astronomical Association* 117 (4): 172–186.

23 Pogson, N. (1856). Magnitudes of thirty-six of the minor planets for the first day of each month of the year 1857. *Monthly Notices of the Royal Astronomical Society* 17 (1): 12–15. https://doi.org/10.1093/mnras/17.1.12.

24 Dufay, J. (1964). *Introduction to Astrophysics: The Stars*. New York: Dover Publications Inc.

25 Brooker, G. (2002). *Modern Classical Optics*. Oxford Master Series in Atomic, Optical and Laser Physics, 1e. Oxford: Oxford University Press.

26 Winston, R., Miñano, J.C., and Benitez, P. (2005). *Nonimaging Optics*. Academic Press.

27 Kervella, P., Thévenin, F., Morel, P. et al. (2003). The interferometric diameter and internal structure of Sirius A. *Astronomy & Astrophysics* 408 (2): 681–688. https://doi.org/10.1051/0004-6361:20030994.

28 Hanbury Brown, R., Davis, J., and Allen, L.R. (1974). The angular diameters of 32 stars. *Monthly Notices of the Royal Astronomical Society* 167 (1): 121–136. https://doi.org/10.1093/mnras/167.1.121.

29 Arvo, J. (1986). Backward ray tracing. In: *Siggraph*. https://courses.cs.washington.edu/courses/cse457/20au/projects/trace/extra/Backward.pdf.

30 Veach, E. and Guibas, L. (1995). Bidirectional estimators for light transport. In: *Photorealistic Rendering Techniques* (ed. G. Sakas, S. Müller, and P. Shirley), 145–167. Berlin, Heidelberg: Springer.

31 Abakumova, A.A., Malinova, T.P., Medennikov, P.A. et al. (2018). Reverse ray tracing method for estimating the illumination from outside the field of view of optical systems. *Journal of Optical Technology* 85 (1): 32–35. https://doi.org/10.1364/JOT.85.000032.

32 Fu, L. (2006). Increasing the brightness of light sources. PhD thesis. Marburg.

33 True, T.T. (1981). Light collection system. US Patent, US4305099A.

34 Li, K. and Inatsugu, S. (2006). Illumination system and method for recycling light to increase the brightness of the light source. US Patent, US20100045937A1.

35 Hottel, H.C. and Sarofim, A.F. (1967). *Radiative Transfer*. McGraw-Hill.

36 Stasenko, A.L. (1967). Self-irradiation coefficient of a Moebius strip of given shape. *Izvestiya Akademii Nauk SSSR, Energetika i Transport*, pp. 104–107.

37 Ballance, J.O. and Donovan, J. (1973). Radiation configuration factors for annular rings and hemispherical sectors. *Journal of Heat Transfer* 95 (2): 275. https://doi.org/10.1115/1.3450045.

38 Cimino, V., Frascarolo, M., Mucciante, M. et al. (2022). Interaction between daylighting and artificial lighting in relation to conservation and perception, according to new illumination system of Sistine Chapel. *Journal of Cultural Heritage* 58: 256–265. https://doi.org/10.1016/j.culher.2022.10.004.

39 Tibbitts, T.W., Morgan, D.C., and Warrington, I.J. (1983). Growth of lettuce, spinach, mustard, and wheat plants under four combinations of high-pressure sodium, metal halide, and tungsten halogen lamps at equal PPFD. *Journal of the American Society for Horticultural Science* 108 (4): 622–630. https://doi.org/10.21273/JASHS.108.4.622.

40 Fylladitakis, E.D. (2023). Controlled LED lighting for horticulture: a review. *Open Journal of Applied Sciences* 13 (2): 175–188.

41 Pinho, P., Jokinen, K., and Halonen, L. (2017). The influence of the LED light spectrum on the growth and nutrient uptake of hydroponically grown lettuce. *Lighting Research & Technology* 49 (7): 866–881. https://doi.org/10.1177/1477153516642269.

42 Ashdown, I. (2023). Comparing PAR and ePAR metrics for horticultural luminaires. https://www.ledsmagazine.com/horticultural-lighting/article/14293724/comparing-par-and-epar-metrics-for-horticultural-luminaires.

43 Craig, D.S. and Runkle, E.S. (2013). A moderate to high red to far-red light ratio from light-emitting diodes controls flowering of short-day plants. *Journal of the American Society for Horticultural Science* 138 (3): 167–172. https://doi.org/10.21273/JASHS.138.3.167.

44 Zhen, S., van Iersel, M., and Bugbee, B. (2021). Why far-red photons should be included in the definition of photosynthetic photons and the measurement of horticultural fixture efficacy. *Frontiers in Plant Science* 12: 693445.

45 Sloper, C. (2013). *The LED Grow Book: Better, Easier, Less Watts*. Createspace Independent Publ.

46 McCree, K.J. (1971). The action spectrum, absorptance and quantum yield of photosynthesis in crop plants. *Agricultural Meteorology* 9: 191–216. https://doi.org/10.1016/0002-1571(71)90022-7.

47 ANSI (2017). Quantities and units of electromagnetic radiation for plants (photosynthetic organisms). https://elibrary.asabe.org/abstract.asp?aid=48303&t=2.

48 ANSI (2018). Recommended methods for measurement and testing of LED products for plant growth and development. https://elibrary.asabe.org/abstract.asp?aid=49804&t=2.

49 DLC (2025). Horticultural lighting resources: terms and definitions. https://www.designlights.org/resources/reports/horticultural-lighting-terms-and-definitions.

50 Calculation of Blackbody Radiance. https://www.spectralcalc.com/blackbody/CalculatingBlackbodyRadianceV2.pdf.

51 Ulbricht, R. (1920). *Das Kugelphotometer (Ulbricht'sche Kugel): Darstellung seiner Theorie, Ausbildung und Anwendung, unter Besonderer Berücksichtigung der Fehlerquellen*. München: Oldenbourg.

52 Lighting Global International Finance Corporation (2014). Integrating sphere measurement. https://www.lightingglobal.org/?s=integrating+sphere.

53 Rowland. A. (1883). On concave gratings for optical purposes. *The London, Edinburgh, and Dublin Philosophical Magazine and Journal of Science* 16 (99): 197–210. https://doi.org/10.1080/14786448308627419.

54 Young, M. (2000). *Optics and Lasers: Including Fibers and Optical Waveguides*. Advanced Texts in Physics. New York, Berlin: Springer.

55 Schulz, G. (1974). *Paradoxa aus der Optik*. Leipzig: Johann Ambrosius Barth.

56 Roberts. D. (1993). Combination lamp and integrating sphere for efficiently coupling radiant energy from a gas discharge to a lightguide. US Patent, USRE34492E. https://patents.google.com/patent/USRE34492E/en.

57 MacLennan, D.A. and Turner, B.P. (2003). Aperture lamp. US Patent, US6509675B2. https://patents.google.com/patent/US6509675B2/en.

58 Leutz, R., Fu, L., and Ries, H. (2006). Carambola optics for recycling of light. *Applied Optics* 45 (12): 2572–2575. https://doi.org/10.1364/AO.45.002572.

59 Li, K. and Inatsugu, S. (2005). Dual paraboloid reflector system configurations forincreasing brightness of an arc lamp. In: *Nonimaging Optics and Efficient Illumination Systems II*, Vol. 5942, 594201. SPIE.

60 Ulffers, H. and Ploke, M. (1962). Beleuchtungseinrichtung für Laufbild- bzw. US Patent, DE1132743B.

61 Ries, H. and Spirkl, W. (1995). A generalized Kirchhoff Law for quantum absorption and luminescence. *Solar Energy Materials and Solar Cells* 38: 39–44.

62 Muschaweck, J. and Rehn, H. (2022). *Designing Illumination Optics*. SPIE Press.

63 Chaves, J. (2016). *Introduction to Nonimaging Optics*, 2e. CRC Press.

64 Breuer, F. (2017). Beleuchtungsmessung mit smartphone-apps. https://publikationen.dguv.de/widgets/pdf/download/article/3391.

65 International Light (2018). What is cosine correction and why is it important in light measurement? https://www.youtube.com/watch?v=NoBG4x4DBmo.

66 European Machine Vision Association (2021). EMVA standard 1288: standard for characterization of image sensors and cameras. https://www.emva.org/wp-content/uploads/EMVA1288General_4.0Release.pdf.

67 Holst, G.C. (2011). *CMOS/CCD Sensors and Camera Systems*, 2e. Bellingham, WA: SPIE.

68 Teledyne Lumenera (2017). How to interpret the Dynamic Range and Signal to Noise Ratio (SNR) in image sensor and industrial camera specifications. https://www.adimec.com/how-to-interpret-the-dynamic-range-and-signal-to-noise-ratio-snr-in-image-sensor-and-industrial-camera-specifications.

69 AZOM (2021). A guide to evaluating camera sensitivity. https://www.azom.com/article.aspx?ArticleID=20215.

70 Teledyne Lumenera (2024). Understanding dynamic range and signal-to-noise ratio when comparing cameras. https://www.lumenera.com/blog/understanding-dynamic-range-and-signal-to-noise-ratio-when-comparing-cameras.

71 ANSI (1997). ANSI/NAPM IT7.228-1997 – Audiovisual systems – electronic projection – fixed resolution projectors (revision and redesignation of ANSI IT7.228-1990). https://www.intertekinform.com/en-gb/standards/ansi-it7-228-1990-95457_saig_ansi_ansi_199792/?srsltid=AfmBOool9X6FkWvflFJbTEd_LhVhUUWsfrWQQGt0xACscMbeuLZ58ib3.

72 Derra, G., Moench, H., Fischer, E. et al. (2005). UHP lamp systems for projection applications. *Journal of Physics D: Applied Physics* 38 (17): 2995. https://doi.org/10.1088/0022-3727/38/17/R01.

73 DLP (2010). Texas Instruments Application Report "DLP System Optics DLPA022" July 2010. Dallas, Texas.

74 Schwaiger, S., Schoemer, R., and Hering, O. (2018). Beleuchtungsvorrichtung mit einer lichtquelle zur emission von beleuchtungslicht. https://patents.google.com/patent/WO2018007117A1/de?oq=+WO+2018007117+A1.

75 Seitz, M. (2017). Device and method for projecting a light pattern. US Patent, US20170205040A1. https://patents.google.com/patent/US20170205040A1/en?oq=DE102016200586A1.

76 Rehn, H., Breidenassel, N., and Grötsch, S. (2006). *Optical Projection Appliance.* Publication of WO2008040278A1.

77 Scott Dewald, D., Penn, S.M., and Davis, M. (2001). Sequential color recapture and dynamic filtering: a method of scrolling color. *SID Symposium Digest of Technical Papers* 32 (1): 1076–1079. https://doi.org/10.1889/1.1831745.

78 Li. K. (2010). Recycling system and method for increasing brightnessusing light pipes with one or more light sources, and aprojector incorporating the same. US Patent, US20100045937A1.

79 Fu, L., Leutz, R., and Ries, H. (2005). Light recycling in solid state devices. In: *Fifth International Conference on Solid State Lighting*, Vol. 5941, 326–330. SPIE.

80 Cunningham, D.W. and Esakoff, G. (1994). Lighting fixture. US Patent, US5446637A.

81 Cunningham, D.W. and Esakoff, G. (1999). Incandescent illumination system. US Patent, USRE36316E.

82 Paßlick, C., Geyer, U., Heßling, T. et al. (2016). Glass light guides for color mixing of high-power LEDs. In: *Polymer Optics and Molded Glass Optics: Design, Fabrication, and Materials 2016*, Vol. 9949, 41–47. SPIE.

83 Rehn, H. (2013). System design of a LED based gobo projector. *LED-based Illumination Systems* 8835: 70–76. https://doi.org/10.1117/12.2022266.

84 Riehemann, S., Grossmann, C., Vogel, U. et al. (2009). Ultra small OLED pico projector: ground-breaking, active illumination-free concept enables new projection applications. *Optik & Photonik* 4 (2): 34–36. https://doi.org/10.1002/opph.201190023.

85 Moonvertising, S.K. (2008). Moonvertising. http://www.nytimes.com/2008/12/14/magazine/14Ideas-Section2-C-t-009.html.

86 ANSI-NEMA FL-1 – Flashlight Wiki. https://flashlightwiki.com/ANSI-NEMA_FL-1.

87 Krieg, J. (2021). Influence of moon and clouds on night illumination in two different spectral ranges. *Scientific Reports* 11 (1): 20642. https://doi.org/10.1038/s41598-021-98060-2.

5

Design

We are now approaching the end of our journey. We started with a refresher on general geometry and optics (Chapter 2), we learned how to understand light for illumination (Chapter 3), and we applied theory to practical problems (Chapter 4). Now we turn to creating initial designs for some typical optical systems and provide some guidance on how to use datasheets and to select light sources.

5.1 Predesign Calculus

As a basis for all design work, we recall the conservation laws for etendue and flux as expressed in Eqs. 3.19 and 3.95, respectively. They provide not only upper limits for design efforts, based on conservation laws and thus independent of optical architectures. They also help greatly to "fence in" the problem, to assess how close an actual design is to the theoretical limit, and they guide us toward promising optical architectures.

5.1.1 Spotlights

Lighting a theater or concert stage, a film set or a TV studio is an art form that is both enabled and limited by the light sources and the optics in the lamp heads. While actors, directors and cinematographers get all the fancy awards, the unsung heroes are the gaffers, without whom no-one could see anything at all. It is the optical designers' job to find technical solutions making the jobs of these heroes easier, and to push those solutions to their respective limits.

To achieve the desired effects, gaffers need a variety of beam profiles. Here is a brief taxonomy [1]:

- Beam light (top hat distribution or hard edge)
- Profile spot (edge intensity larger than 2/3 of center intensity)
- Spotlight (edge intensity larger than 1/3 of center intensity) (Figure 5.1)
- Soft spot (the edge becoming diffuse)
- Wash- or floodlight (beam angle 100° or larger, completely diffuse)

Engineering Illumination Optics: Case Studies and Exercises, First Edition. Henning Rehn and Julius Muschaweck.
Published 2026 by Wiley-VCH GmbH.

frimufilms/Adobe Stock Photos

Figure 5.1 Spotlight.

This is the general idea, although different manufacturers will have slightly different names and specifications.

Exercise 5.1: LED spotlight (1)

Design a spotlight (cf. Figure 4.75) with a narrow circular beam based on the LED from Exercise 4.7, with a diameter of 20 mm and $\Phi_S = 1200\,\text{lm}$.

The spotlight (more precisely, its exit pupil) is circular and has a diameter of 130 mm. For a start, assume ideal efficiency of the optical system and a Lambertian source.

(a) What is the narrowest angular beam extent with a sharp cutoff that can be achieved?

(b) What maximum luminous intensity can be achieved with this spotlight?

(c) What on-axis illuminance is possible at 5 m distance?

(d) What is the FWHM of a $\cos^n$ beam with the same peak intensity and the same flux (just different shape)?

(e) Create a first design with a soft beam, to corroborate your theoretical results. What is the simplest design you can think of?

Solution 5.1

(a) Our light-emitting surface (LES) has a radius $r_S = 10\,\text{mm}$. We assume we can collect rays from all angles from the LED ($\theta_S = 90°$) – which we could, if we had a perfectly matching compound collimator, or, more realistically, a plano-convex condenser lens with a tiny air gap. The inverse problem is studied in Exercise 5.3, where you will have to find a source for a given fixture.

Etendue conservation means that

$$10\,\text{mm} \cdot \sin\theta_S = 65\,\text{mm} \cdot \sin\theta \tag{5.1}$$

which gives a far-field angle of $\theta = \pm 8.8°$.

(b) The corresponding solid angle (Eq. 2.11) is

$$\Omega = 2\pi\,\text{sr}(1 - \cos 8.8°) = 0.074\,\text{sr} \tag{5.2}$$

Assuming a top-hat distribution, our intensity will be

$$I = \frac{\Phi}{\Omega} = \frac{1200\,\text{lm}}{0.074\,\text{sr}} \approx 16.2\,\text{kcd} \tag{5.3}$$

Another way to get the same result would be to compute the source luminance L_S from the etendue and multiply this with the aperture area, as we did in several exercises in Chapter 4.

(c) The illuminance follows the inverse square law (Eq. 3.142),

$$E_0 = \frac{I_0}{D^2}\text{sr} = \frac{16.2\,\text{kcd}}{(5\,\text{m})^2}\text{sr} \approx 650\,\text{klx} \tag{5.4}$$

At this distance and aperture diameter, we do not expect near-field effects.

(d) We estimate the cosine exponent using Eq. 3.131,

$$n = 2\pi\,\text{sr}\frac{16.2\,\text{kcd}}{1200\,\text{lm}} - 1 = 85 \tag{5.5}$$

We obtain the FWHM angle $2\theta_{\text{FWHM}}$ via

$$0.5 = (\cos\theta_{\text{FWHM}})^{85} \quad \Rightarrow \quad \theta_{\text{FWHM}} = 8.1° \tag{5.6}$$

This angle is a little smaller than above because of the $\cos^n$ tail.

(e) We have no way of knowing what possible solution you found, but the simplest we can think of is a parabolic reflector (see Figure 5.2). It will need a diameter of 130 mm, 1/4 of which is its focal length, i. e. 32.5 mm (see Section 2.5.1). The coma, as is common for all simple conic sections, will create soft beam edges, albeit with a central peak that is too sharp. The inevitable self-blocking in the center is less than 3%; for a real source with a PCB, heat sink and a spider to hold it, the loss would be larger, but not necessarily too large to be practical. The peak intensity and beam width will be just about right. For a real design, the awkward placement of LED, heat sink and wiring would not be acceptable, though. Hollow CPC-like primary optics work well only with larger CoB LEDs, in our experience. As the next-simpler system, we would imagine an aspheric primary condenser glass lens and a PMMA Fresnel lens, with enough defocus and maybe slight scattering to remove the source inhomogeneities. If you like, try your hand with this approach, for example using the ray file for the ams-OSRAM Soleriq® S9 (because ams-OSRAM provides separate yellow and blue ray files, thus also capturing color inhomogeneities) as a real source.

(Continued)

Solution 5.1 (Continued)

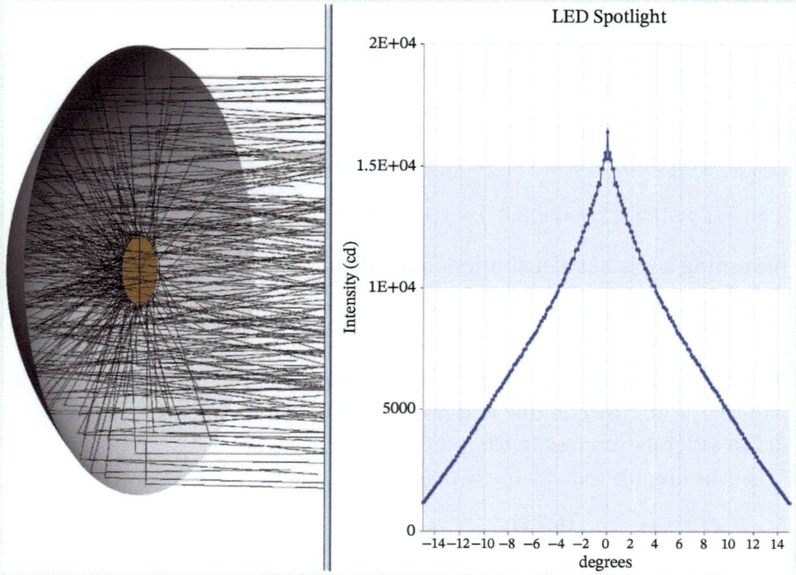

Figure 5.2 A simple parabolic reflector corroborates our estimates. The peak intensity it provides is 16.2 kcd, including self-blocking, and its FWHM angle is ±8.8°.

Dr. Wisenheimer:

$\theta_S = 90°$ is often unrealistic, for example when you use a lens system. You better start with $\theta_S = 60°$. Of course, this implies a loss with a remaining efficiency of

$$\eta = \sin^2 60° = 75\% \tag{5.7}$$

5.1.2 Solar Simulators

Solar simulators produce essential, but not all, aspects of sunlight. Their purpose is to study plants, test architectural installations and satellites (see Figure 5.3), or measure solar cells under well-defined and stable laboratory conditions; details are specified in the IEC 60904 standard.

In most cases, simulators provide solar irradiance, but not solar radiance, and approximate the solar spectrum more or less well. Typically, a larger beam angle than the solar angular subtense is used, and must be used when the source radiance is smaller than the Sun's huge value of 2 Gnit (see Knowledge Corner Topic 3.19). Solar simulators are often based on discharge lamps, sometimes also halogen lamps, or nowadays a combination of LEDs. The target area can be as small as a single solar cell (20 cm^2) or as large as one or more modules (several square meters).

Solar simulator designs include solar cell testers [2], high-flux solar simulators (Figure 5.3) [3], and devices for testing CPV (concentrated photovoltaics, see Figure 5.4) [4, 5].

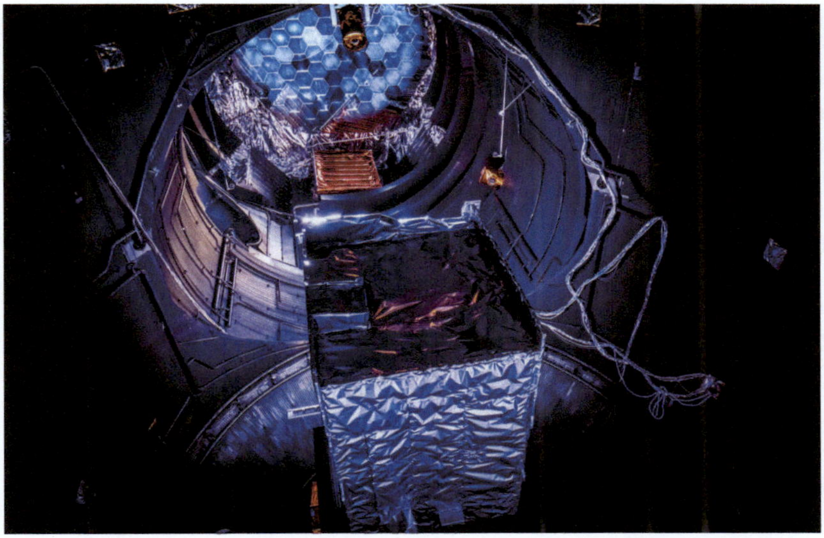

Figure 5.3 Juice (Jupiter Icy Moon Explorer) thermal development model inside ESA's Large Space Simulator, which features a powerful solar simulator. *Source:* Courtesy of ESA–M.Cowan, ESA standard license.

Figure 5.4 A solar simulator used for solar thermal experiments. *Source:* Courtesy of Ole Jørgen Nydal, NTNU, Norway.

Exercise 5.2: Solar simulator (1)

Is it possible to design a solar cell tester with a 120…150 W halogen lamp? (This was quite an intelligent question on researchgate.)

Solution 5.2

We discuss the problem in terms of photometry, as our intention is to produce a solar simulator with reasonable spectral quality.

A 120 W halogen lamp with a filament temperature of 2700 K will produce approximately 1400 lm of visible light (and, in addition, a lot of IR), and we can assume a luminance of 12 Mnit (Figure 3.52). The *effective luminance* L_{eff} of a filament coil model should be set to half of that value.

The target illuminance for the solar simulator is $E_{\text{T}} = 111$ klx (Eq. 3.225). The beam half angle θ follows from the projected solid angle needed to illuminate the target,

$$\Omega_{\text{p}} = \frac{E_{\text{T}}}{L_{\text{eff}}} = \frac{111\ \text{klx}}{6\ \text{Mnit}} = \pi\ \text{sr} \sin^2\theta \quad \Rightarrow \quad \theta \approx \pm 4° \tag{5.8}$$

Consequently, we have to design an optical system that collimates the light of the filament accordingly.

There are various types of 120 W halogen lamps:

(a) For a linear bulb, we assume a cylindrical filament of a length of $l = 100$ mm and a diameter of $2R = 1$ mm inside an extruded parabolic (trougher) reflector. According to Eq. 2.120, the angle in the far field of Figure 4.1 will be (θ in rad!)

$$\frac{R}{2f} < \theta < \frac{R}{f} \tag{5.9}$$

and thus

$$\frac{R}{2\theta} < f < \frac{R}{\theta} \tag{5.10}$$

For $R = 0.5$ mm, we guess and try $f = 7$ mm and set up the source and reflector in our illumination software (see simulation below).

(b) There are reflectorized 150 W halogen lamps for medical purposes (Figure 5.5, the term "medical" means they operate at a higher filament temperature and a higher radiance at the cost of a lower lifetime; for a solar simulator, short lifetime is undesirable) in a smooth, non-faceted elliptical reflector, medical halogen lamp HLX 64634.

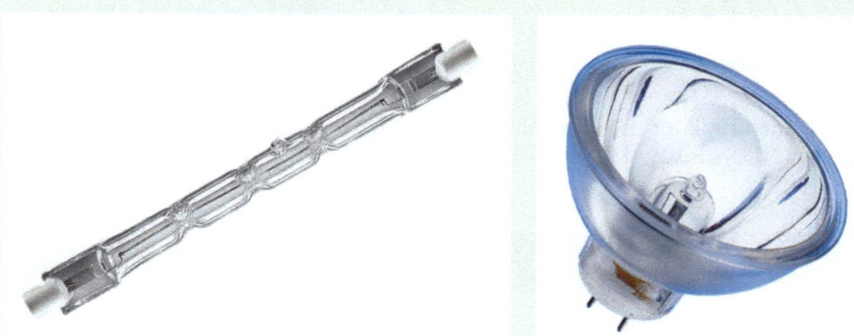

Figure 5.5 Halogen lamp types. *Source:* Courtesy of ams-OSRAM.

Looking at a family datasheet, we find a *working distance* of 32 mm and a front diameter of 50 mm, i.e. an MR16-compatible size. On this basis, we can set up a reasonable elliptical reflector model. For simplicity, we identify the working distance with the linear eccentricity c and the front diameter with $2b$. This gives us a major half axis $a = 40.6$ mm, a focal length $f = 8.6$ mm, and an eccentricity $e = 0.79$. However, optical design software requires other, equivalent values: the conic constant $k = -e^2 = -0.624$ and the vertex radius $R = b^2/a = 15.4$ mm (cf. Section 2.5.2).

The filament may be modeled as an axially oriented cylinder of $L = 4$ mm and $R = 1$ mm , set to 700 lm (a realistic efficacy of 5 lm/W in the reflector, the software showing a luminance of about 7 Mnit). This is a conservative approach. The luminance of a real tungsten surface should be twice that value, which is correct for a filament temperature of 2700 K (cf. Figure 3.52).

Simulation 5.1: Halogen sources in reflector

Linear bulb

We set up a parabolic trougher reflector with $f = 5$ mm (see above) and insert a cylindrical source of length $l = 100$ mm and $R = 0.5$ mm with a flux of 1400 lm, yielding a luminance of only 1.4 Mnit (smaller than above but perhaps good enough), see Figure 5.6.

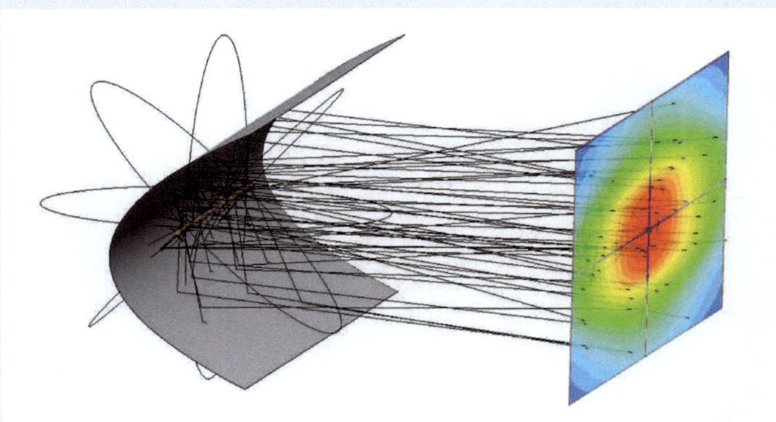

Figure 5.6 Extruded parabolic reflector with linear source: illuminance distribution.

(Continued)

Simulation 5.1: (Continued)

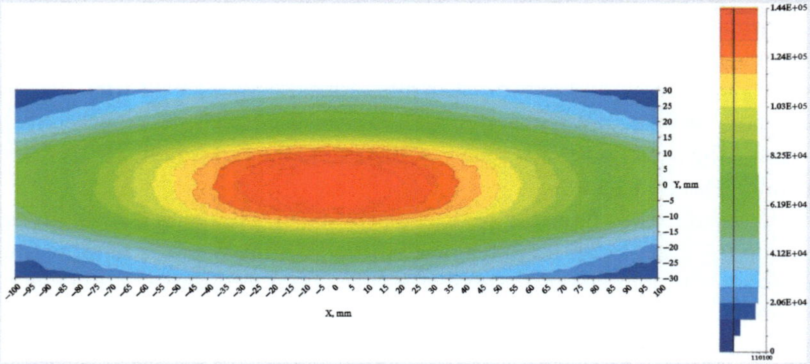

Figure 5.6 (Cont'd)

There is a large area of $E > 140$ klx (Figure 5.6) – task fulfilled. However, the angular distribution is not at all circular, and, of course, the spectrum for any halogen lamp without spectral filtering is quite different from the Sun's. We provide the trougher system simulation file.

Reflector lamp

For the MR16 reflector and the values determined above, we can set up a model as illustrated in Figure 5.7.

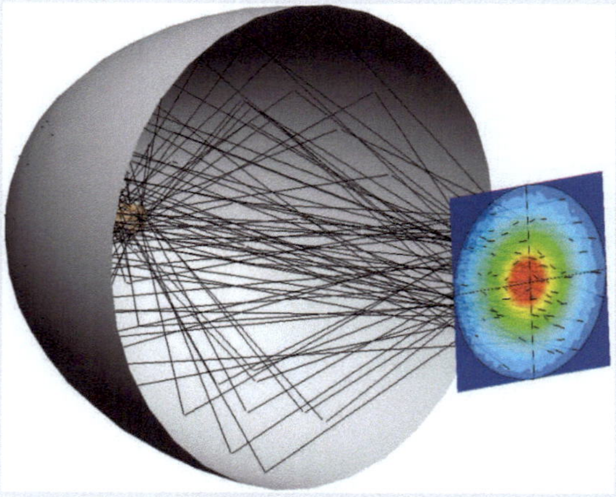

Figure 5.7 Simulated illuminance distribution in the secondary focal plane of a reflectorized medical halogen lamp.

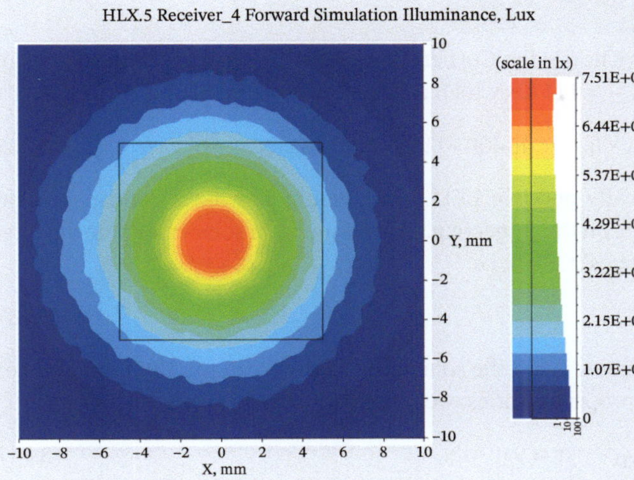

Figure 5.7 **(Cont'd)**

In a $10 \cdot 10\,\text{mm}^2$ region, we obtain an average illuminance of about 3 Mlx, by far higher than the target value (Figure 5.7). Consequently, this area may be imaged with an optical system of magnification

$$\beta = \sqrt{\frac{3\ \text{Mlx}}{111\ \text{klx}}} \approx 27 \tag{5.11}$$

and we would obtain an area of $(270\,\text{mm})^2$ illuminated with sufficient illuminance. We provide the MR16 system.

Dr. Wisenheimer:
You may add a $10\,\text{mm} \cdot 10\,\text{mm}$ integrating rod and image its exit area for better homogeneity.

Exercise 5.3: Solar simulator (2)

You are given a $L = 300$ Mnit white LED (that's the highest LED luminance that could be obtained at the time when this text was written) as a source for a solar simulator. How many suns can you achieve?

Dr. Wisenheimer:
The irradiance or illuminance of solar collectors or solar simulators is usually compared to the natural Sun irradiance or illuminance. The ratio is called "suns".

Solution 5.3

From the Sun data (Knowledge Corner Topic 3.19) we take the standard solar illuminance value $E_{v,E} = 111$ klx which can be concentrated to up to (cf. Eq. 4.80)

$$E_{v,E,max} = C_{max} \cdot E_{v,E} = 46\,400 \cdot 111\,\text{klx} = 5.15\,\text{Glx} \tag{5.12}$$

The best theoretically possible LED optics would provide a hemispheric or Lambertian incidence on the target (e.g. a multi-junction solar cell made for concentrating photovoltaics (CPV)), i.e.

$$U_T = A_T \cdot \pi\,\text{sr} \tag{5.13}$$

In other words, we would flood the solar cell etendue with the LED luminance. Doing so with an appropriate optics, we obtain

$$E_{LED,max} = L_{LED} \cdot \pi\,\text{sr} \approx 940\,\text{Mlx}, \tag{5.14}$$

leading to a ratio of

$$C_{LED} = \frac{E_{LED,max}}{E_{v,E}} \approx 8470 \tag{5.15}$$

A solar simulator for testing CPV devices with this LED can theoretically produce 8470 suns. In practice, CPV cells should rarely receive more than 1000 suns to avoid overheating and to accommodate some tolerance; the LED considered here would easily allow this value.

Dr. Wisenheimer:

The ratio of maximum solar concentration to the "maximum LED suns" is equal to the luminance ratio of the Sun (corrected for the atmosphere, Eq. 3.223) and the LED:

$$\frac{C_{max}}{C_{LED}} = \frac{46\,400}{8470} = \frac{L_{v,E}}{L_{LED}} = \frac{1.63\,\text{Gnit}}{300\,\text{Mnit}} \approx 5.5 \tag{5.16}$$

For solid-state white light, created with blue light and phosphor conversion, so-called LARP devices (laser-activated remote phosphor) achieve even higher luminances, albeit with less efficiency and not very homogeneous.

Exercise 5.4: Solar simulator (3)

Low-budget LED solar cell tester: Sketch an optical system that can provide an illuminance of 130 klx (one sun plus some reserve) in a $r_T = 50$ mm circle. Use and identify a single lens. Calculate the necessary LED flux. How can you secure a good illuminance uniformity?

Solution 5.4

A simple, yet promising approach is to choose a lens and simply set up a $2f$ configuration (Figure 5.8). We will discuss details in Knowledge Corner Topic 5.1.

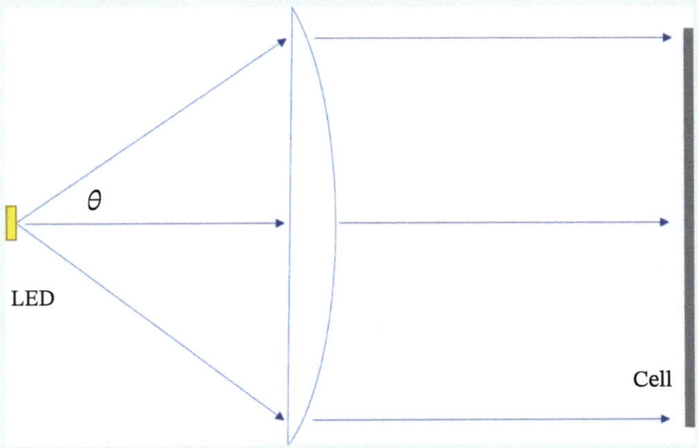

Figure 5.8 Basic LED solar simulator setup.

For simplicity, we look for a Fresnel lens; we found an off-the-shelf $f = 51$ mm $d = 109$ mm lens in this Fresnel lens catalog.

We are aware that we will not achieve aplanatic imaging (free of spherical aberration and coma) in this way – this would require a principal *shell*, whereas the Fresnel lens can only offer a principal plane; the same is true for planar metalenses, although this is not always recognized in the metalens world [6]. We ignore that for simplicity, and see how far we can get.

At the test area, we need

$$\Phi_\mathrm{T} = E_\mathrm{T} A_\mathrm{T} = 130\,\mathrm{klx} \cdot \pi(50\,\mathrm{mm})^2 \approx 1000\,\mathrm{lm} \tag{5.17}$$

The collection efficiency of the lens is a function of the cone half angle θ from the source (in air, $n = 1$, Eq. 3.11),

$$\eta = \sin^2\theta = \frac{\tan^2\theta}{1 + \tan^2\theta} = \frac{d^2}{4f^2 + d^2} = \frac{109^2}{4 \cdot 51^2 + 109^2} = 53\% \tag{5.18}$$

As a consequence, we need a 2000 lm-plus-LED for this design. A CoB LED with a LES diameter of 9 mm should be suitable.

Before we simulate (below), we calculate the illuminance distribution in the target plane (y') by investigating a ring $\mathrm{d}y'$.

For the lens, we have

$$y' \approx \tan\theta \cdot f \tag{5.19}$$

and

$$\mathrm{d}y' = \frac{f\,\mathrm{d}\theta}{\cos^2\theta} \tag{5.20}$$

(Continued)

Solution 5.4 (Continued)

The flux into a cone of $\pm\theta$ is (using Eq. 3.109 and Eq. 2.12)

$$\Phi(\theta) = I_0 \cdot \pi \,\mathrm{sr} \sin^2\theta = \Phi_S \sin^2\theta \tag{5.21}$$

hence

$$\mathrm{d}\Phi(\theta) = \Phi_S \cdot 2\sin\theta\cos\theta\,\mathrm{d}\theta = \Phi_S \cdot 2\frac{1}{(1+y'^2/f^2)^2}\frac{y'\mathrm{d}y'}{f^2} \tag{5.22}$$

and

$$E_T(y') = \frac{\mathrm{d}\Phi}{2\pi y'\mathrm{d}y'} = \frac{\Phi_S}{\pi}\frac{f^2}{(f^2+y'^2)^2} \tag{5.23}$$

which is closely related to the $\cos^4$ law (Exercise 3.17) and results in a considerable decrease of illuminance at the edge of the target (Figure 5.9).

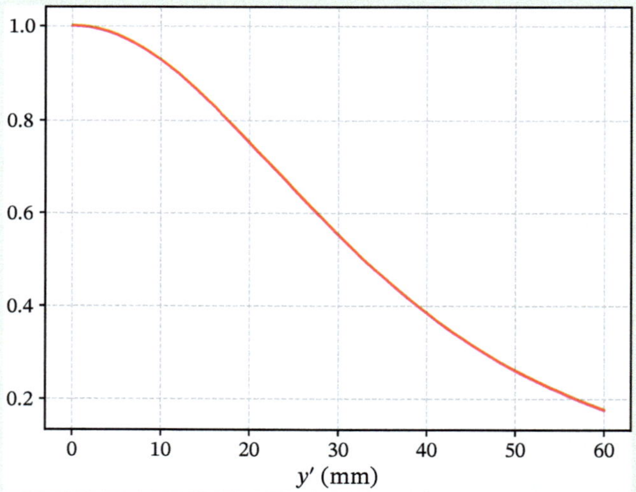

Figure 5.9 Target (relative) illuminance distribution according to Eq. 5.23.

I know why!

When I float around in the Fresnel lens, I can see how the inner edge of each tooth casts a shadow on the next sloped flank. When I then fly to the target and look back, I see, on the same sloped flank, an outer bright ring and an inner dark ring. But if what I see is not fully flashed, I cannot get full irradiance! You could tilt the vertical flank to avoid shadowing, but then I would see that tilted flank dark. I guess Dr. Wisenheimer would say: it's a matter of etendue conservation.

Any effects of distortion come on top of this. Still, this is a "minimum viable" product. See, for example, this Solar cell tester.

In Knowledge Corner Topic 5.1, we show how this happens and how to improve uniformity.

Dr. Wisenheimer:
For a more Sun-like spectrum, you may use a Multi Chip LED.

Simulation 5.2: Solar cell tester simulation

The simulation in Figure 5.10 shows that we obtain indeed an almost triangular shaped illuminance distribution at the target.

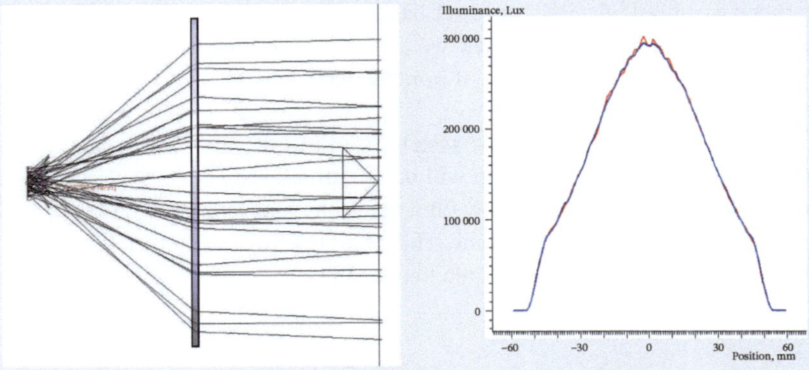

Figure 5.10 Solar cell tester comprising a 9 mm-LES CoB LED (set to $\Phi_S = 2500\,\mathrm{lm}$) and a Fresnel lens: illuminance distribution.

See Simulation file.

Can we improve on this?

Knowledge Corner Topic 5.1: Köhler design

We analyze the simple LED solar simulator design shown in Figure 5.8; the result is displayed in Figure 5.11.

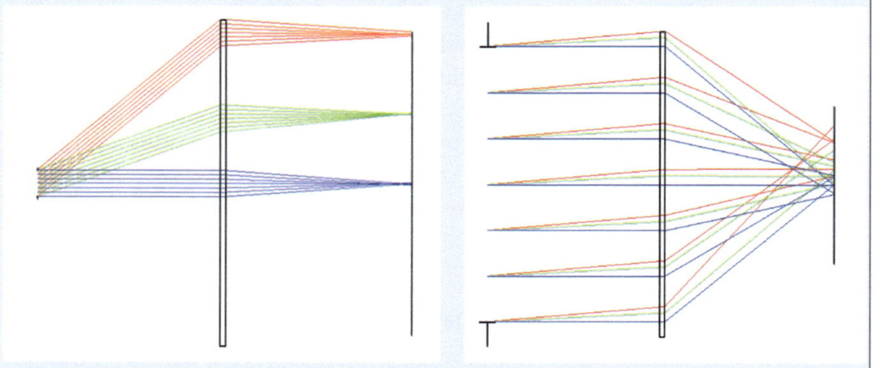

Figure 5.11 Fresnel lens in a 2 r_T = 50 mm configuration. Left: optical design to focus the LED far field at the target. Right: the same design set up from target to LED.

The 2f configuration seems to work properly in the forward direction. Full field distortion is at about 20% only. But if we look at the design from the target side (Figure 5.11, right), we can identify considerable performance problems and a 50% pincushion distortion at the LED.

Perhaps we should design our tester backwards, swapping source and target? And there is another aspect: When we are using the target area as the pupil, the optical design software will fill it homogeneously with rays. Does this cause homogeneous illuminance, too?

Let us try. We set up an imaging system from infinity with a ±5° field and a front stop of d = 100 mm (the stop will be the target; see Figure 5.12, left). The image is the LED area and we allow for a three-element lens system to focus the rays from infinity onto the LED plane. The result is a Köhler illumination system with a telecentric pupil as a conjugate to the LED area.

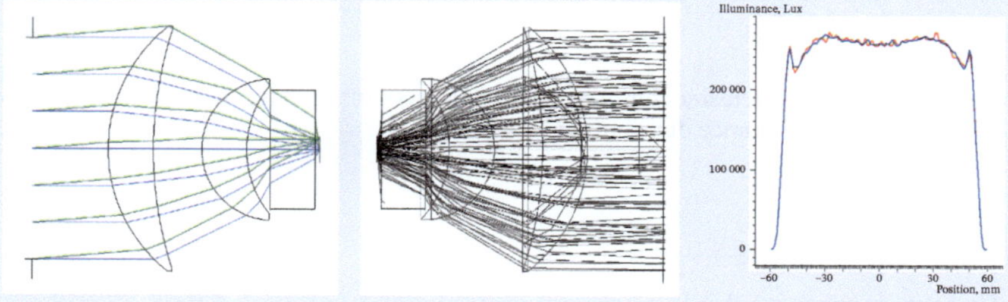

Figure 5.12 Left: Köhler design for the solar simulator task. Center: a non-sequential simulation of the design featuring the same source and target as in Figure 5.10 shows that it does indeed produce the desired homogeneous illuminance distribution (right).

We provide the Zemax design file and the LightTools simulation file.

The design does even realize an AOI of 64° onto the LED plane (which will increase the efficiency considerably), and with nearly zero distortion, too (Figure 5.12).

We call our approach "Köhler design", because for a finite pupil distance it would produce the classical Köhler arrangement [7]. Especially for Köhler illumination, it is a promising approach to play with objects, images, fields and pupils to "motivate" imaging optics software, which knows little about illumination, to create systems for homogeneous illuminance. We discuss this approach more thoroughly in [8].

The next exercise will illustrate how to use the same approach for a related solar simulator problem.

Exercise 5.5: Solar simulator (4)

Distant target: Sketch an optical system that can provide an illuminance of 130 klx in a $r_T = 50\,\text{mm}$ circle *at a distance of* $D = 1\,\text{m}$. Can the LES $d = 9\,\text{mm}$ CoB LED from the previous exercise still be used? If so, estimate the necessary exit pupil size of the optical system.

Solution 5.5

The necessary target flux is $\Phi_T = 1000\,\text{lm}$ as calculated above. Does the LED we chose provide sufficient luminance? This will depend on the exit aperture.

The source etendue is

$$U_S = (\pi \cdot 4.5\,\text{mm})^2 \approx 200\,\text{mm}^2\text{sr} \tag{5.24}$$

The target etendue is calculated using the far-field approximation (Eq. 3.42), which connects distance, target and exit pupil sizes. Setting target etendue equal to that of the source, we obtain an exit pupil semidiameter (or reflector exit aperture) of

$$R = \sqrt{\frac{U_S}{\text{sr}}} \cdot \frac{D}{\pi r} = \frac{\sqrt{200\,\text{mm}^2} \cdot 1\,\text{m}}{\pi \cdot 0.05\,\text{m}} = 90\,\text{mm} \tag{5.25}$$

which is a little large but probably feasible. For longer distances, we would need an LED of higher luminance to maintain the front size of the device.

Optical design

How can we design optics to homogeneously illuminate a distant circular area? We try a similar backward approach. An object of $R = 90$ mm (located in the *center* of the top drawing in Figure 5.13) has a rear entrance pupil *on the left*, at the specified distance of 1 m. The ray pencils are drawn back to the pupil from which they travel forward through the object, passing a lens system and forming an image on the right which is just the size of the CoB LED (cf. [8]).

(Continued)

Solution 5.5 (Continued)

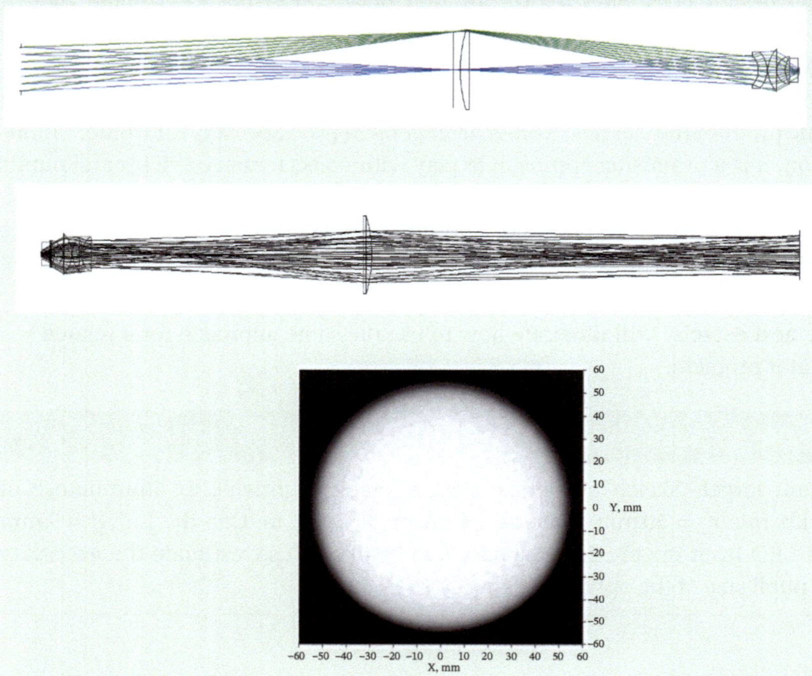

Figure 5.13 Top: Inverse imaging Köhler design to homogeneously illuminate a small distant circular area (far right), which will later become the CoB LED. Center: Simulation in reverse direction, compared to the imaging design. Now, rays travel from the CoB LED on the left to the target on the right, like in reality. Bottom: The real target illuminance distribution which we obtain is quite homogeneous.

We provide the imaging optical design file and the corresponding illumination system simulation file.

We have achieved a homogeneous illuminance distribution on a distant target without any use of an integrator. As usual in Köhler illumination, this is possible because the directional distribution of the LED is homogeneous, too. Imaging performance is not important here as it would affect only the virtual image of the LED near the last lens.

Dr. Wisenheimer:

The design process often leaves the lens which is closest to the LED (Figure 5.12) as a plane parallel plate made of some heavy flint glass instead of a "real" lens [9].

The effect of the plate is to produce a virtual image of the source that can be imaged well by just a few simple other lenses (cf. Exercise 2.8).

5.1.3 Medical Lighting

We now leave the world of solar simulators behind us and enter the world of *medical lighting* with all its special needs, starting with surgical lights (Figures 5.14 and 5.15).

Surgical lights

For any standard surgery that does not require microscopes or endoscopes, surgical lights illuminate the surgery area. They can also be used during diagnosis.

Surgical lights are standardized in DIN EN IEC 60601, which essentially describes a light distribution at a distance of $D = 1$ m. The customer, or the optical designer, can choose a couple of parameters within certain limits. For example, width and center illuminance of such a distribution can be chosen, but the way to measure them is standardized (Figure 5.16).

Exercise 5.6: Surgical light

A customer specifies a distribution with the following parameters:

Parameter	Value
center illuminance	$E_C = 150\,\text{klx}$
Spot diameter at $E_{10} = 10\% \cdot E_C$	$d_{10} = 25\,\text{cm}$
Spot diameter at $E_{50} = 50\% \cdot E_C$	$d_{50} = 18\,\text{cm}$

Calculate the required source flux. Calculate the necessary lamp head diameter for a halogen lamp ($L_S = 10\,\text{Mnit}$) and a white LED ($L_S = 100\,\text{Mnit}$), neglecting all other requirements.

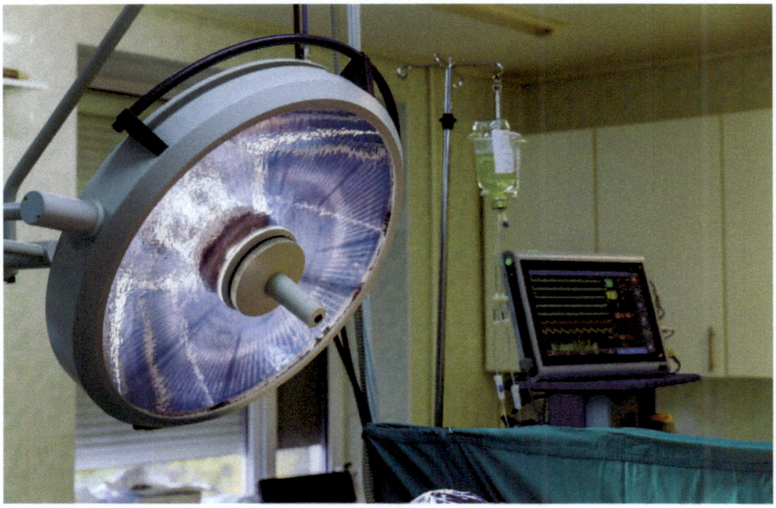

ungvar/Adobe Stock Photos

Figure 5.14 Legacy surgical light with center source in a large head.

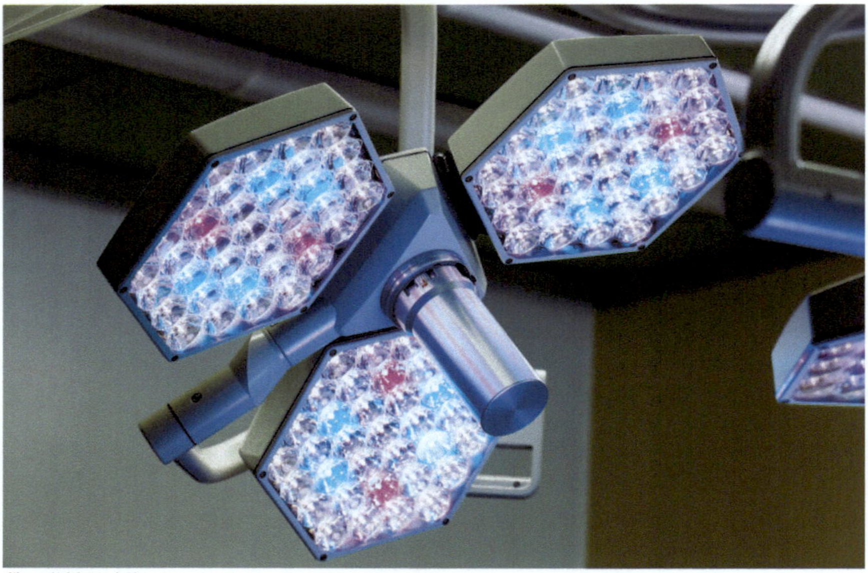

zlíkovec/Adobe stock Photo

Figure 5.15 Surgical light on the basis of LEDs with individual TIR lenses. They look colorful, showing how high-quality (high-CRI) white is achieved by mixing light from several LEDs. In particular, single-color cyan LEDs help to increase the CRI by filling the gap between blue pump light and yellow phospor light.

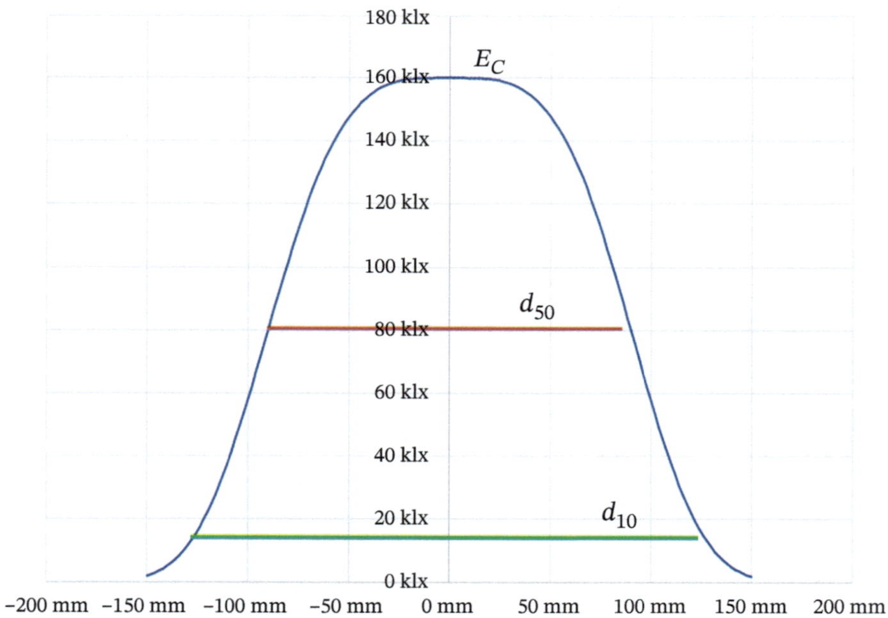

Figure 5.16 Supergauss shaped illuminance distribution of a surgical light.

Solution 5.6

We first check whether a Gaussian distribution such as

$$E(r) = E_C \exp\left[-\frac{1}{2}\left(\frac{r}{\sigma}\right)^2\right] \tag{5.26}$$

can meet the specification. The equation

$$E_{10} = E_C \exp\left[-\frac{1}{2}\left(\frac{r_{10}}{\sigma}\right)^2\right] \tag{5.27}$$

yields

$$r_{10} = \sigma\sqrt{-2\ln 0.1} = 2.15\sigma \tag{5.28}$$

and the same way

$$r_{50} = \sigma\sqrt{-2\ln 0.5} = 1.18\sigma \tag{5.29}$$

Thus, a Gaussian illuminance distribution will not work: It features a ratio of

$$\frac{r_{10}}{r_{50}} = 1.82 \neq \frac{25\text{ cm}}{18\text{ cm}} = 1.39 \tag{5.30}$$

that violates the specification.

It is a common approach to aim for a so-called *supergauss distribution*

$$E(r) = E_C \exp\left[-\frac{1}{2}\left(\frac{r}{\sigma}\right)^s\right] \tag{5.31}$$

with an additional parameter s (which would be $s = 2$ for standard Gaussian) that provides the desired degree of freedom to adjust the softness of the beam. We can easily calculate the parameters. The radii

$$r_{10} = \sigma\sqrt[s]{-2\ln 0.1} \tag{5.32}$$

and

$$r_{50} = \sigma\sqrt[s]{-2\ln 0.5} \tag{5.33}$$

yield

$$\frac{r_{10}}{r_{50}} = \sqrt[s]{\frac{\ln 0.1}{\ln 0.5}} \tag{5.34}$$

hence

$$s = \frac{\ln\left(\frac{\ln 0.1}{\ln 0.5}\right)}{\ln r_{10} - \ln r_{50}} = 3.65 \tag{5.35}$$

(Continued)

Solution 5.6 (Continued)

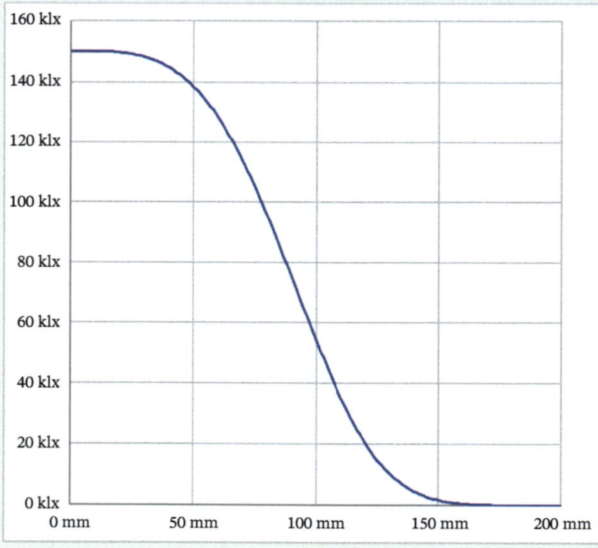

Figure 5.17 Supergauss distribution meeting the specification of the task.

The total flux (whole target plane) of supergaussian distributions with $s > 2$ is smaller than that of a Gaussian with the same standard deviation due to their steeper slopes and smaller tails (Figure 5.17).

Then we find the desired standard deviation

$$\sigma = \frac{r_{10}}{\sqrt[3.65]{-2\ln 0.1}} = \frac{r_{50}}{\sqrt[3.65]{-2\ln 0.5}} \approx 82\,\text{mm} \tag{5.36}$$

and calculate the total contained flux numerically, obtaining

$$\Phi_\text{T} = 4146\,\text{lm} \tag{5.37}$$

with 4000 lm within $r_{10} = 125$ mm.

We provide the file Supergauss.xlsx for these calculations.

Next, we calculate the system etendue U_T, defined by the light tube formed by the head aperture with R_H and both

- r_{10}, and
- a small spot in the center (radius $r \ll r_{50}$).

This is basically a check for the First and the Second Law.

For both light tube etendues, we use the far-field approximation (Eq. 3.42) because of $r \ll r_{10} \ll D$:

$$U_\text{T} = \left(\frac{\pi\, r\, R_\text{H}}{D}\right)^2 \text{sr} \tag{5.38}$$

We check for the flux within r_{10}

$$\Phi_{10} = L_S U_T(r_{10}) \tag{5.39}$$

and obtain

$$R_H = \frac{D}{\pi\, r_{10}} \sqrt{\frac{\Phi_{10}}{L_S\,\mathrm{sr}}} \tag{5.40}$$

(stating "high luminance enables small optics").

The on-axis illuminance (in the center spot of a small radius r) is

$$E_C = \frac{\Phi}{A} = \frac{L_S U_T(r)}{\pi\, r^2} = L_S \frac{\pi\, R_H^2}{D^2}\,\mathrm{sr} \tag{5.41}$$

which requires a head semi-diameter of

$$R_H = \sqrt{\frac{E_C D^2}{\pi\, L_S\,\mathrm{sr}}} \tag{5.42}$$

These formulas define the head size as a function of source luminance.

Realize	$\mathbf{\Phi_{10}}$	$\mathbf{E_C}$
L_S	R_H	R_H
10 Mnit	51 mm	69 mm
100 Mnit	16 mm	22 mm

Consequently, the center illuminance E_C is the feature that is harder to achieve, requiring a larger aperture than needed for Φ_{10}.

Small apertures, however, cause undesirable strong shadows. The standard contains shadow requirements, which may cause a much larger actual head diameter than we calculated here. Still, our calculations are extremely useful, because (i) we know which source luminance we need, and (ii) they tell us by how much, for a given source luminance, we must *dilute* the light in the lamp head, e.g. by controlled scattering with microfacets, and we can choose our optical architecture accordingly – just another example of how etendue/flux/luminance calculations guide us toward viable solutions.

Dr. Wisenheimer:

A Gaussian distribution in the spatial domain is convenient because it can be integrated analytically (cf. Knowledge Corner Topic 2.12).

$$\begin{aligned}\Phi &= \int_0^\infty E(r)\,\mathrm{d}A \\ &= E_C \int_0^\infty \exp\left[-\frac{1}{2}\left(\frac{r}{\sigma}\right)^2 \cdot 2\pi\, r\right] \mathrm{d}r = 2\pi\,\sigma^2 E_C\end{aligned} \tag{5.43}$$

(Continued)

Solution 5.6 (Continued)

We could achieve the same flexibility with a super-cos^n distribution,

$$E = \cos^n(\gamma(s)), \quad \gamma(s) = \frac{\theta^s}{(\pi/2)^{s-1}} \tag{5.44}$$

but there would be little advantage, and supergauss distributions are common and well known.

The specification and calculation state a *prescribed illuminance* problem, corresponding to a *fuzzy target* in the phase-space framework.

The fill of the target phase space can be described by a corresponding luminance diagram similar to that of a fuzzy source.

When microinvasive surgery is planned, we need to illuminate the surgery area through the same thin endoscopes that the surgeons use for the camera and to insert the instruments.

Fiber illumination

Optical fibers come in various flavors: Glass or polymer or even liquid-filled, with and without cladding, as single fibers or fiber bundles. But they all share two properties: They are round (rotationally symmetric), and they have a high-index core and a low-index cladding (the lowest-index cladding is air). See Knowledge Corner Topic 2.8 for more details on their optical function.

It is very common for an optical fiber to be *the* bottleneck in an illumination system. The typical goal is to squeeze as much light as possible into the small fiber *etendue*, while overall efficiency is a secondary consideration: luminance is king!

Exercise 5.7: LED to fiber coupling

(a) A standard fiber ($d = 1$ mm, NA = 0.5) needs to be be filled with green LED light. What radiant flux is possible?

(b) Your customer needs 1000 lm white light to be emitted from a light guide as thin as possible, while he or she does not care about efficiency. Choose a suitable light guide and LED. Do you need a lens for coupling?

Solution 5.7

(a) At the time this text was written, there were high-current LEDs in the blue and green operated at 3 A/mm^2 and producing up to $\Phi = 3$ W in the blue or 500 lm in the green from a chip size of 1 mm^2. So, green luminance can reach

$$L = \frac{500\ \text{lm}}{\pi\ \text{mm}^2\text{sr}} = 160\ \text{Mnit} \tag{5.45}$$

The etendue of the suggested fiber is (Eq. 3.42)

$$U = A \cdot X = A \cdot \mathrm{NA}^2\pi\,\mathrm{sr} = (0.5\,\mathrm{mm})^2\pi \cdot 0.5^2\pi\,\mathrm{sr} = 0.6\,\mathrm{mm}^2\mathrm{sr} \tag{5.46}$$

so we can expect up to

$$\Phi = L \cdot U = 160\,\mathrm{Mnit} \cdot 0.6\,\mathrm{mm}^2\mathrm{sr} = 100\,\mathrm{lm} \tag{5.47}$$

which corresponds (at ≈ 550 lm/W) to about $\Phi = 180$ mW.

As we mentioned, there are similar commercial products available, such as a fiber-coupled LED source.

(b) There are several types of light guides available.

- Plastic optical fibers (POF) or polymer optical fibers
- Liquid light guides
- Fiber bundles
- Numerical aperture between 0.4 and 0.8

The transmission of such light guides is on the order of 60% (POF), 75% (liquid), and 50% (bundle) for a length of $l = 2$ m. Taking into account the light guide transmission and some potential geometrical losses, we need an LED of about 1500–2000 lm. It is pretty obvious that we need a high-luminance LED and not a CoB type. We end up choosing something like the

OSLON Boost HX with up to 2000 lm out of 2 mm^2. This chip area has an etendue of

$$U_\mathrm{S} = \pi\,\mathrm{sr}\,A = 6.3\,\mathrm{mm}^2\,\mathrm{sr} \tag{5.48}$$

We look for a light guide slightly larger than that, and with high transmission, for example, this liquid light guide with $d = 3$ mm, $\mathrm{NA} = 0.55$, and thus an etendue of $1.5^2\mathrm{mm}^2\pi \cdot 0.55^2\pi\,\mathrm{sr} \approx 6.7\,\mathrm{mm}^2\,\mathrm{sr}$.

The "to lens or not to lens" question is discussed in [8]. Here, we could use *direct attach*, also known as "butt coupling" (the chip diagonal is smaller that the fiber diameter). The coupling efficiency is then given by the NA ratio

$$\eta = \left(\frac{0.55}{1}\right)^2 \approx 30\% \tag{5.49}$$

which violates the specification, despite the fact that we have not accounted for the small air gap required air. Can we beat that by using coupling optics?

Imagine a truncated cone that connects the inscribed circle of the chip area ($1.25 \cdot 1.59\,\mathrm{mm}^2$) to the fiber entrance. The inner circle covers 61.5% of the full area. The NA (angular acceptance) of the cone is then found from

$$\frac{3\,\mathrm{mm}}{1.25\,\mathrm{mm}} = \frac{NA}{0.55} \tag{5.50}$$

(Continued)

Solution 5.7 (Continued)

which yields an NA > 1, suggesting that the device is simply transmitting all angles. So the use of some (imaging or non-imaging) optics already provides an efficiency gain by a factor of two. In total, we have

$$\Phi_T = 70\% \cdot 61.5\% \cdot 2000\,\text{lm} \approx 850\,\text{lm} \quad (5.51)$$

which is slightly below the specified value and calls for better optics or a larger fiber and/or LEDs. But there should be some room for improvement, because the angular magnification (collimation) was quite strong. We proceed by means of a simulation.

Simulation 5.3: LED-to-fiber coupling

We set up a Lambertian source of $1.25 \cdot 1.59\,\text{mm}^2$ area and 1400 lm (to take into account the light guide transmission of 70%). The receiver size is $r = 1.5$ mm and NA = 0.55 (the fiber, see Figure 5.18).

We vary the front diameter of the light guide

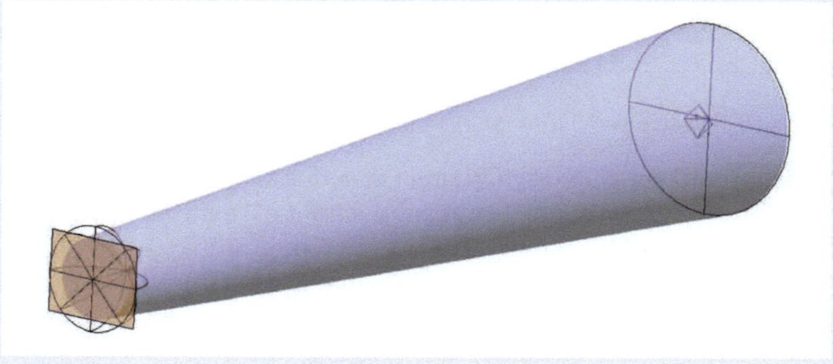

Figure 5.18 LED-to-fiber coupling by a tapered cylinder.

and compare, searching for a maximum (see Figure 5.19).

Direct attachment (no light guide)	420 lm
Inscribed circle	875 lm
Circumscribed circle	948 lm
best $r = 0.83$ mm	1200 lm

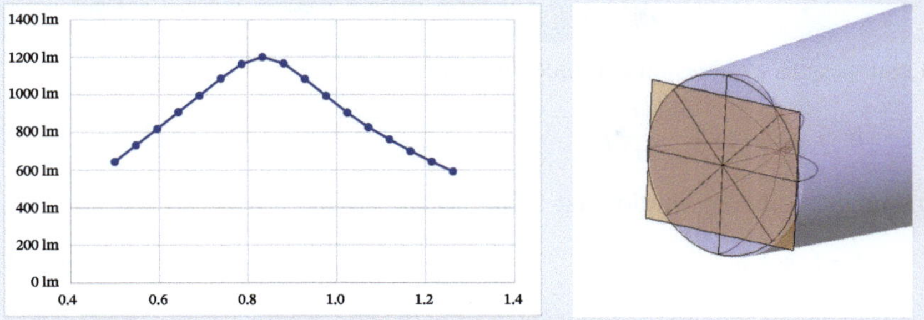

Figure 5.19 Left: Collected flux as a function of LED-side semi-diameter (mm). Right: Best choice.

The NA corresponding to maximum efficiency

$$\mathrm{NA} = 0.55 \cdot \frac{3\,\mathrm{mm}}{2 \cdot 0.83\,\mathrm{mm}} \approx 1 \tag{5.52}$$

is easy to understand as we collect the maximum area and all angles.

We provide the simulation file.

If the luminance of a white LED is not sufficient for an application, we may pump the phosphor with a blue laser diode. This technology is called LARP (laser-activated remote phosphor). The Gaussian irradiance distribution of the laser beam is inscribed to the white-light luminance distribution (see Figure 5.20). The mathematics of such a source distribution has been introduced in Knowledge Corner Topic 3.12.

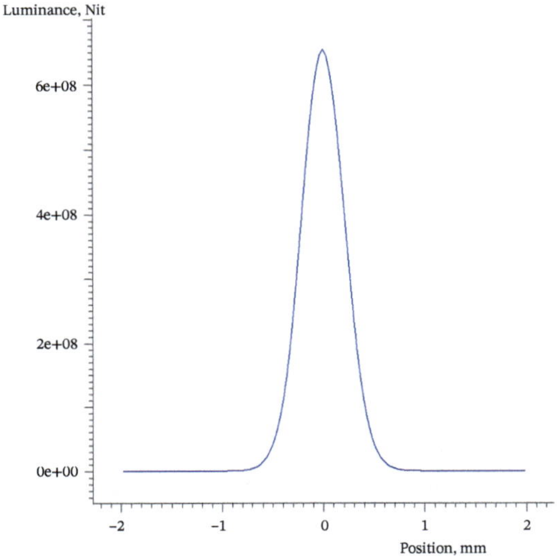

Figure 5.20 Luminance distribution of a LARP source, derived from ray data.

Exercise 5.8: Optimum magnification of a LARP source

A source with a Gaussian exitance distribution

$$E(r) = E_0 \exp\left(-\frac{r^2}{2\sigma^2}\right) \tag{5.53}$$

and an intensity distribution like a Lambertian emitter

$$I = I_0 \cos\vartheta \tag{5.54}$$

needs to be imaged (not collimated) to an optical fiber with a semi-diameter r_F = 1 mm and NA = 0.5.

Here, the magnification is the target-to-source coordinate ratio

$$M = \frac{r_F}{R} = \frac{\sin\theta}{\text{NA}} \tag{5.55}$$

where R is the coordinate on the source that is imaged to the fiber radius, and θ is the source angle that is imaged to the NA limit angle arcsin NA. A small magnification means that all the source area is coupled to the fiber, but the acceptance angle for source rays is small. For large magnification, on the other hand, a large angular range is covered but only light from a small area at the source is collected. So there might be an efficiency maximum in between both situations.

Calculate the optimum magnification for a source with a $\sigma = 0.3$ mm spatial distribution.

Solution 5.8

The source flux within a circle of radius R is

$$\Phi(R) = \int E(r)\,\mathrm{d}A = E_0 \int_0^R \exp\left(-\frac{r^2}{2\sigma^2}\right) \cdot 2\pi\, r\,\mathrm{d}r \tag{5.56}$$

which can be integrated analytically:

$$\Phi(R) = 2\pi\, E_0 \left[-\sigma^2 \exp\left(-\frac{r^2}{2\sigma^2}\right)\right]_0^R = 2\pi\, E_0\sigma^2 \left[1 - \exp\left(-\frac{R^2}{2\sigma^2}\right)\right] \tag{5.57}$$

The total source flux is ($R \to \infty$)

$$\Phi_S = 2\pi\, E_0\sigma^2 \tag{5.58}$$

Including the acceptance angle limitation, we obtain

$$\Phi(R, \theta) = \Phi_S \left[1 - \exp\left(-\frac{R^2}{2\sigma^2}\right)\right] \sin^2\theta \tag{5.59}$$

We switch to coordinates in the image plane and insert the fiber parameters:

$$\Phi(M) = \Phi_S \left[1 - \exp\left(-\frac{r_F^2}{2\sigma^2 M^2}\right)\right] \text{NA}^2 \cdot M^2 \tag{5.60}$$

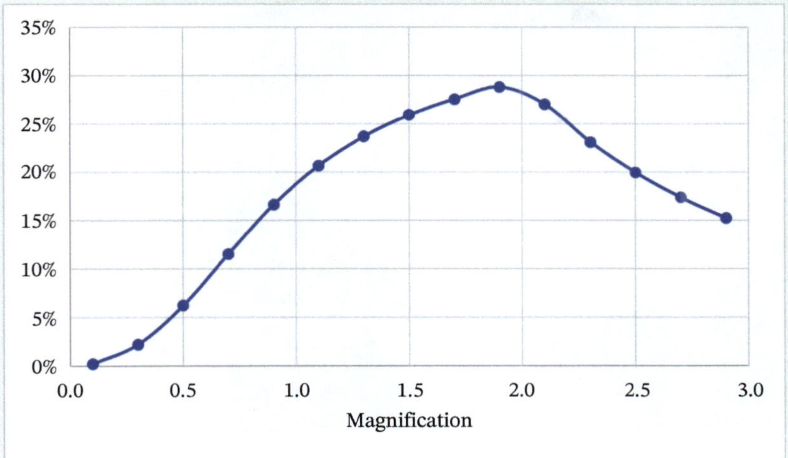

Figure 5.21 Collection efficiency as a function of magnification.

Here, the left slope is determined by angular collection efficiency, the right slope by the spatial one. The peak is approximately at $M = 1/\text{NA}$ (Figure 5.21). This is no surprise: At that magnification, $\theta = 90°$, and the captured source area is minimal while filling the fiber completely, i.e. the captured source area is maximally concentrated at the hot spot.

Provided file Magnification.xlsx.

Intense Pulsed Light

Intense pulsed light (IPL) describes a cosmetic technology for hair removal by light. Such IPL devices (Figure 5.22) operate with a Xe flash lamp with a red-band pass filter (to avoid burning the skin by light that does not act on the hair follicles). Literature [10] shows that a typical IPL device produces a radiant exposure (fluence) of $H = 1500\,\text{J/cm}^2$ within 20 ms onto an area of $A_\text{T} = 5\,\text{cm}^2$.

Exercise 5.9: IPL hair removal

Order from above:

This is a much too expensive way to produce a few red photons. Redesign the device using red LEDs!

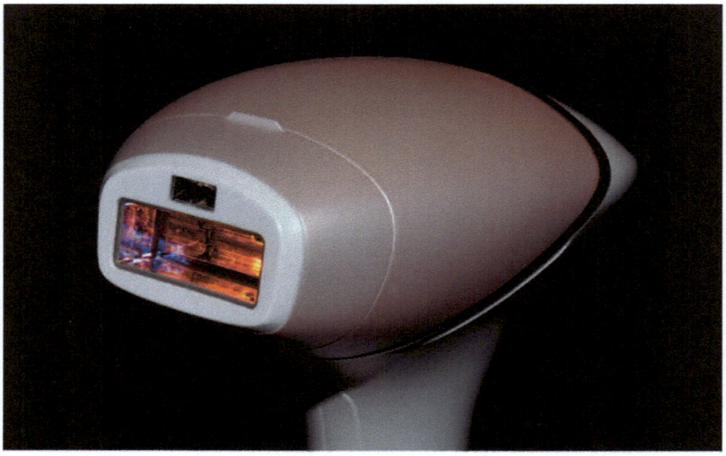

Figure 5.22 Home IPL device.

Solution 5.9

The "brightest" (highest-luminance) red LED (at the time this book was written) can produce about $\Phi_v = 600\,\mathrm{lm}$ from a chip area $A_{\mathrm{LED}} = 2\,\mathrm{mm}^2$. We assume a spectral luminous efficacy (see Knowledge Corner Topic 3.24) for red LEDs of $K = 180\,\mathrm{lm/W}$. The corresponding radiant flux of the LED is then

$$\Phi = \frac{\Phi_v}{K} = 3.3\,\mathrm{W} \tag{5.61}$$

which means a radiant exitance of $E = 1.65\,\mathrm{W/mm}^2$.

In contrast, the IPL fluence is then

$$E_{\mathrm{Xe}} = \frac{H}{\tau} = \frac{1500\,\mathrm{W\,s/cm}^2}{0.02\,\mathrm{s}} = 75\,\frac{\mathrm{kW}}{\mathrm{cm}^2} = 750\,\frac{\mathrm{W}}{\mathrm{mm}^2} \tag{5.62}$$

This exceeds the LED upper limit by a factor of 450 and explains why such devices are still using flash lamps. Alas, no chance for red LEDs.

5.2 Evaluation of Datasheets and Product Specifications

There is more information in datasheets and product specifications than meets the eye. Since many quantities are related by the laws of physics, there can be contradictions, and it is possible to compute additional quantities by applying these laws, reverse-engineering them from the specifications.

Let us learn how to read a product specification and derive important details.

Exercise 5.10: LED datasheet

Estimate an LED's chip area and luminance from what is stated in a typical LED datasheet.

Solution 5.10

Chip area is often not specified in the datasheet but is important for optical design tasks. Sometimes you can get it from a drawing (usually close to the end of the document). Alternatively, you can estimate it by simulation from a ray file (see below).

Voltage for green/blue/white LEDs: a value of around 3 V indicates one chip in a row, multiples of that value suggest more than one chip. The maximum current also gives some indication, and thermal resistance as well because it is roughly correlated with chip area through

$$A \cdot R_{\text{th}} \approx 5 \,\frac{\text{K}}{\text{W}}\text{mm}^2 \tag{5.63}$$

For example, have a look at this LED.

Forward voltages of 2.8, 5.6, 11.2 V indicate 4 chips, available with different wiring. A thermal resistance of 1.25 K/W indicates an area of 4 mm^2. The product picture and the drawing in the datasheet show 4 chips, each with an area of about $(1.5\ \text{mm})^2$. We download a ray file and simulate.

Simulation 5.4: LED investigation

Creating a good source model from vendor-provided ray files is notoriously difficult. Linking rays correctly to actual package geometry and making the rays start just outside that geometry is something that, in our experience, only ams-OSRAM is doing really well. (Disclaimer: We have introduced this ray file generation workflow at OSRAM almost twenty (!) years ago. Other vendors had ample time to catch up, but unfortunately did not. The reason they didn't is probably that the ones who suffer – you, the optical designers – don't get to make the purchasing decisions often enough.)

Anyway, what you can do in any ray tracing software is tracing the rays to a distant (say, one meter) sphere, and then back to a plane. That plane may well be inside the CAD geometry and/or the original ray file bounding box – which is why you need to trace to the sphere in the first place.

Then, you can vary the plane location to find a sharp image of source structures. Specifically, we are looking for chip boundaries. The result is shown in Figure 5.23).

(Continued)

Simulation 5.4: (Continued)

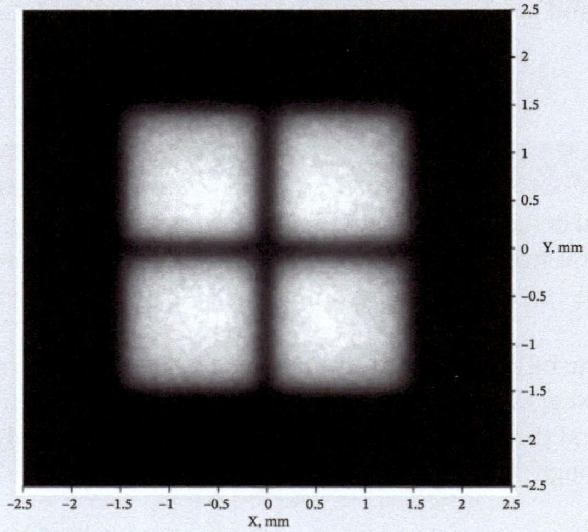

Figure 5.23 LED chip area by simulation. We estimate chips with an area of 4 x 2 mm^2, arranged with gaps of 0.2 mm.

And here is the simulation file.

Knowing the chip area including the gaps, we can estimate the luminance of the LEDs by using Eq. 3.9

$$L_S = \frac{\Phi}{A \cdot \pi\,\mathrm{sr}} = \frac{1000\,\mathrm{lm}}{9\,\mathrm{mm}^2 \cdot \pi\,\mathrm{sr}} \approx 35\,\mathrm{Mnit} \tag{5.64}$$

which is not very much, but suitable for many applications.

Dr. Wisenheimer:

The etendue of the four single chips of Figure 5.23 is

$$U = \pi\,\mathrm{sr} \cdot A = 8\pi\,\mathrm{mm}^2\,\mathrm{sr} \approx 25\,\mathrm{mm}^2\,\mathrm{sr} \tag{5.65}$$

The *effective etendue* of the arrangement is

$$\begin{aligned} U_{\mathrm{eff}} &= \pi\,\mathrm{sr} \cdot (1.4\,\mathrm{mm} + 0.2\,\mathrm{mm} + 1.4\,\mathrm{mm})^2 \\ &= 9\pi\,\mathrm{mm}^2\mathrm{sr} \approx 28\,\mathrm{mm}^2\mathrm{sr} \end{aligned} \tag{5.66}$$

We face some dilution because we have to include the dark cross in the center, which we most likely cannot "cut out" by optical means.

If we want to go the imaging way: the whole chip area is inside a circumscribed circle of $R = 2.1$ mm. Then the effective etendue is

$$U_{\mathrm{eff}} = (\pi R)^2\mathrm{sr} \approx 43.5\,\mathrm{mm}^2\mathrm{sr} \tag{5.67}$$

which is much larger that the above figure, but realistic. When we want to create a circular target with rotationally symmetric optics (lenses), e.g. in a typical Köhler system, we will most likely end up with the source, or an intermediate image of the source, underfilling a circumscribed circle.

In the case of a sparse array of such chips, we would image them separately, have a circumscribed circle of $r = 1$ mm and would obtain a combined etendue of

$$U_{\text{sparse}} = 4(\pi\, r)^2\ \text{sr} \approx 39.5\,\text{mm}^2\text{sr} \tag{5.68}$$

We conclude that the effective etendue of LED arrays is increased with respect to the actual source etendue when taking into account the subsequent optics.

An expensive and large workaround could be to combine the individual sources with the help of light guides as shown in Figure 4.76.

Exercise 5.11: A PAR64 lamp

PAR64 tungsten halogen lamps comprise a halogen burner and a parabolic glass (or metal) reflector and a front "lens", which can be clear, scattering, or have some optical function such as a lens array. For the naming, see Knowledge Corner Topic 4.1.

The halogen burner is usually arranged with the filament oriented along the lamp axis. The filament often consists of several strings arranged in a rectangle. As a result, the far field may be a little asymmetric. Mainstream beam sizes are NSP (narrow spot) and MFL (medium flood).

"Par cans" were often used with attached gel filters for a variety of spectral effects such as a change of color temperature (Figure 5.24).

Roman Yastrebinsky/Shutterstock Liveshot/Shutterstock

Figure 5.24 Par64 lamp for installation in a *PAR Can*.

Check the datasheet of an arbitrary 1 kW PAR64 NSP lamp. For example, PAR64 64738/4 with a total flux of $\Phi = 15\,800$ lm, peak intensity $I_0 = 255$ kcd and CCT = 3000 K.

(a) Calculate the beam angle of the lamp from its photometric data.

(b) Can you realize an LED version of such a fixture? Choose a CoB LED with sufficient flux. What beam angle can be achieved?

Solution 5.11

(a) We use the $\cos^n$ beam model. Equation 3.130 allows us to calculate the beam exponent

$$n = 2\pi \mathrm{sr}\frac{I_0}{\Phi} = 2\pi \mathrm{sr}\frac{255\,\mathrm{kcd}}{15\,800\,\mathrm{lm}} = 101 \tag{5.69}$$

Then, Eq. 3.137 yields $\theta_{0.5} = 6.7°$, so we have a beam angle of 13.5°.

Dr. Wisenheimer:

PAR64 means a front diameter of 200 mm (cf. Knowledge Corner Topic 4.1), hence the front area is 31 400 mm^2. The peak luminance can be estimated according to

$$L = \frac{I_0}{A} = \frac{270\,\mathrm{kcd}}{31\,400\,\mathrm{mm}^2} \approx 10\,\mathrm{Mnit} \tag{5.70}$$

We compare this to the theoretical filament luminance of $L = 40$ Mnit at 3100 K (calculated via Eq. 3.199) and note a considerable dilution – the source luminance is spread over a much larger etendue. This is not surprising considering the incandescent technology (see Figure 5.25).

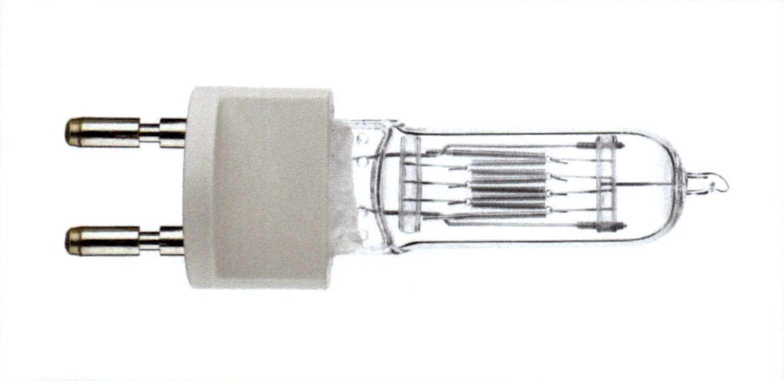

Figure 5.25 1kW tungsten halogen bulb. *Source:* Courtesy of ams-OSRAM.

(b) LED retrofit: To be on the safe side regarding flux, we should choose a $d > 30$ mm CoB LED, for example this CoB LED, and can calculate with $\Phi =$ 20 klm at a reasonable color.

We conceive an optical system that collects rays from the LED from angles of up to 60°. Etendue conservation then states (cf. Eq. 3.73)

$$16\,\mathrm{mm} \cdot \sin 60° = 100\,\mathrm{mm} \cdot \sin\theta \tag{5.71}$$

which yields $\theta = \pm 8°$. Its performance is thus close to that of the original fixture and a retrofit should be technically feasible.

Exercise 5.12: Skylight

Skylights or light fingers are collimated beams for architectural or entertainment purposes. A typical skylight is based on a high-power Xe lamp (basically a cinema projection lamp) in a parabolic reflector. Technically, these are cousins of search lights.

A popular example is the Sky Beam on top of the Luxor hotel in Las Vegas (Figure 5.26). We can retrieve basic photometric data via public information, such as a peak intensity of "42.3 billion candlepower" (which seems to imply "cd").

Figure 5.26 Sky Beam on top of the Luxor in Las Vegas. *Source:* AI rendering by the authors.

The setup consists of 39 pieces of a 7kW Xe fixture with a specified "Peak Beam Candle Power: 795 million" (which means 795 Mcd).

The published photometric data claim that such a fixture in the spot mode produces 7500 fc an a distance of $100' \approx 30$ m) and the beam diameter in this distance is $d = 18''$.

Calculate the luminance, luminous flux and luminous efficacy (Eq. 3.244) of such a fixture. Discuss.

Solution 5.12

According to the website, the fixture produces an on-axis illuminance of 7500 fc corresponding to 80 klx at a distance of $100' \approx 30$ m.

The inverse square law (Eq. 3.142) indicates

$$I_0 = 80\,\text{klx} \cdot \frac{(30\,\text{m})^2}{\text{sr}} = 72\,\text{Mcd} \tag{5.72}$$

(we apply the law and neglect our considerations from Knowledge Corner Topic 3.17).

The beam has a diameter of $d = 18'' = 457$ mm. From Eq. 3.107, we deduce an effective luminance of

$$L_0 = \frac{I_0}{A} = \frac{72\,\text{Mcd}}{\pi \cdot (228.5\,\text{mm})^2} \approx 440\,\text{Mnit} \tag{5.73}$$

which is more or less realistic for a Xe discharge lamp (with an average luminance in the order of 200 Mnit and a peak of about 1 Gnit [11]).

The cone angle is defined by the beam diameter of 42″ at 100′

$$\tan\theta = \frac{21''}{100'} \approx \frac{0.53\,\text{m}}{30\,\text{m}} \tag{5.74}$$

yielding a beam angle (FWHM) of $2\theta = 2°$ and a solid angle of

$$\Omega = 2\pi(1 - \cos 1°) \approx 1\,\text{msr} \tag{5.75}$$

If this were a top-hat beam, its flux would be

$$\Phi = I\Omega = 72\,\text{Mcd} \cdot 1\,\text{msr} \approx 72\,000\,\text{lm} \tag{5.76}$$

If this were a $\cos^n$ beam instead (Knowledge Corner Topic 3.15), we would have

$$\cos^n 1° = 0.5 \tag{5.77}$$

$$n = \frac{\ln 0.5}{\ln(\cos 1°)} = 4550 \tag{5.78}$$

The luminous flux of such a light can be estimated via Eq. 3.130

$$\Phi = \frac{2\pi\,\text{sr}}{n+1} I_0 = 72\,\text{Mcd} \cdot \frac{2\pi\,\text{sr}}{4551} \approx 100\,\text{klm} \tag{5.79}$$

which is higher than above because there is some flux outside the FWHM by its definition. We proceed with this value.

The luminous efficacy of the fixture is

$$\eta = \frac{100\,\text{klm}}{7\,\text{kW}} = 14\,\text{lm/W} \tag{5.80}$$

which is a somewhat small value, suggesting an optical efficiency of the fixture in the order of 50% (as typical Xe cinema lamp efficacy is about or slightly above 30 lm/W [11, Ch. 10], [12]).

In summary, these are the specifications of the Xe fixture:

Item	Datasheet	Calculated
Electrical power	7 kW	
Illuminance in a distance of 30 m	7500 fc	80 klx
On-axis luminous intensity per fixture "central beam candle power"	795 Mcd	72 Mcd
On-axis luminous intensity for 39 fixtures	42.3 Gcd	
Luminous flux per fixture		100 klm
Luminous efficacy	30…40 lm/W (lamp) [12]	14 lm/W (fixture)
Beam diameter in a distance of 30 m	42″	1.1 m
Beam angle (FWHM)		2°
Beam diameter	18″	457 mm
Luminance		440 Mnit

Discussion: Besides some minor inconsistencies, we face a big mystery regarding the beam intensity, as there is a factor of ten between some newspaper values and the photometry that seemingly corresponds well to reasonable luminous flux and luminance data.

Dr. Wisenheimer:
The Dark Sky Association would probably have quite an opinion about these sky lights, and the light pollution they create.

Exercise 5.13: LED optics

An LED business leader claims enthusiastically "With our LED, we can achieve 15 000 candela from a 50 mm diameter optic". Is this a realistic claim or simply marketing talk? Check this datasheet.

Solution 5.13

The highest flux value for the 9 mm LES is $\Phi_S = 2000$ lm. The corresponding luminance is

$$L = \frac{\Phi_S}{(\pi\, r_S)^2\,\text{sr}} = \frac{2000\,\text{lm}}{200\,\text{mm}^2\,\text{sr}} \approx 10\,\text{Mnit} \tag{5.81}$$

The maximum on-axis luminous intensity with a lens of the given size $r_F =$ 25 mm is

$$I_0 = L\pi r_F^2 = 10\,\text{Mnit} \cdot 2000\,\text{mm}^2 = 20\,\text{kcd} \tag{5.82}$$

So the intensity claim may well be correct, but it is still marketing talk: 10 Mnit is simply not that much.

Dr. Wisenheimer:

Never rely on the brightest LED bin when planning for a project. Everybody wants to get it, so it may not be available when you need it.

If there are a several bins offered, choose a flux value from the middle range for typical, the minimum flux of the second brightest bin for best case, and the minimum flux of the second darkest bin for worst case (the extreme bins tend to be sparsely populated in production).

Exercise 5.14: Diode laser radiometry

Check the data sheet of this 5 W blue laser diode. In the specification, we find the following angular data: fast axis ±25°, slow axis ±5° at $1/\text{e}^2$. Calculate its radiant exitance and intensity!

Solution 5.14

We guess, based on experience, $M_\perp^2 \approx 2$ for the fast axis and $M_\parallel^2 \approx 12$ for the slow axis. From Eq. 2.89, we find

$$w = M^2 \frac{\lambda}{\pi \sin\theta} \tag{5.83}$$

and obtain $w_\perp \approx 0.7\,\mu\text{m}$, $w_\parallel \approx 22\,\mu\text{m}$. The area of the facet is

$$A = 2w_\perp \cdot 2w_\parallel \approx 62\,\mu\text{m}^2 \tag{5.84}$$

(extent at the $1/\text{e}^2$ level) and the average radiant exitance there is

$$E = \frac{5\,\text{W}}{62\,(\mu\text{m})^2} \approx 80\,\frac{\text{kW}}{\text{mm}^2} \tag{5.85}$$

For a solid angle estimate, we proceed as in Eq. 5.98 and obtain

$$\Omega \approx 4\pi\,\mathrm{sr}\,\sin(25^\circ/2)\,\sin(5^\circ/2) = 0.12\,\mathrm{sr} \tag{5.86}$$

which gives the average intensity in the $1/e^2$ region as

$$I = \frac{5\,\mathrm{W}}{0.12\,\mathrm{sr}} = 42\,\frac{\mathrm{W}}{\mathrm{sr}} \tag{5.87}$$

Peak values may be derived by extending the Gaussian beam calculus (Knowledge Corner Topic 2.12) to *xy* symmetry.

The radiance is (using Eq. 3.21)

$$\begin{aligned} L &= \frac{\Phi}{U} = \frac{\Phi}{(M^2\lambda)^2\,\mathrm{sr}} \approx \frac{\Phi}{M_\perp^2 M_\parallel^2 \lambda^2\,\mathrm{sr}} \\ &= \frac{5\,\mathrm{W}}{2\cdot 12\cdot (0.45\,\mu\mathrm{m})^2\,\mathrm{sr}} \approx 1\,\frac{\mathrm{MW}}{\mathrm{mm}^2\,\mathrm{sr}} \end{aligned} \tag{5.88}$$

and the radiant intensity

$$I = L\cdot A = 1\,\frac{\mathrm{MW}}{\mathrm{mm}^2\,\mathrm{sr}}\cdot 62\,\mu\mathrm{m}^2 = 62\,\frac{\mathrm{W}}{\mathrm{sr}} \tag{5.89}$$

comes out slightly larger than calculated above. The reason for this deviation lies in the etendue calculation with the applied approximations such as for the solid angle. We check this using a simulation.

Dr. Wisenheimer:

Take care when using the term "intensity" in discussions with laser folks. And take even more care regarding "brightness" or "brilliance". This topic is also discussed in [13, p. 271] with a perfect confusion of terms and units.

The terms "brighter" and "brightness" are more colloquial than technical. Sometimes they refer to luminance, sometimes to flux. Avoid them in technical statements as this can lead to misconceptions [14–18].

Simulation 5.5: Laser diode simulation

Lasers produce coherent light and should be simulated by Gaussian beam propagation (BSP in CodeV or POP in Zemax, not to forget FRED or Virtual Lab). Surprisingly, a simulation on the basis of *ray files* works rather well in many cases, at least for diode lasers.

We set up a system consisting of laser diode and receiver. The receiver is positioned in front of the package and then *defocused* by 0.95 mm into the plane of the laser facet (Figure 5.27). This feature allows for the light measurement on a *virtual screen*.

(Continued)

Simulation 5.5: (Continued)

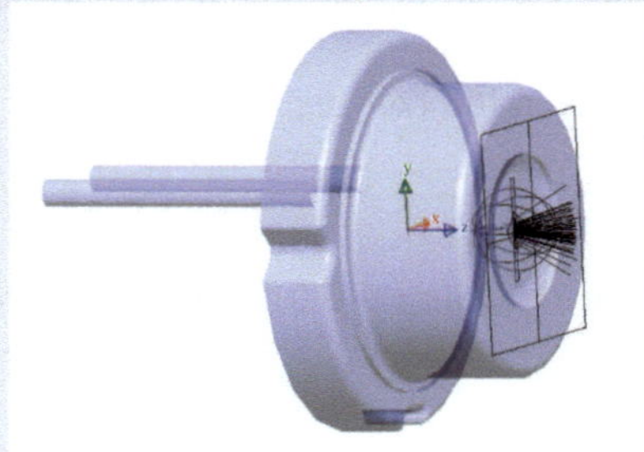

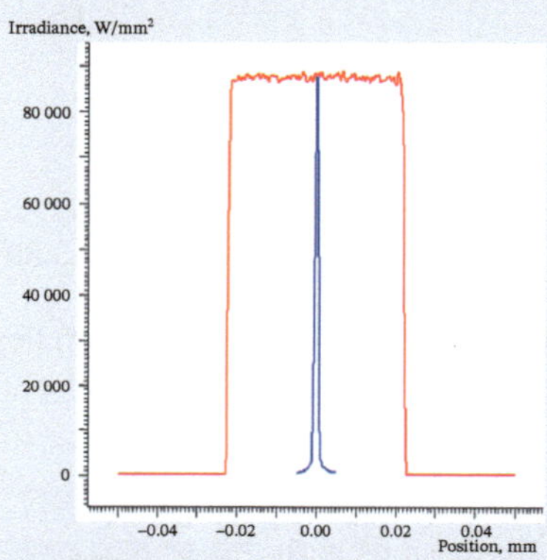

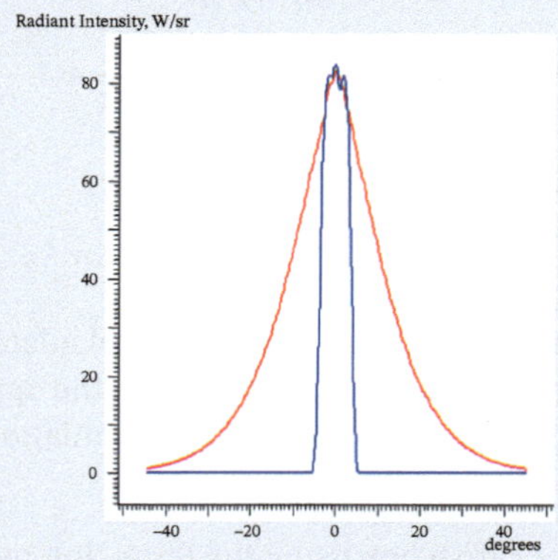

Figure 5.27 Laser diode simulation (left). The spatial extent is approximately 1.6 μm · 45 μm. Being set to 5 W, the source shows a peak radiant exitance of 85 kW/mm² (left). The angular distribution confirms the angles given in the data sheet and yields a peak intensity of 80 W/sr (right).

We provide the simulation file.

The results from calculation (average) and simulation (peak) agree rather well. Typically, our analytic estimates are useful in the beginning of a project. Later, we can get more accuracy from the simulation.

We see that the shape of the beam in both cross-sections is not actually Gaussian, thus challenging various assumptions. But we may also question the validity or accuracy of the ray data as their measurement or generation by simulation is not always straightforward.

Dr. Wisenheimer:

There is an additional small difference of the waist's z locations of the fast and slow axes noticeable during the ray data analysis. This *astigmatism* should be taken into account for accurate simulation and design if generic sources are used.

Usually, diode laser ray files provide more information about facet size and angular distribution than datasheets.

A comparison with an ideal Gaussian beam would allow for the estimation of the M^2 factors for both axes (cf. Eq. 5.83).

On the other hand, be careful and do not trust such ray files too much. They may not be based on measurement but generated by a simulation, e.g. using a perfect point source with measured intensity, and thus might not represent all aspects of a real laser diode. Moreover, the ray file might correspond to the design status, not the *typical* product.

In addition, take care to account for the significant tolerances of the divergence angles and the position and orientation of the laser chip in the package.

Broadband NIR LEDs became popular as light sources for hyperspectral imaging and spectral sensing. They feature a NIR (600…1000 nm) phosphor pumped by a blue LED. The appearance of many objects in this wavelength range gives useful information on their composition or maturity level. In the internet, you can find more information about NIR instruments and NIR LEDs.

Exercise 5.15: IR LED vs. incandescent emitter

Calculate the spectral radiance of a wide-spectrum NIR-phosphor LED and compare the result to a black-body emitter of 3000 K, in the 500…1000 nm spectral range.

Solution 5.15

We start with the LED: The datasheet claims a radiant flux of 50 mW within the range 600…1050 nm and 150 μW/nm from 750…1000 nm. The maximum DC current would allow an increase by a factor of 1.3. The spectral radiant flux as specified is then

$$\Phi_\lambda = \frac{65\,\text{mW}}{450\,\text{nm}} \approx 144\,\frac{\mu\text{W}}{\text{nm}} \tag{5.90}$$

The radiance from the effective area is

$$L = \frac{\Phi_S}{U} = \frac{65\,\text{mW}}{\pi\,\text{sr}\cdot 4\,\text{mm}^2} \approx 5\,\frac{\text{mW}}{\text{mm}^2\text{sr}} \tag{5.91}$$

The spectral radiance is

$$L_\lambda = \frac{L}{\Delta\lambda} = \frac{5\,\text{mW/mm}^2\text{sr}}{450\,\text{nm}} = 10\,\frac{\mu\text{W}}{\text{mm}^2\text{sr nm}} \tag{5.92}$$

We do the comparison: A black-body emitter at 3000 K (Eq. 3.194, Figure 5.31) has a spectral radiance of

$$L_\lambda = 1\,\frac{\text{mW}}{\text{mm}^2\,\text{sr}\,\text{nm}} \tag{5.93}$$

(Continued)

Solution 5.15 (Continued)

at 950 nm, which is 100 times that of the LED. Of course, radiance is not the only criterion for the selection of a light source, and LEDs may well have other advantages in a sensor system.

Dr. Wisenheimer:
Discharge lamps hardly have a suitable spectrum in the NIR below 1000 nm, but laser-driven plasma sources provide a spectrally homogeneous broadband NIR emission with a much higher spectral radiance of up to 100 mW/ (mm^{2}sr nm), albeit at a high cost and bulky dimensions.

There are exciting innovations in automotive front lighting. Advanced headlight technologies make it possible to drive with a continuous high beam with portions of it masked out to prevent glare for oncoming traffic [19]. This can be realized on the basis of projectors [20, 21], scanners [22] or pixelated light sources [23].

Exercise 5.16: Pixelated LED

The datasheet of a pixelated LED states:

- 320 columns times 80 rows of pixels,
- pixel pitch 40 μm,
- minimum luminance 75 Mnit.

Calculate the luminous flux of the device.

Solution 5.16

We calculate the flux from the area and the etendue:

$$A = 25\,600 \cdot (0.04\,\text{mm})^2 = 41\,\text{mm}^2 \tag{5.94}$$

$$U = \pi\,\text{sr} \cdot A \approx 130\,\text{mm}^2\text{sr} \tag{5.95}$$

$$\Phi = L \cdot U = 75\,\text{Mnit} \cdot 130\,\text{mm}^2\text{sr} \approx 10\,000\,\text{lm} \tag{5.96}$$

Thermal engineers will require that the device must not run continuously at full load.

Dr. Wisenheimer:
Pixelated sources of this kind, micro-LEDs [24], or the OLED from Exercise 4.47 pose the challenge of imaging an object with a Lambertian characteristics. A wise choice of the $F/\#$ determines both efficiency as (cf. Eq. 3.14)

$$\eta \propto \left(\frac{1}{2F}\right)^2 \tag{5.97}$$

and, as every imaging optical designer will confirm, the necessary effort for an imaging design with sufficient performance. This tradeoff calls for a reduction of the emission angle, for example by a microcollimator per pixel (similar to the microlenses on CMOS image sensors) or by using projection microdisplays which feature smaller acceptance angles.

We continue the pixel headlight story in Exercise 5.19.

5.3 Selecting a Light Source

In many instances, illumination design is asking for some optics to connect a given source and a target. Sometimes, there is just a target specification and the designer is free to select a source. We discuss this in detail in [8].

Here, we show how to derive the properties of a source to be chosen. We start by touching an interesting field of illumination optics.

Airfield lighting

Aviation safety requires several lamp types with precisely specified performance. Accordingly, a variety of optical solutions is needed. The standards are publicly available (we refer to the CS ADR-DSN.U.940 "Aeronautical ground light characteristics" section).

There is approach lighting (see Figure 5.28), landing systems (e.g. runway threshold illumination), runway and taxiway lighting, and many more. These amazing posters give a good overview.

Figure 5.28 Airport lights as seen from an approaching aircraft.

An approach centerline light (Figure 5.29) produces a specified beam to guide an aircraft during landing – with a distribution as shown in Figure 5.30. Such fixtures are embedded in the runway and called *runway inset lights*.

Figure 5.29 Runway inset light.

Exercise 5.17: Airfield lighting

Consider a runway inset light with a front slit of $A_T = 10\,\text{mm} \cdot 50\,\text{mm}$. The far-field specification is shown in Figure 5.30. What luminous flux is included in this distribution?

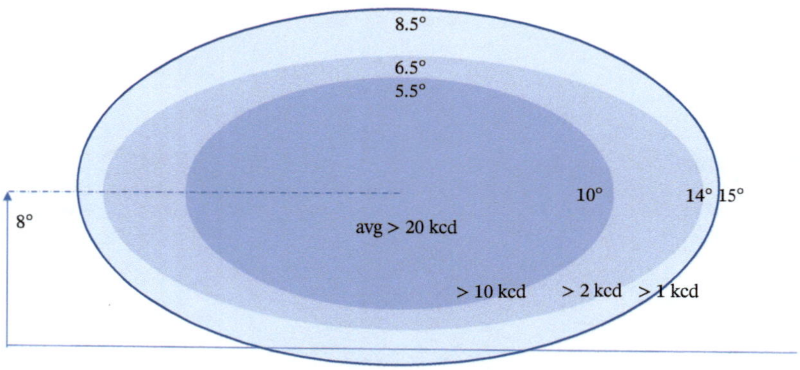

Figure 5.30 Specification (isocandela plot) for an approach centerline inset light.

White LEDs (whose color point tolerances are also restricted by the standard) are to be used to operate the device. Which source luminance is required?

Solution 5.17

We roughly estimate the solid angle (Eq. 2.11) of an elliptical cone. We have now two angles ($\pm\theta_a$ and $\pm\theta_b$) instead of one and use HANDLE WITH CARE

$$\Omega = 4\pi \sin\frac{\theta_a}{2}\sin\frac{\theta_b}{2} \tag{5.98}$$

We calculate the solid angle of the center region, then of the outer ones and subtract the already used inner regions:

Region	θ_a	θ_b	Ω	$\Delta\Omega$	I_T	Φ_T
Inner ellipse	10°	5.5°	0.05 sr		20 000 cd	1050 lm
Mid ellipse	14°	6.5°	0.09 sr	0.03 sr	5000 cd	170 lm
Outer Ellipse	15°	8.5°	0.12 sr	0.03 sr	1500 cd	50 lm

The target etendue is hard to estimate because we have a target *distribution*. But we can estimate the luminance for the center region which likely needs the highest luminance,

$$L = \frac{\Phi_T}{U_T} = \frac{I_T}{A_T} = \frac{20\,000\,\text{cd}}{500\,\text{mm}^2} = 40\,\text{Mnit} \tag{5.99}$$

The system etendue is

$$U_S = \frac{\Phi_T}{L} = \frac{1050\,\text{lm}}{40\,\text{Mnit}} \approx 26\,\text{mm}^2\,\text{sr} \tag{5.100}$$

which indicates an LED area of up to $A_S \approx 8\,\text{mm}^2$.

Several LEDs are available that can produce the required flux from that area, even if we had to take into account an optical and geometrical efficiency of less than 100%.

Dr. Wisenheimer:
Looking at the shape of the exit aperture, we may want to try a design with a 5x1 shaped automotive headlight LED.

A selection of special systems

Exercise 5.18: NIR heater

A customer is asking for 4 kW of incoherent NIR light (750…1200 nm) into a $d = 30$ mm target aperture. For practical reasons, we add a $\pm 30°$ maximum acceptance angle to the specification. Is there any incandescent source which can meet the specification?

Solution 5.18

First, we calculate the target etendue (Eq. 3.42):

$$U_T = (\pi \cdot 15\,\text{mm} \cdot \sin 30°)^2\,\text{sr} = 555\,\text{mm}^2\,\text{sr} \tag{5.101}$$

which means we need at least a source radiance of

$$L = \frac{4000\,\text{W}}{555\,\text{mm}^2\text{sr}} = 7.2\,\frac{\text{W}}{\text{mm}^2\text{sr}} \tag{5.102}$$

We plot the spectral radiance of a black body (Eq. 3.194) at 3000 K (Figure 5.31).

The average value may be on the order of 900 W/(m^2 sr nm), and multiplied with a spectral width of 450 nm, we get $L = 0.4\,\text{W/mm}^2\text{sr}$, which is insufficient by more than a factor of ten. Of course, we may still increase the filament temperature a little and gain a factor of two, but the conclusion remains valid.

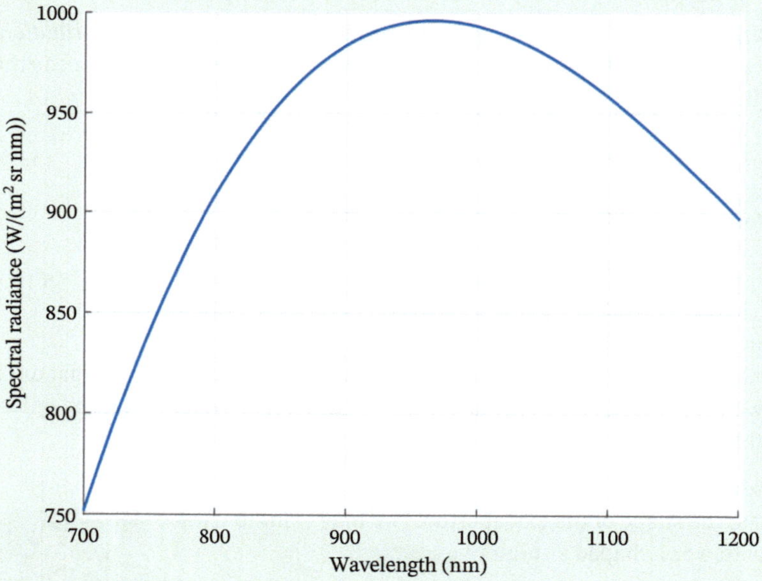

Figure 5.31 Spectral radiance of a black-body surface at 3000 K.

You may use the provided file CIE_ Planck.xlsx to calculate total radiance in the given wavelength range (Figure 5.31).

We would need a filament temperature of 6500 K for the required radiance, which can be hardly provided by any solid. Instead, we may want to try a discharge lamp, for example some kind of Xe cinema lamp, since Xe has strong emission lines in the near infrared. It is only a matter of overall efficiency that at 6500 K, the emission peak is not in the infrared anymore: Higher temperature means higher black-body radiance, for all wavelengths.

In the next exercise, we are tasked with selecting a suitable light source for a projection headlamp.

Exercise 5.19: Pixel beam with DLP

It is planned to use a DMD to implement a pixelated headlight [25].

- The idea is to use a DMD with a diagonal of 0.46″ and ±12° tilt angle.
- The luminous flux to be projected is $\Phi_T = 1500\,\text{lm}$.
- The DMD efficiency is $\eta_D = 70\%$.
- Transmission efficiency of the projection optics is $\eta_O = 80\%$.

Suggest a suitable light source.

Solution 5.19

The DMD tilt angle determines the cone angle leaving the DMD and thus the etendue [26]:

$$\begin{aligned} U_{\text{DMD}} &= A\Omega_P = 960 \cdot 480 \cdot (0.0076\,\text{mm})^2 \cdot \pi\,\text{sr}\,\sin^2(12°) \\ &= 26.6\,\text{mm}^2 \cdot 0.136\,\text{sr} = 3.6\,\text{mm}^2\text{sr} \end{aligned} \tag{5.103}$$

We need a source with a luminance of at least

$$L = \frac{\Phi_T}{\eta_D \eta_O U} = \frac{1500\,\text{lm}}{0.56 \cdot 3.6\,\text{mm}^2\text{sr}} = 744\,\text{Mnit} \tag{5.104}$$

which is almost three times the luminance of the brightest available white LED. Consequently, such a projection headlamp requires a laser-pumped phosphor source [27] (LARP) or a high-intensity discharge [28] (HID) source.

We stay in the automotive world and design innovative displays.

Exercise 5.20: HUD projector

Suggest a light source for an automotive head-up display (HUD, see Figure 5.32) [29] with the following set of parameters:

Quantity	Value
Distance between virtual image and the eye box	2.4 m
H/V field of view (angular size of the virtual image)	$\pm 5° \cdot \pm 2°$
Eye box radius	100 mm
Windshield reflectivity	10%
DMD efficiency	$\eta_D = 70\%$
Illumination optics efficiency	$\eta_O = 70\%$
Overfill	10% of area

(Continued)

Exercise 5.20: (Continued)

In addition, a DMD with a diagonal of 0.3″ and 12° tilt angle is to be used. The luminance of the virtual image should be 15 kcd/m^2.

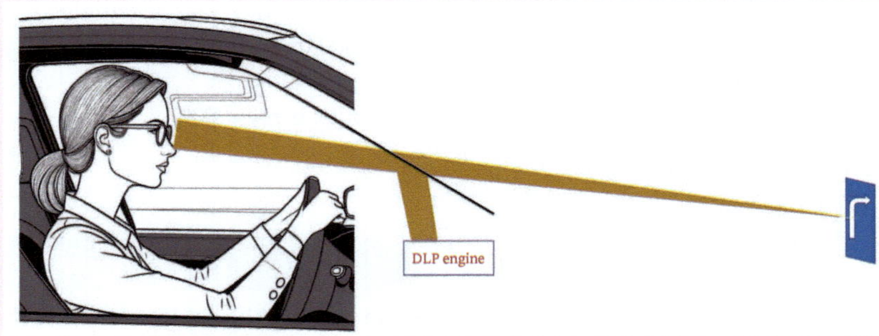

Figure 5.32 Automotive HUD. *Source:* AI rendering and graphics by the authors.

Which type of LED(s) can be used (if any)? Calculate a suitable screen gain to fill the HUD etendue.

Before attempting a solution, however, we need to introduce the concept of *screen gain*.

Knowledge Corner Topic 5.2: Screen gain

Consider a projector illuminating a screen. For simplicity, assume the light is incident within a cone of a half angle θ, constant for the whole screen, with perpendicular chief rays, i.e. a telecentric incidence, and an on-axis intensity of the projector of

$$I_{0,\mathrm{S}} = L_\mathrm{S} A \tag{5.105}$$

Adding a Lambertian scatterer to the screen would distribute the incident light from an etendue of

$$U_\mathrm{S} = A \sin^2 \theta \pi \,\mathrm{sr} \tag{5.106}$$

to an etendue of

$$U_\mathrm{L} = A\pi \,\mathrm{sr} \tag{5.107}$$

(the technical term is *dilution*), reducing the effective luminance to

$$L_\mathrm{L} = \frac{\Phi}{U_\mathrm{L}} = \frac{\Phi}{U_\mathrm{S}} \sin^2 \theta = L_\mathrm{S} \sin^2 \theta \tag{5.108}$$

The intensity after Lambertian scattering would then be

$$I_{0,\mathrm{L}} = L_\mathrm{L} A = I_{0,\mathrm{S}} \sin^2 \theta \tag{5.109}$$

However, Lambertian scattering is both hard to achieve and undesirable: We do want some scattering, but not so strong. For an arbitrary scatterer, the *screen gain* is defined as the ratio of the actual on-axis intensity I_0 to that of a Lambertian emitter of the same luminous exitance

$$g = \frac{I_0}{I_{0,\mathrm{L}}} \tag{5.110}$$

With Eq. 5.108, we have

$$g = \frac{L}{L_\mathrm{L}} = \frac{U_\mathrm{L}}{U} \tag{5.111}$$

which describes the etendue increase due to the scattering.

Screen "gain" does not increase luminance. It simply describes the fact that the luminance *reduction* by the screen is less than that of a Lambertian scatterer. The maximum luminance we can obtain is that of the incident beam.

The effective luminance reflected by the screen is

$$L = L_\mathrm{S}\, g \sin^2 \theta \tag{5.112}$$

Any claims of higher concentration ratios such as for certain LEDs [30] would rely on other effects than just scattering (such as recycling light from higher angles, similar to deglaring films).

The Second Law prevents us to achieve gain from a true Lambertian source ($\theta = 90°$), where our calculus just results in $g = 1$. Screen gain is always limited by

$$g \leq \frac{1}{\sin^2\theta} \tag{5.113}$$

Solution 5.20

It is quite normal for HUDs to have larger etendue than typical displays. The reason is that a sufficiently large virtual image at a finite distance must be seen from a relatively large eyebox: Drivers aren't equally tall, and they don't sit still.

Eye box and virtual image form a (non-rotationally symmetric) light tube, whose etendue we now estimate. This is the HUD system etendue U_T.

$$\begin{aligned} U_\mathrm{T} &= A\Omega_\mathrm{P} = \pi \cdot (100\,\mathrm{mm})^2 \cdot \pi\,\mathrm{sr}\ \sin 5° \sin 2° \\ &= 3.14 \cdot 10^4\,\mathrm{mm}^2 \cdot 0.0095\,\mathrm{sr} = 300\,\mathrm{mm}^2\,\mathrm{sr} \end{aligned} \tag{5.114}$$

which is obviously much larger than the DMD etendue of

$$\begin{aligned} U_\mathrm{DMD} &= A\Omega_\mathrm{P} = 848 \cdot 480 \cdot (0.0076\,\mathrm{mm})^2 \cdot \pi\,\mathrm{sr}\ \sin^2 12° \\ &= 23.5\,\mathrm{mm}^2 \cdot 0.136\,\mathrm{sr} = 3.2\,\mathrm{mm}^2\,\mathrm{sr} \end{aligned} \tag{5.115}$$

(Continued)

Solution 5.20 (Continued)

This etendue ratio requires an *exit pupil expander* [31] to spread the projector etendue to the HUD system etendue without affecting the image.

The exit pupil expander is often realized by a scattering screen placed in an intermediate image of the projector. There are also realizations with a microlens array [32] or by diffraction [33]. The expander area is then imaged to the virtual image with the eye box as the pupil, maybe with an optical correction that compensates the windshield aberrations.

(a) For the given virtual image luminance, we need

$$\Phi_\mathrm{T} = LU_\mathrm{T} = 15\,\mathrm{kcd/m^2} \cdot 300\,\mathrm{mm^2\,sr} = 4.5\,\mathrm{lm} \tag{5.116}$$

reflected from the windshield to the eyebox.

Thus, we need a source of about

$$\Phi_\mathrm{S} = \frac{4.5\,\mathrm{lm} \cdot 110\%}{10\% \cdot 70\% \cdot 70\%} \approx 100\,\mathrm{lm} \tag{5.117}$$

and with a luminance of

$$L_\mathrm{S} = \frac{\Phi_\mathrm{S}}{U_\mathrm{DMD}} = \frac{100\,\mathrm{lm}}{3.2\,\mathrm{mm^2\,sr}} = 31\,\mathrm{Mnit} \tag{5.118}$$

Perhaps we need a little more as the screen gain will scatter the light and fill an etendue slightly larger than that of the HUD.

The luminance diagram in Figure 5.33 [8] explains the development of flux and luminance during propagation through the projector.

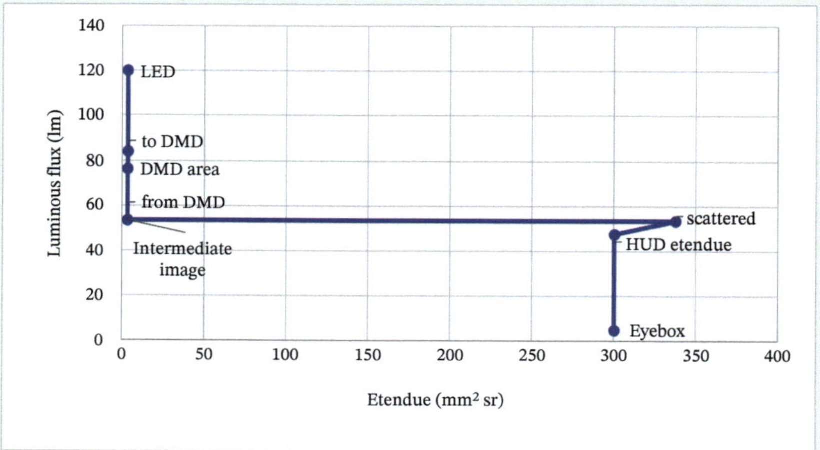

Figure 5.33 Luminance diagram (cf. Figure 4.52) of the HUD. The long horizontal step is due to the exit pupil expander.

With Eq. 5.111, we can calculate a matching screen gain:

$$g = \frac{U_\mathrm{T}}{U_\mathrm{DMD}} = \frac{300\,\mathrm{mm^2\,sr}}{3.2\,\mathrm{mm^2\,sr}} \approx 94 \tag{5.119}$$

There is a helpful Calculator provided by the DMD maker.

We provide the spreadsheet file luminance_HUD.xlsx for playing with the parameters, and hopefully a few insights.

Exercise 5.21: LED spotlight (2)

The datasheet of an LED spotlight shows (among other information)

- a 70 W LED source,
- a $2\theta = 13...43°$ zoom range,
- $\Phi_T = 3500\,\text{lm}$ output,
- a front lens diameter $2r_F = 60\,\text{mm}$.

Assume an optical efficiency of $\eta = 70\%$. Guess what the LED source might be (this is a sequel to Exercise 5.1).

Solution 5.21

The etendue of the device in the tele position is

$$U_T = (30\,\text{mm})^2\pi \cdot (\sin 6.5°)^2\pi\,\text{sr} \approx 114\,\text{mm}^2\,\text{sr} \tag{5.120}$$

(the specification shows the full angle, but as usual, we need the half angle).

The source flux is

$$\Phi_S = \frac{\Phi_T}{\eta} = \frac{3500\,\text{lm}}{70\%} \approx 5000\,\text{lm} \tag{5.121}$$

which indicates an efficacy of

$$\eta = \frac{5000\,\text{lm}}{70\,\text{W}} \approx 70\,\text{lm/W}, \tag{5.122}$$

quite a reasonable value. The necessary source luminance is at least

$$L_S = \frac{\Phi_S}{U_T} = \frac{5000\,\text{lm}}{114\,\text{mm}^2\,\text{sr}} \approx 44\,\text{Mnit} \tag{5.123}$$

With the given power range, one may consider, for example, the only circular LED we know with up to 5000 lm for a good bin at maximum current. Its actual luminance is more than twice as high as needed in this example, so it would be a reasonable candidate for such a fixture.

Figure 5.34 A beam light.

Exercise 5.22: Search light

A search light (or beam light, flash light, see Figure 5.34) is announced with the following key parameters:

- Flux $\Phi = 12$ klm
- Front diameter $2r_{\mathrm{F}} = 195$ mm
- Zoom range $2\theta = 3\ldots22°$
- Beam distance 4000 m (for 0.25 lx, ANSI NEMA FL-1 [34])
- Output up to 4 Mcd
- Power 378 W

The supplier claims a "homogeneous beam profile". Is a *top-hat beam* realistic? What kind of source could the supplier be using?

Solution 5.22

In the tele position (±1.5°), the solid angle is

$$\Omega = 2\pi(1 - \cos\theta) \approx 0.002\ \mathrm{sr} \tag{5.124}$$

and for a top-hat beam a flux of more than

$$\Phi = I_0\Omega = 4\,\mathrm{Mcd} \cdot 0.002\,\mathrm{sr} = 8000\,\mathrm{lm} \tag{5.125}$$

is required. The above specification supports the top-hat claim; it could actually be a top-hat beam.

We use Eq. 3.107 to estimate the luminance

$$L = \frac{I_0}{A} = \frac{4\,\text{Mcd}}{\pi r_F^2} = 134\,\text{Mnit} \tag{5.126}$$

This is a large, but not outlandish value, even though $\eta \approx 70\%$ requires an actual source luminance of $L_S \approx 200$ Mnit. Cold-white automotive LEDs with this luminance are available, but a single LED will not be sufficient. Accordingly, we guess that there is a sparse array of relatively small LED chips inside, with individual collimators and subsequent homogenization optics; in [35], we present possible homogenizing elements and how and why to choose them.

Exercise 5.23: Surgical headlight

Surgical headlights are like hats with a directional light (Figure 5.35). Formerly operated with Xe lamps via light guides, these devices are today mostly LED-based and driven by batteries. Here is an overview.

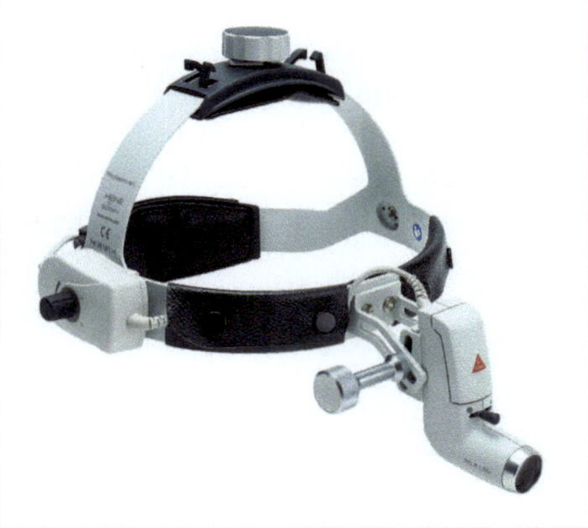

Figure 5.35 Surgical headlight. *Source:* Courtesy of Heine.

As most of the devices on the market are essentially similar, we continue with a generic specification based on reasonable datasheets:

- Working distance is defined by an industry standard and measures 16 ″,
- 100 klx is required there,
- 20…100 mm adjustable spot size diameter,
- 5 W LED power,
- 20 mm lens diameter.

The spot size is often adjustable by a zoom factor of 5. There is only one illuminance value provided; it seems that the zoom does not change illuminance (of course, the headlight is electronically dimmable).

(Continued)

Exercise 5.23: (Continued)

We conclude that the size setting is not really a zoom. Instead, it is realized by an adjustable aperture which is then imaged to the working distance. This seems reasonable given that illuminance is much more important for the surgeon than total flux, and an actual fivefold optical zoom is simply unrealistic within the given dimensions and cost.

(a) Estimate the necessary source luminance from the above data.
(b) Suggest a source and an optical design.
(c) Does the inverse square law (Eq. 3.142) apply for your design?

Solution 5.23

(a) To estimate the luminance, we start with Eq. 3.112 and calculate the projected solid angle from the exit pupil of the lens,

$$L = \frac{E}{\Omega_\mathrm{p}} \approx \frac{E \cdot D^2}{\pi\,\mathrm{sr}\,r^2} = \frac{100\,\mathrm{klx} \cdot (406\,\mathrm{mm})^2}{\pi\,\mathrm{sr} \cdot (10\,\mathrm{mm})^2} = 52\,\mathrm{Mnit} \tag{5.127}$$

This value is quite realistic with current white LEDs, although warm-white, high-CRI, high-luminance LEDs are still hard to find.

(b) We would suggest an optical design based on imaging an iris diaphragm with a single lens as shown in Figure 5.36. The LED may be collimated with a TIR lens (which is simpler and cheaper than at least two condenser lenses).

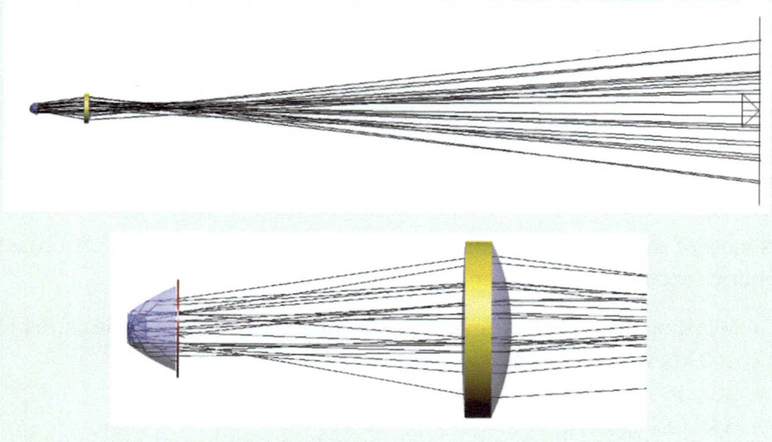

Figure 5.36 Optical design for a surgical headlight: LED, TIR lens, adjustable Iris, projection lens. The system is producing a peak illuminance of 100 klx for an LED flux of 600 lm from 1 mm^2 chip area. The optical efficiency is about two thirds, average illuminance is half of the peak.

We provide the system file.

A small white projection LED would be suitable from an optical design point of view; color temperature and color rendering may be an issue.

(c) Inverse square law: in Knowledge Corner Topic 3.16, we have shown that the law applies to any source, for a sufficiently large distance. Does it apply to an image as well? We double-check by means of a simulation, similar to Exercise 4.43.

Simulation 5.6: Inverse square law (2)

We vary the distance of the receiver to the imaging lens and plot $1/\sqrt{E_0}$; the result is shown in Figure 5.37.

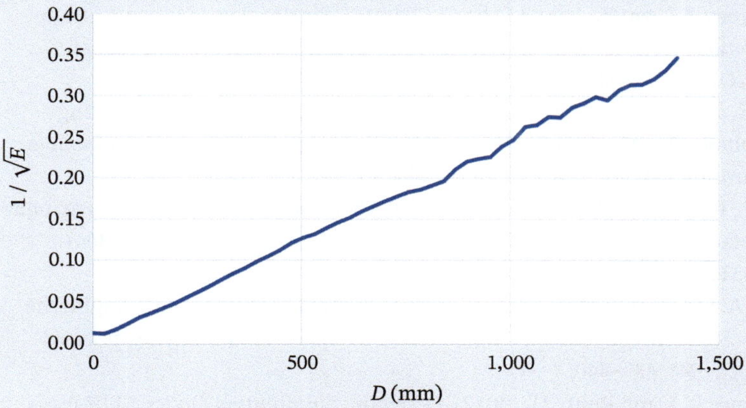

Figure 5.37 Plot of $1/\sqrt{E_0}$ (in $1/(\sqrt{\text{klx}})$) vs. distance D.

Yes, the inverse square law also applies to this type of projector, and the effective source location is right at the exit aperture, from where the distance was measured. This is no surprise, since on-axis, all you see, starting from a short distance, is the fully flashed exit aperture. A fully flashed, constant luminance aperture is indistinguishable from a Lambertian source, and that is what we see here.

We provide the spreadsheet file.

Dear illumination optics professional – you made it up to this point, so we praise your perseverance and the level of sophistication you have achieved: all the desired light seems to have reached the receiver. The simulations are completed, designs have converged.

It certainly was a tour de force for you (we readily admit that writing the book was also a tour de force for us), but we do hope you also had some fun, and we also do hope that some of the things you learned while reading the book will find their way into the applications and products you help create.

It is now time to summarize some highlights in the following final chapter.

References

1 Rehn, H. (2021). An unusual zoom design for a variable edge beam. In: *International Optical Design Conference*, Vol. 12078. Optica Publishing Group.

2 Bliss, M., Betts, T.R., and Gottschalg, R. (2008). Advantages in using LEDs as the main light sourcein solar simulators for measuring PV device characteristics. In: *Reliability of Photovoltaic Cells, Modules, Components, and Systems,* Vol. 7048. 45–55. SPIE Press.

3 Gallo, A., Marzo, A., Fuentealba, E. et al. (2017). High flux solar simulators for concentrated solar thermal research: a review. *Renewable and Sustainable Energy Reviews* 77: 1385–1402. https://doi.org/10.1016/j.rser.2017.01.056.

4 Domínguez, C., Antón, I., and Sala, G. (2008). Solar simulator for concentrator photovoltaic systems. *Optics Express* 16 (19): 14894–14901. https://doi.org/10.1364/OE.16.014894.

5 Rehn, H. and Hartwig, U. (2010). A solar simulator design for concentrating photovoltaics. In: *Nonimaging Optics: Efficient Design for Illumination and Solar Concentration VII,* Vol. 7785, 123–129. SPIE Press.

6 Aieta, F., Genevet, P., Kats, M. et al. (2013). Aberrations of flat lenses and aplanatic metasurfaces. *Optics Express* 21 (25): 31530–31539. https://doi.org/10.1364/OE.21.031530.

7 Köhler, A. (1893). Ein neues Beleuchtungsverfahren für mikrophotographische Zwecke. *Zeitschrift für wissenschaftliche Mikroskopie und für Mikroskopische Technik* 10 (04): 433–440.

8 Muschaweck, J. and Rehn, H. (2022). *Designing Illumination Optics.* SPIE Press.

9 Brick, P. and Rehn, H. (2018). Illumination device and method for producing an illumination device. https://patents.google.com/patent/WO2018197636A3/de?oq=WO+2018%2f197636.

10 Town, G., Ash, C., Eadie, E. et al. (2007). Measuring key parameters of intense pulsed light (IPL) devices. *Journal of Cosmetic and Laser Therapy* 9 (3): 148–160. https://doi.org/10.1080/14764170701435297.

11 Muschaweck, J. and Rehn, H. *Design Elements of Illumination Optics. Components and Systems.* In preparation.

12 UXL Xenon Short Arc | Ushio America, Inc. (2017). https://www.ushio.com/product/uxl-xenon-short-arc.

13 Miller, J.L., Schwertz, K., Friedman, E., et al. (2020). *Photonics Rules of Thumb*, 3e. SPIE.

14 Alman, D.H. (1977). Errors of the standard photometric system when measuring the brightness of general illumination light sources. *Journal of the Illuminating Engineering Society* 7 (1): 55–62. https://doi.org/10.1080/00994480.1977.10748961.

15 de Vries, A., Heynderickx, I., and de Kort, Y.A. (2022). From luminance to brightness: a data-driven approach to support brightness assessments in open plan offices. *Lighting Research & Technology* 54 (8): 798–818. https://doi.org/10.1177/14771535221117365.

16 Dupuis, R.D. and Krames, M.R. (2008). History, development, and applications of high-brightness visible light-emitting diodes. *Journal of Lightwave Technology* 26 (9): 1154–1171. https://doi.org/10.1109/JLT.2008.923628.

17 Khanh, T.Q., Bodrogi, P., Zandi, B. et al. (2023). Brightness perception under photopic conditions: experiments and modeling with contributions of S-cone and ipRGC. *Scientific Reports* 13 (1): 14542. https://doi.org/10.21203/rs.3.rs-2559514/v1.

18 Waters, C.E., Mistrick, R.G., and Bernecker, C.A. (1998). Brightness perception from sources of non-uniform luminance. *Journal of the Illuminating Engineering Society* 27 (2): 89–101. https://doi.org/10.1080/00994480.1998.10748237.

19 Wood, M. (2019). Pixelated headights. https://www.mikewoodconsulting.com/articles/Protocol%20Summer%202019%20-%20Pixelated%20Headights.pdf.

20 Bhakta, V.R. (2015). High resolution adaptive headlight using Texas Instruments DLP® technology. In: *ISAL Proceedings.*

21 Günther, A. (2010). *Entwicklung eines aktiven Kraftfahrzeugscheinwerferkonzepts mit DMD-Array*. PhD thesis. Ilmenau.

22 Asari, T., Miyachi, M., Oda, Y. et al. (2021). Adaptive driving beam system with MEMS optical scanner for reconfigurable vehicle headlight. *Journal of Optical Microsystems* 1 (01): 014501. https://doi.org/10.1117/1.JOM.1.1.014501.

23 Brick, P., Günther, A., and Grötsch, S. (2021). Automotive headlamp optics for a pixelated μLED source. *Proc. SPIE* 11874: 20–23.

24 Vögl, F., Avramescu, A., Lex, A. et al. (2024). Enhanced forward emission by backside mirror design in micron-sized LEDs. *Optics Letters* 49 (18): 5095–5098. https://doi.org/10.1364/OL.537525.

25 Knöchelmann, M., Held, M.P., Kloppenburg, G. et al. (2019). High-resolution headlamps – technology analysis and system design. *Advanced Optical Technologies* 8 (1): 33–46. https://doi.org/10.1515/aot-2018-0060.

26 Texas Instruments (2010). DLP system optics. https://download.kamami.pl/p584882-dlpa022.pdf.

27 Erdl, H. and Hanafi, A. (2014). Lighting device for a motor vehicle. https://patents.google.com/patent/US20140226352/en.

28 Weichmann, U., Cromwijk, J.W., Heusler, G. et al. (2005). Light-sources for small-etendue applications: a comparison of xenon-and UHP lamps. In: *Projection Displays XI,* 13–26. SPIE.

29 Firth, M. (2019). Introduction to automotive augmented reality head-up displays using TI DLP technology. https://www.ti.com/lit/wp/dlpy009/dlpy009.pdf.

30 New Multi-Color PhlatLight LEDs. LED professional (2010). https://www.led-professional.com/products/led-modules-led-light-engines/new-multi-color-phlatlight-leds.

31 Chen, F. and Miller, J.O. (2021). Systems and methods of increasing pupil size in a display system, September 2019. US Patent, US20190278076A1. https://patents.google.com/patent/US20190278076A1/en?oq=US20190278076A1.

32 Urey, H. and Powell, K.D. (2005). Microlens-array-based exit-pupil expander for full-color displays. *Applied Optics* 44 (23): 4930–4936. https://doi.org/10.1364/AO.44.004930.

33 Urey, H. (2001). Diffractive exit-pupil expander for display applications. *Applied Optics* 40 (32): 5840–5851. https://doi.org/10.1364/AO.40.005840.

34 ANSI-NEMA FL-1 - Flashlight Wiki. https://flashlightwiki.com/ANSI-NEMA_FL-1.

35 Muschaweck, J. and Rehn, H. (2019). Illumination design patterns for homogenization and color mixing. *Advanced Optical Technologies* 8 (1): 13–32. https://doi.org/10.1515/aot-2018-0051.

6

Conclusions

You started this journey, dear reader, with the first paragraph in the introduction:

> To become a good optical designer for illumination applications, an engineer must acquire a very specific set of technical skills: he or she must understand the underlying theory, must command the toolbox of optical elements and know how to combine them, and must be a proficient user of (at least one) optical design software. In addition to reading books, managing projects, and learning from senior colleagues, as an engineer on your way to expanding your skills, you must also think through problems on your own, use your own mind, stumble over obstacles, and find your way to a solution – all by yourself. This may be hard, sometimes exhausting or even embarrassing. However, solving problems by yourself is an indispensable part of the journey.

More than one hundred exercises later, we hope that the material in this book did indeed help you to acquire this set of skills. We hope that you did think through the exercises yourself, resisting the temptation to peek ahead at the solutions. When we wrote this book, starting from a selection of exercises from our live courses, we ran into a large number of fascinating side topics, many of which we discussed with you in the Knowledge Corner café, sometimes showing you simulations. We hope that these more textbook-like parts help to convey the deep fascination about light for illumination that we feel even more strongly now than when we started our careers in illumination optics design. It amazes us every day how things fit together like a giant jigsaw puzzle, because the conservation laws (flux, *etendue*, luminance, skewness) perfectly connect everything. Knowing the conservation laws and the underlying ideas thoroughly, we – and you – can just go with the flow instead of blindly searching for solutions that may be impossible to find.

In this spirit, we now present what we consider the most important formulas, takeaways and recommendations from this book.

Engineering Illumination Optics: Case Studies and Exercises, First Edition. Henning Rehn and Julius Muschaweck.
Published 2026 by Wiley-VCH GmbH.

6.1 Formulas to Remember

In our opinion, all the equations in this book are valuable, but those that are collected and repeated in this chapter really stand out. You might want to commit the simpler ones to memory – for the others, it is most likely sufficient if you just remember that they exist and where you can find them.

6.1.1 Snells Law, Expressed as a Conservation Law

Refraction at an interface is described as the conservation of the *angular extent* product

$$X = n^2\Omega_\mathrm{p} = \text{const} \tag{6.1}$$

before and after the interface (Eq. 2.45). This is equivalent to the common formulation of Snell's law, $n \sin\theta = \text{const}$, plus the fact that both the rays and the surface normal are coplanar.

6.1.2 Conic Sections

You have learned how to use conic sections in optical design, with pencil, paper and pocket calculator, or with the help of a spreadsheet. The two key conic section formulas are the vertex equation (Eq. 2.101)

$$(1 + k)z^2 - 2\rho z + r^2 = 0 \tag{6.2}$$

and its polar form (Eq. 2.103)

$$r_1(\delta) = \frac{\rho}{1 + e\cos\delta} \tag{6.3}$$

6.1.3 Etendue Calculations

It is quite common for illumination optics engineers to know that etendue is somehow very important. However, its details often remain mysterious. In Section 3.2, we lifted the mystery veil, allowing us to actually use etendue as an indispensable design tool.

Frequently, we think of etendue as the "size" of a *light tube*: one aperture sends light into another aperture, with the light tube consisting of all rays involved. The etendue U is then the 4D volume of this ray set in appropriate coordinates (Eq. 3.7) on a screen (Knowledge Corner Topic 3.1):

$$U = \iint \mathrm{d}X\,\mathrm{d}A = \iint n^2 \cos\theta \,\mathrm{d}\Omega\,\mathrm{d}A \tag{6.4}$$

In the common case of a circular source radiating into a circular, finite-distance pupil on the optical axis, all in air, *Hottel's formula* (Eq. 3.36) gives a simple analytical expression for that light tube's etendue:

$$U = \frac{\pi^2}{4}\left(\sqrt{(r+R)^2 + D^2} - \sqrt{(r-R)^2 + D^2}\right)^2 \text{ sr} \tag{6.5}$$

An important special case of Hottel's formula, and indeed of the general definition in Eq. 3.7, is that of a telecentric, rotationally symmetric beam in a medium with index n:

$$U = AX = An^2\Omega_p = r^2\pi(n \sin\theta)^2\pi \text{ sr} = (n\,\pi\, r \sin\theta)^2 \text{ sr} \tag{6.6}$$

derived as Eq. 3.11, which is not too far away from the paraxial definition $U = A\Omega$. However, its preconditions are now better understood.

In this book, we also brought the case for *angular extent* $X = n^2\Omega_p$ before you, the jury. The quantity

$$n^2\Omega_p = \int n^2 \cos\theta \,\sin\theta \,d\theta \,d\phi \tag{6.7}$$

keeps popping up everywhere in the calculations. Moreover, it appears in a conservation law (see Snell's law above). The relation $U = AX$ tells us something, too: Area and angular extent enter phase-space considerations on an equal footing. Such quantities deserve a dedicated symbol, and a dedicated name as well. But symbols and names, such as A for area, become commonplace only when sufficiently many engineers – the jury – understand and use them.

6.1.4 Etendue Estimation by Simulation

If your system is too complex to be described by a simple equation, you can still (and often quite easily) estimate its etendue by means of ray-tracing software.

To do this, send light of constant source luminance L_S through a system of unknown etendue U and measure the collected target flux Φ_T. You will then obtain the etendue of the rays that actually make it to the target as

$$U = \frac{\Phi_T}{L_S} \tag{6.8}$$

derived as Eq. 3.40. Be careful to follow the guidelines in Knowledge Corner Topic 3.7 on how to set up your system for this purpose!

6.1.5 Etendue and Coherence

A system etendue of

$$U \leq \left(\frac{\pi}{2}\lambda\right)^2 \tag{6.9}$$

ensures a certain spatial coherence at the target (Eq. 3.78), revealing an almost magical connection between ray optics and quantum mechanics.

6.1.6 Luminance and Etendue

Etendue connects flux and luminance: luminance[1] is the density of flux in phase space,

$$L = \frac{d\Phi}{dU} \quad \Rightarrow \quad L_{avg} = \frac{\Phi}{U} \tag{6.10}$$

(Eqs. 3.91 and 3.108).

---

1 Strictly speaking, "our" luminance is termed *basic luminance*, since it includes a factor of n^2 in the angular extent. We take the liberty of omitting "basic" in our writing, because it is basic luminance that is conserved along a ray even under refraction, and we believe that the short, undecorated name should be reserved for the conserved quantity, and not the other way around.

6.1.7 The Radiometry Diamond

Four quantities, each as a triplet of photometric/radiometric/spectral siblings, form the foundation of radiometry: flux Φ_v, intensity I_v, illuminance E_v, and luminance L_v. These quantities are geometrically connected with each other. In Figure 3.28, reproduced here as Figure 6.1, we display these connections in an intuitive, diamond-shaped diagram, which every illumination optics designer should be able to draw when awakened at three o'clock in the morning.

6.1.8 Efficiency Limit

The conservation laws for etendue and luminance impose an upper limit for the achievable flux transfer from a homogeneous source to a target, given by their etendue ratio

$$\eta\text{max} = \min\left(\frac{U_T}{U_S}, 1\right) \tag{6.11}$$

derived as Eq. 3.19, or expressed in radiometric terms as Eq. 3.95,

$$\Phi_T \leq \min\left(L_S \cdot U_T, \Phi_S\right) \tag{6.12}$$

6.1.9 Lambertian Emitters

Lambertian emitters have been introduced in Knowledge Corner Topic 3.12 as a source of spatially and directionally constant luminance. For Lambertian emitters, many general formulas take simple and beautiful forms, such as the source exitance[2] (flux per area)

$$E = n^2 \pi\, L \tag{6.13}$$

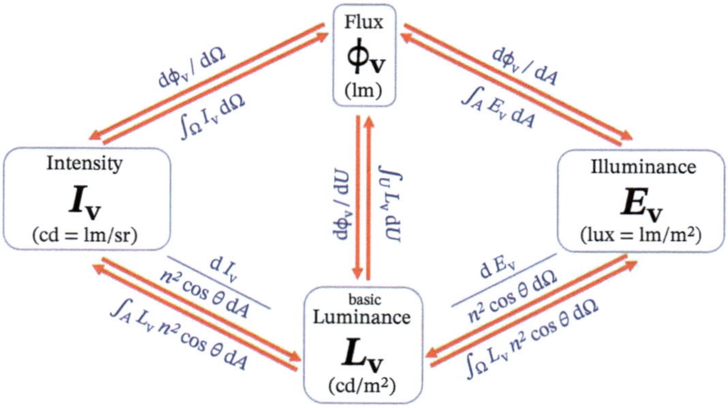

Figure 6.1 The radiometry diamond (reproduced from Figure 3.28).

2 Strictly speaking, luminous or radiant flux per area is termed *illuminance* E_v or *irradiance* E when the light comes, and *luminous exitance* M_v or *radiant exitance* M when the light leaves. Since we tend to think of radiometric quantities as living on virtual *screens* (see Knowledge Corner Topic 3.1) as opposed to real surfaces, we tend to ignore this distinction and just use E or E_v.

and the intensity profile

$$I(\theta) = I_0 \cos\theta \tag{6.14}$$

with

$$I_0 = L \cdot A \tag{6.15}$$

We remind you that these relations describe the important special Lambertian case only, and are not valid in general.

6.1.10 Inverse Square Law

When a beam with intensity I_v illuminates a perpendicular target at distance d, the target illuminance is given by

$$E_v = \frac{d\Phi_v}{dA} = \frac{I_v}{d^2}\,\text{sr} \tag{6.16}$$

derived as Eq. 3.142. (When the target has a non-perpendicular orientation, an additional factor of $\cos\theta$ applies.)

We have learned that the inverse square law is valid in the far field of any source (not just Lambertian ones). In fact, the law *defines* where the far field begins – if it is violated, you are simply not far enough away. It also helps us to define the effective center of the source.

6.1.11 cos^4 Law

Taking a camera picture of a Lambertian emitter, or illuminating a distant parallel plane with a Lambertian emitter, will approximately result in an inhomogeneous image or target illuminance (bright center and darker edges) in the form

$$E_T(\theta) = E_0 \cos^4\theta \tag{6.17}$$

derived as Eq. 3.190. The actual distribution will depend on imaging aberrations and $F/\#$.

6.1.12 cos^n Beams

We introduced this powerful and handy tool to describe soft collimated beams in Eq. 3.128:

$$I(\theta) = \frac{n+1}{2\pi\,\text{sr}}\Phi\cos^n\theta \tag{6.18}$$

where Φ is the total flux of the beam, and n is the exponent defining the beam shape. For small angles, a $\cos^n$ beam is nearly identical to a Gaussian beam, but much easier to handle analytically.

6.1.13 Stefan–Boltzmann Law

Thermal sources such as incandescent lamps or the Sun approximately emit black-body radiation. The radiant exitance of a black-body emitter, embedded in refractive index n, as a function of absolute temperature T, is given by[3]

$$E(T) = n^2 \sigma T^4 \tag{6.19}$$

with the Stefan–Boltzmann constant σ as defined in Eq. 3.197. Review Section 3.5 for a background on why the term "black body" makes sense, and for how to model non-black sources.

6.1.14 Color Weight

Additive mixing of light from different sources is governed by their respective *color weights* (cf. Eq. 3.270), a very useful quantity in colorimetric calculations, given by Eq. 3.264:

$$W = X + Y + Z = \int \bar{w}(\lambda) S_\lambda(\lambda) \mathrm{d}\lambda \tag{6.20}$$

We obtain the CIE x, y color coordinates of the additive mix by computing the average, weighted with the color weights W_i:

$$x_{\text{mix}} = \frac{\sum x_i W_i}{\sum W_i} \qquad y_{\text{mix}} = \frac{\sum y_i W_i}{\sum W_i} \tag{6.21}$$

in perfect analogy to computing the coordinates of the center of mass for a particle distribution. It is the color weight, not the luminous flux, that plays the role of mass in this analogy.

6.1.15 Maximum Concentration

Applying the universal conservation laws (Knowledge Corner Topic 3.11) to the concentration of a beam of light, collimated into the projected solid angle Ω_p, yields a concentration limit (cf. Eq. 4.80) of

$$C_{\text{max}} = \frac{\pi \text{ sr}}{\Omega_\text{p}} \tag{6.22}$$

For our Sun, we obtain a concentration limit of

$$C_{\text{max}} = \frac{1}{\sin^2 \theta_\text{S}} \approx 46\,000 \tag{6.23}$$

6.2 Takeaways and Recommendations

- It is incredible how much can be calculated analytically, how well these estimates (sometimes ) match corresponding simulations (see, e. g. Exercise 4.3), and how far they guide us toward viable solutions.

3 Many textbooks erroneously omit the factor of n^2, which is required to ensure consistency with the Second Law.

- Do not hesitate to use "dirty" approximations where useful and appropriate (with the proper amount of care, of course)!
- Mastering a collection of a few basic formulas and the theory behind them renders you business-fluent in the world of illumination optics.
- Frequently, you may model reflectors using formulas and numerical techniques.
- It is no big deal to handle a mix of international (SI) and US customary units. If possible, stick to SI, though.
- Illumination design often differs substantially from paraxial estimates and from imaging pre-design. If possible, use the power of fine-tuning by means of optimization!
- Select the right tools for your task:
  - Equations and analytic calculations are just one option. Playing with simple simulations is a complementary, equally valuable approach. However, while formulas are free, simulations require commercial software.
  - Use simulations to verify your calculations, and use calculations to confirm your simulations.
  - Let spreadsheets and/or scripting languages assist your work. One of us prefers Excel spreadsheets, the other prefers Mathlab and Python. Both approaches are equally valid and have their pros and cons.
  - Color calculator, the supplier's LED tools, Julius' spectrum library and more: There is helpful software for free out there!
- Do not merely read datasheets: there is much information between the lines. Calculate and draw more information!
- Always specify (e.g. by using an index) which luminance, flux, *etendue* etc. you mean. Source or system or target? Confusion lets you or others fail – thinking of *screens* as entities on which these quantities live can help to avoid confusion.
- Use the internet for complementary information (e.g. to find the spectrum of a comparable source if you have only a minimum datasheet).
- Strive for simple systems.
- Always think first. Optical design happens in the mind of the optical designer.

And now the final hint: Throughout your career, you may be challenged by quotes such as

> Everyone knew it was impossible, until a fool who didn't know came along and did it. (*origin unknown.*)

While this can be true in some cases, more often than not it is wrong, because what we do – and can do – in illumination optics is governed by conservation laws. Stay calm and patient when you have to handle a contradiction between a *conservation law* and somebody with superficial knowledge [1]. As you know, conservation laws will win. Always.

Reference

1 Kruger, J. and Dunning, D. (1999). Unskilled and unaware of it: how difficulties in recognizing one's own incompetence lead to inflated self-assessments. *Journal of Personality and Social Psychology* 77 (6): 1121–1134. https://doi.org/10.1037/0022-3514.77.6.1121.

Index

Engineering Illumination Optics: Case Studies and Exercises, First Edition. Henning Rehn and Julius Muschaweck.

Published 2026 by Wiley-VCH GmbH.